Biomarkers of Environmental Contamination

Edited by

John F. McCarthy
Lee R. Shugart

CRC Press
Taylor & Francis Group
Boca Raton London New York

CRC Press is an imprint of the
Taylor & Francis Group, an **informa** business

First published 1990 by CRC Press
Taylor & Francis Group
6000 Broken Sound Parkway NW, Suite 300
Boca Raton, FL 33487-2742

Reissued 2018 by CRC Press

This book contains information obtained from authentic and highly regarded sources. Reasonable efforts have been made to publish reliable data and information, but the author and publisher cannot assume responsibility for the validity of all materials or the consequences of their use. The authors and publishers have attempted to trace the copyright holders of all material reproduced in this publication and apologize to copyright holders if permission to publish in this form has not been obtained. If any copyright material has not been acknowledged please write and let us know so we may rectify in any future reprint.

Library of Congress Cataloging-in-Publication Data

Biomarkers of environmental contamination / edited by John F. McCarthy, Lee R. Shugart.
     p.   cm.
    Includes bibliographical references and index.
    ISBN 0-87371-284-6
    1. Indicators (Biology).   2. Pollution—Environmental aspects.   3. Environmental monitoring.   I. McCarthy, John F.   II. Shugart, Lee R.
    QH541.15.I5B594   1990
    363.73'63—dc20   90-38072

A Library of Congress record exists under LC control number: 90038072

Publisher's Note
The publisher has gone to great lengths to ensure the quality of this reprint but points out that some imperfections in the original copies may be apparent.

Disclaimer
The publisher has made every effort to trace copyright holders and welcomes correspondence from those they have been unable to contact.

ISBN 13: 978-1-315-89116-3 (hbk)
ISBN 13: 978-1-351-07026-3 (ebk)

Visit the Taylor & Francis Web site at http://www.taylorandfrancis.com and the
CRC Press Web site at http://www.crcpress.com

Preface

The purpose of this book is to provide an introduction to and review of the state-of-science concerning the use of biological markers in animals and plants as an innovative approach to evaluating the ecological and health effects of environmental contamination. Specifically, the book examines the status of research on the development, application, and validation of biological markers, either as indicators of exposure to toxic chemicals in the environment or as predictors of the adverse consequences of that exposure.

Biological markers are measurements at the molecular, biochemical, or cellular level in either wild populations from contaminated habitats or in organisms experimentally exposed to pollutants that indicate that the organism has been exposed to toxic chemicals, and the magnitude of the organism's response to the contaminant. Biological markers measured in wild animals can directly contribute to detecting, quantifying, and understanding the significance of exposure to chemicals in the environment. These measurements in environmental species may also help assess the potential for human exposure to environmental pollutants and for predicting the human health risks.

The concept of using biological markers to evaluate biological hazards has attracted considerable attention from regulatory agencies and is currently under evaluation at a number of research facilities. NOAA is testing biological markers as a component of their National Trends and Status Program and EPA is implementing programs at their research laboratories for the development and application of biological markers at contaminated sites. The National Marine Fisheries Service Research Laboratory is exploring this approach as well as agencies in other countries including Canada, The Netherlands, Germany, and several Scandinavian Countries. The goal of research by these agencies is to validate the use of biological markers as sensitive, cost-effect tools for evaluating risk of environmental hazards and to implement biological monitoring as a component of their regulatory plans; for example, EPA's Environmental Research Lab-Narragansett has begun a preliminary investigation using biological markers to document pollution at Black Rock Harbor and New Bedford Harbor, EPA's Environmental Monitoring Systems Laboratory in Cincinnati has a biomarker research program evaluating biomarkers in contaminated streams, and biological markers are a component of the National Pollution Discharge Elimination System (NPDES) permit for the Department of Energy facilities in Tennessee.

The key to effective implementation of biomarker-based biomonitoring is the need to integrate the insights and expertise of specialists in a number of technical specialties. The purpose for this symposium, and the contribution of this book, is to provide a forum to bring together the insights of specialists in a range of technical fields that permits them to focus their expertise on questions of environmental contamination. Experts in the areas of histology, immunology, enzymology, and nucleic acid/protein chemistry gathered at the symposium to discuss the application of their methods and approaches to this problem. In addition, scientists involved in the more integrated field studies also discussed the advantages of biological monitoring and consideration in the interpretation of data from animals collected in the field. This book represents a unique and timely synthesis of state-of-the-science approaches to biological monitoring of environmental contamination.

The book is intended to be of interest to a wide range of specialists. Different chapters provide overviews and recent research progress on the application and validation of biomarkers within a number of technical specialties such as molecular biology, biochemistry, histology, and immunology are detailed. For toxicologists, the book provides an introduction to an innovative approach to assessing environmental contamination that also permits the toxicologist to advance fundamental understanding of the expression of toxicity on a molecular and cellular level. The chapter on the application of biomarker measurements to ecological risk analysis is, to our knowledge, the first attempt to place biomarkers within a risk analysis paradigm. More generally, though, scientists in government agencies, both national and international, will find this synthesis of biomarker methodologies applicable to their missions. Regulators will find this volume particularly interesting as a comprehensive and up-to-date discussion of this promising approach to assessing the potential for environmental risks associated with contamination.

The editors wish to acknowledge and thank the Exploratory Studies Program of the Oak Ridge National Laboratory who provided funds to develop a research program in biomarkers at our facility. We also thank the senior management of the Environmental Sciences Division, David E. Reichle and Carl W. Gehrs, for their strong support and assistance as we brought together the interdisciplinary team needed to address the complex issues surrounding the development and application of biomarker-based biomonitoring. The ideas and approaches that underlie the organization and philosophy expressed in this book grew out of discussions and research among a team of individuals, including Marshall Adams, Larry Barnthouse, Rhonda Epler, Mark Greeley, Kitty Gustin, Braulio Jimenez, Deborah Millsap, Fred Sloop, and Glenn Suter.

The Oak Ridge National Laboratory is operated by Martin Marietta Energy Systems, Inc., under Contract DE-AC05-840R21400 with the U.S. Department of Energy. This is Publication No. 3518 of the Environmental Sciences Division, Oak Ridge National Laboratory.

John F. McCarthy and Lee R. Shugart
Environmental Sciences Division
Oak Ridge National Laboratory
Oak Ridge, Tennessee

John F. McCarthy is Group Leader for Biological Chemistry at the Environmental Sciences Division of the Oak Ridge National Laboratory (ORNL). He received a B.S. at Fordham University in New York, a Ph.D. in biological oceanography at the Graduate School of Oceanography of the University of Rhode Island, and did postdoctoral research at the Biology Division of ORNL before transferring to the Environmental Sciences Division in 1980. His principal research interest is in the transport, fate, and biological effects of contaminants. His research on the role of environmental and physiological factors affecting contaminant accumulation by aquatic organisms underscored the difficulties in understanding and predicting uptake and effects of toxicants in complex environmental exposures. Recognizing the advantages of biological marker measurements for evaluating environmental contamination, Dr. McCarthy redirected his efforts to coordinating an interdisciplinary team of colleagues at ORNL, and in 1986 obtained funding from the ORNL Exploratory Studies Program to develop a program in biological markers in animal species. This has grown to a diversified program of laboratory and field studies on biomarker responses in aquatic and terrestrial organisms.

Dr. McCarthy also maintains an active research program on the environmental transport of contaminants, particularly with respect to the role of organic and inorganic colloidal particles on the subsurface transport of pollutants. He serves as the coordinator of the Geochemical Transport Processes/Colloids Subprogram of the Subsurface Science Program of the U.S. Department of Energy. Dr. McCarthy is a member of the American Chemical Society, the Society for Environmental Toxicology and Chemistry, and the International Humic Substances Society.

Lee R. Shugart is a Senior Staff Member in the Environmental Toxicology Section of the Environmental Sciences Division of the Oak Ridge National Laboratory and an adjunct faculty member of the University of Tennessee Graduate School of Biomedical Sciences. His professional training is in biochemistry and his research interests are concerned with molecular mechanisms of environmental genotoxicity and the role of nucleic acids. He has conducted extensive research on the chemical modification of cellular macromolecules by environmental contaminants on fish, rodents, and humans. His recent studies have focused on the development of new methodologies for quantifying the interaction of genotoxic agents with deoxyribonucleic acids and proteins. He has published over 60 articles in the scientific literature on such topics as protein biosynthesis, mechanisms of enzyme action, nucleic acid biochemistry, and the effect of genotoxic agents on environmental species. Professional society membership includes the American Chemical Society; Sigma Xi; American Society of Biochemistry and Molecular Biology; and Society of Environmental Toxicology and Chemistry.

Contents

SECTION FOUR
GENOTOXIC RESPONSES

SECTION FIVE
METAL METABOLISM

SECTION SIX
APPLICATION OF BIOMARKERS IN FIELD EVALUATION

CONCLUDING REMARKS

SECTION ONE

Overview

CHAPTER 1

Biological Markers of Environmental Contamination

John F. McCarthy and Lee R. Shugart

The lone voice of Rachel Carson warning of impending environmental disaster has been joined over the last 25 years by a ground swell of concern about the effects of man's activities on our planet. Key among the issues concerning environmentalists and regulators are the health and ecological effects of toxic chemicals being released into the environment. Unfortunately, assessing either exposure to or effects of environmental contaminants is fraught with uncertainties. There are a number of difficulties in estimating the extent of exposure to toxic chemicals in the environment, and even greater difficulty in attributing, let alone predicting, the adverse health or ecological effects that result from the exposure.

Exposure is difficult to assess because of the wide diversity of potential routes of exposure (air, soil, water, and food chain), the large differences in the biological availability of contaminants associated with the different environmental media, and individual and species-specific differences in the pharmacodynamic disposition of the toxicants. All these processes affect the amount of toxicant that enters the animal and reaches critical molecular targets. For example, many organic contaminants are readily taken up and accumulated by aquatic organisms if the chemicals are dissolved in water, but the bioavailability is much lower if the same compounds are sorbed to soil or sediment particles. Seasonal changes in habitat, feeding habits, or metabolic activity can modify the nature and extent of exposure. Exposure to environmental chemicals cannot be readily quantified by measuring body burdens because, for example, many toxic chemicals do not

3

bioaccumulate, but rather are metabolized; in many cases, the metabolites are more toxic than the parent compound present in the environment. Furthermore, the relationship between body burden and toxic response is complex and not fully understood. The presence of complex mixtures of contaminants, the more realistic environmental scenario, creates further uncertainty. In fact, we understand the toxicity of relatively few of the tens of thousands of chemicals released into the environment, and have almost no information on the action of even well-characterized chemicals when they are components of a complex mixture.

The adverse health or ecological effects resulting from environmental exposures are even more difficult to describe than is exposure. The nature and extent of adverse effects will depend on the magnitude and duration of exposure, the mode of action of the toxicant, length of time required to manifest a disease state, and the susceptibility of the organism. Furthermore, susceptibility is a function of individual and species-specific factors as well as age, nutritional status, and the cumulative impacts of other chemical, environmental, or physiological stresses. In addition, the often long latent period between exposure and manifestation of an adverse effect makes it difficult to attribute a disease or other impairment to an environmental exposure.

The above statements are equally valid for understanding toxic effects in either humans or animals in nature. Faced with these complications and uncertainties, there is growing interest in an approach that evaluates exposure and effects of environmental chemicals by the use of biological markers, which can be defined as measurements of body fluids, cells, or tissues that indicate in biochemical or cellular terms the presence of contaminants or the magnitude of the host response.

OBJECTIVES OF THE SYMPOSIUM

It is the objective of this symposium volume to review the current state of science as it pertains to the scientific basis, current state of development, validation, and use of biological markers (biomarkers) in environmental research. The emphasis is on identifying and evaluating exposure of environmental species and effects on the health of environmental species and the integrity of their ecosystem. Biological markers are also being very actively developed and evaluated as indicators of exposure and health effects in humans, but this symposium will not focus on this research; interested readers are referred to the report of the National Research Council's Committee on Biological Markers (NRC 1987).

The various chapters of this volume describe different types of biomarkers that offer promise for environmental monitoring. The biomarkers are arranged in categories defined by the nature of the toxic endpoint being probed. Chapters that introduce each biomarker response include an introductory tutorial to provide the reader with a background review of the biomarker, including the scientific basis and rationale for the endpoint being used as a biomarker, and a brief history of its application to environmental or health problems. In most chapters, the authors illustrate the utility of their biomarker measurements by describing specific results of studies with organisms from contaminated environments.

Anatomical and cytological abnormalities are classic endpoints that have long been used as an indicator of deleterious exposure to pollutants in the environment. Dr. Hinton provides an overview of this area and Drs. Cormier and Yamashita describe additional studies which apply these endpoints in assessments of environmental pollution.

Several different types of adaptive biochemical and immunological responses have been useful as biomarkers. Drs. Spies and Stegeman and their collaborators introduce one of the best characterized biochemical responses—the induction of the cytochrome P-450 mixed function oxidase (MFO) system. Their paper describes an intensive characterization of responses of MFO parameters in marine fish, and the interactions between this contaminant detoxication system and other critical endpoints such as reproductive competence and genotoxicity. Drs. Jimenez and Fossi not only provide additional examples of the application of this biomarker, but also emphasize the need for understanding the influence of normal physiological functions on the activity of this enzyme system. Dr. Halbrook reviews and provides a field demonstration of an alternate and noninvasive technique for assessing enzyme activity in vivo by measuring the sleep time of animals dosed with a narcotic metabolized by the MFO system. Another adaptive response of animals to stress that has recently attracted attention as a biomarker is the induction of a class of proteins collectively characterized as stress proteins. This promising class of biomarkers is reviewed by Dr. Sanders, who also provides data on responses of animals exposed to pollutants. A vital adaptive response of organisms is the competence of the immune system. Immunotoxic endpoints are powerful integrators of the effects of chemical exposure, physiological status, and disease challenge. Dr. Weeks provides an introduction to this class of biomarkers and illustrates the use of macrophage responses as a biomarker in fish from a contaminated estuary.

Dr. Shugart provides an overview of another class of biomarkers that monitor the effects of environmental pollutants on DNA. Dr. Kurlec presents an additional illustration of application of genotoxic biomarkers in evaluating contamination, and makes the point that proper interpretation of these endpoints as biomarkers requires an understanding of the normal physiological processes affecting these responses. Dr. McMahon reviews and presents data utilizing sophisticated analysis of oncogenes in fish from polluted harbors as a measure of genotoxic effect.

The next section of the symposium dealt with endpoints that measure aberrations in metal metabolism as biomarkers indicating that metals are exerting toxic effects. Dr. Petering discusses the general background for this area of research and suggests the utility of measurement of metals bound to the metal-binding protein, metallothionein, as a biomarker. Drs. Benson, Garvey, and Veldhuizen-Tsoerkan take a slightly different tack in that they advocate quantifying increases in the levels of metallothionein itself as a biomarker of metal exposure.

Applications of biomarkers in field evaluations are illustrated in many of the chapters dealing with specific biomarkers; however, studies in the last section of the symposium focus on evaluation of environmental contamination rather than on the biomarker methodology. Dr. Lower introduces this section with a review of the use of animals and plants as sentinels for evaluating environmental pollution. Three field studies present different approaches to using biomarker-based biomonitoring in aquatic and marine systems. Dr. Adams summarizes some of the results of an ongoing evaluation of the effects of environmental stress on fish through the use of a suite of biomarker responses; this study is a component of an EPA-mandated permit-to-discharge industrial effluents under the National Pollution Discharge Elimination System. Dr. Long presents a summary of studies of San Francisco Bay by the Status and Trends Program of the National Oceanic and Atmospheric Administration, and compares responses of biomarkers with several endpoints such as toxicity tests and concentrations of chemicals in the environment. Finally, Dr. Lambertson presents results of his health evaluation of North Atlantic whales.

One of the limitations encountered in attempts to use biomarker responses for field evaluations is the lack of a conceptual paradigm that incorporates biomarkers in a formal hazard assessment or risk analysis. Dr. Suter reviews this critical area and presents an approach for using biomarkers in ecological assessments.

Throughout this volume, the point is made that biomarkers offer great potential for evaluating environmental contamination, but that there are a number of hurdles that must be overcome before that full potential is realized. In the conclusion to this volume, Dr. McCarthy proposes a research plan for focusing and coordinating resources necessary to develop and implement a biomarker-based environmental monitoring program.

BIOMARKER-BASED BIOMONITORING

Before addressing the specific biomarkers described in this volume, it may be instructive to consider how biomarkers could be used to evaluate exposure or effects of environmental contaminants. In an environmental context, biomarkers offer promise as sensitive indicators demonstrating that toxicants have entered organisms, been distributed within the tissues, and are eliciting a toxicological effect at critical targets. Due to these attributes, biomarkers measured in animals from sites of suspected contamination can be an important and informative component of an environmental monitoring program. In such a biomonitoring program, the biomarker responses of animals or plants from a suspect site would be compared to those of the same species collected from pristine "reference" sites. Ideally, reference sites are ecologically identical to the suspect sites, except with respect to the presence of contaminants. The reference organisms provide a baseline against which to compare perturbations caused by the chemical exposure and helps to account for species-specific differences in biochemistry or toxicology as well as effects of seasonal or other environmental factors on the endpoint being used as a biomarker. Based on the magnitude and pattern of the biomarker responses, the environmental species offer the potential of serving as

- *sentinels* demonstrating the presence of bioavailable contaminants and the extent of exposure,
- *surrogates* indicating potential human exposure and effects, and
- *predictors* of long-term effects on the health of populations or the integrity of the ecosystem.

There are two elements of this approach which should be discussed: (a) Why monitor animals in the environment? Why not rely on measurements of chemical residues in the environment, or on standardized toxicity bioassays as a basis for evaluating potential exposure to contaminants? (b) Why measure biological markers in the animals? Why not measure body burdens of priority pollutants or rely on community-level responses, such as decreased diversity or loss of sensitive populations, as the key endpoints for environmental protection?

Monitoring of Environmental Species

Why monitor organisms in the environment? Data from the biological system that is the target of toxicant action provide important information that is not readily available from chemical analyses of air, water, or soil or from bioassays.

Temporally and Spatially Integrated Measure of Bioavailable Pollutants— Chemical analyses are a necessary and informative component of any evaluation of environmental contamination because they unequivocally demonstrate the presence of particular toxicants. However, chemical analyses have significant limitations because they provide data that not only are difficult to relate to a biological effect but may not even accurately reflect the real status of chemical concentrations over time and space. Thus, chemical sampling is like a snapshot. Changes can result from storm events, shifts in winds, or intermittent releases from industrial plants. Furthermore, contamination is often geographically patchy; a quiet pool in a stream may accumulate highly contaminated silt, while the gravel bottom a few feet away may have only trace levels of contaminants. Animal responses integrate the relative concentration and the bioavailability of contaminants temporally and over the organism's spatial range, thus providing a measure that is more relevant to evaluating ecological or health risks.

Attributing Exposure and Risks to Environmental Pollutants—Environmental legislation often addresses problems due specifically to exposure to environmental sources of hazardous chemicals. Attributing human health effects to chemical exposure from the environment, such as from toxic waste sites, is problematic because humans are exposed to a wide variety of life style and workplace exposures. While wild animals share our exposure to chemicals in the environment, they are relatively protected from these extraneous exposures. Monitoring environmental species may offer a means of teasing apart those exposures (and health risks) due to environmental toxicants from those due to other sources of deleterious exposure.

Significance of Different Routes of Exposure as Indicated by Habitat—Comparison of responses of species with different habitats or at different trophic levels provides information on the significance of different routes of exposure. Benthic vs pelagic fish, foliage- vs soil-associated insectivores, and herbivores vs top carnivores provide contrasts in the extent of their exposure to bioavailable contaminants in water, soil, or food chain pathways. Information that exposure occurs primarily through the food chain, for example, or through contact with the sediment can help in prioritizing additional monitoring and suggest strategies for intervention or remediation.

Limitations of Toxicity Bioassays—Bioassays offer many advantages for comparing the relative toxicity of specific chemicals or of specific effluents. However, toxicity tests also have serious limitations for biological monitoring because most do not account for the effect of (1) chemical speciation in the environment, (2) kinetics and hysteresis in sorption of chemicals to sediment, (3) accumulation through foodchains, and (4) modes of toxic action which are not readily measured as short-term (7–21 day) effects on survival, growth or reproduction. Furthermore, methods for realistically exposing animals to some environmental media are limited and routes and extent of exposure often fail to mimic those in nature. Most bioassays used for environmental protection use aquatic organisms and aqueous phase contaminants. Sediment toxicity tests often rely on equilibration of chemicals in a sediment-water mixture, or employ benthic organisms whose survival, growth, or reproductive success may be dependent not only on the presence of toxicants, but also on physical characteristics of the sediment (sand, silt, organic content). Other than in plant systems and bird toxicity tests, there are few useful bioassays for evaluating the effects of environmental contaminants on terrestrial organisms.

Even sensitive and well-characterized bioassays are not sufficient to warn of danger to ecological integrity. This is illustrated by the results of a three year biomonitoring program in a stream affected by discharges from the U.S. Department of Energy's Y-12 Plant in Oak Ridge, Tennessee. The monitoring program, which is a component of the facility's National Pollution Discharge Elimination System permit to discharge, includes biomarker measurements of fish, population and community surveys of fish and benthic invertebrates, and toxicity bioassays of water discharged from the plant into the stream (*Ceriodaphnia* survival and reproduction and fathead minnow larval survival and growth[2]). The monthly bioassay results indicated that the effluents caused no observed effect in either bioassay even though no fish were present near the industrial discharge, and community surveys demonstrated clear downstream effects on diversity and species richness (Loar et al.[2] and unpublished data). While the toxicity tests have been very useful in testing and controlling release of effluents from specific process streams within the plant, they were not sensitive enough to warn of observed ecological effects of the discharges from the plant. In contrast, biomarker measurements showed significant differences along a gradient going downstream from the industrial discharge, and correlated well with community level effects.[1,2]

Advantages of Biomarker Measurements

Measurement of biomarker responses to exposure offers the potential of providing information that cannot be obtained from measurements of chemical concentrations in environmental media or in body burdens. For example, measurement of biomarker responses can provide evidence that organisms have been exposed to toxicants at levels that exceed normal detoxication and repair capacity and that have induced responses within molecular and cellular targets. This can provide crucial evidence for establishing a scientifically defensible link between toxicant exposure and ecologically relevant effects at a population or community level. As will be discussed throughout this volume, however, it is important to recognize that limitations in our current understanding of the molecular and biochemical mechanisms of toxic action often prevent unequivocal interpretation of biomarker responses, especially in relating them to a specific consequence, such as eventual development of cancer, increased susceptibility to disease, or decreased reproduction. While a coordinated research program is needed to address these limitations comprehensively (see concluding chapter of this book), biomarker responses have a number of advantages even in their present state of development.

Exposure to Rapidly Metabolized Contaminants—Biomarkers provide evidence of exposure to compounds that do not bioaccumulate or are rapidly metabolized and eliminated, such as organophosphates or polynuclear aromatic hydrocarbons. The biomarkers provide information that integrates the toxic action of the parent compound as well as any toxic metabolites.

Integrate Pharmacodynamic and Toxicological Interactions—Biomarkers integrate the toxicological and pharmacokinetic interactions resulting from exposure to complex mixtures of contaminants, and they present a biologically relevant measure of toxicant interactions in target tissues. Direct and indirect interactions of multiple contaminants upon the uptake and internal distribution of chemicals, as well as synergism or antagonism of the toxicants' action, are integrated within the organism. The biomarkers express the cumulative effect of toxicant interactions in molecular or cellular targets.

Exposure vs Ecological Effects—The results of population monitoring are the ultimate indicators of ecological effects. However, population responses such as occurrence, abundance, and reproduction do not provide an indication of their cause. Furthermore, population monitoring tends to be a rather insensitive indicator of effects because of the variability of animal populations and the imprecision of field monitoring techniques.

Therefore, it is useful to monitor biomarkers to provide a more sensitive and

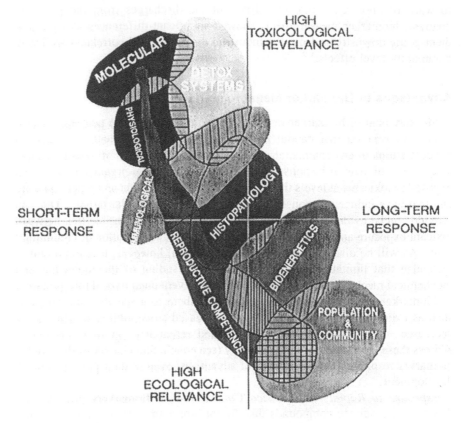

Figure 1. The relationship between responses at levels of biological organization and the relevance and time scales of responses. (From Adams, S. M., K. L. Shepard, M. S. Greeley, Jr., M. G. Ryon, B. D. Jimenez, L. R. Shugart, J. F. McCarthy, and D. E. Hinton, "The Use of Bioindicators for Assessing the Effects of Pollutant Stress in Fish," *Marine Environmental Research*, 28:459–464 (1989).)

precise indicator of the nature and magnitude of effects, and to gain insights into what caused the effect. The correlations between the biomarkers of exposure and the indicators of ecological effects are expected to be better than those between the population monitoring parameters and the indicators of exposure.

Conceptually, a suite of biomarkers and higher level response indicators can provide evidence to statistically test hypotheses about the linkage between exposure to toxic chemicals and ecologically relevant effects. The rationale for this approach is indicated in Figure 1, which illustrates the relationship between responses of different levels of biological organization and the relevance and time scales of the responses. Responses at each level provide information that helps us to understand and interpret the relationship between exposure and adverse effects. Responses measured at the lower levels of biological organization, such as DNA damage or enzyme activity, often provide sensitive and specific responses to particular toxicants (Figure 1). These biomarkers offer advantages as *measures of exposure*, and may be diagnostic of the type of contaminant to which the organism is exposed, but the biological significance to the overall structure and function of a population or ecosystem is unclear. Conversely, responses at higher levels of biological organization, such as changes in population abundance or species diversity, are directly relevant to concerns about *ecological effects*, but cannot, by themselves, prove whether differences among sites are due to pollutants or to natural ecological factors. A comprehensive and integrated monitoring program needs to consider responses at several levels of organization, including biomarkers measured at the organismic or suborganismic level as well as population and community level indicators. The goal in examining responses at these different levels of organization is to answer two critical questions:

1. Are organisms exposed to levels of toxicants that exceed the capacity of normal detoxication and repair systems?
2. If there is evidence of exposure, then is the chemical stress impacting the integrity of the populations or communities?

Evidence of exposure from body burdens or from responses of lower levels of biological organization provide an answer to the first question. In particular, the biomarkers of exposure indicate the biological significance of chemicals which may have entered the animal; i.e., did the chemical reach molecular and biochemical targets and cause detectable damage or induce a protective response? The second question can be addressed by determining whether the responses to the toxicants are propagated up through successively higher levels of biological organization (biomarkers of effects and population parameters). If chemical exposure is responsible for a high level ecological effect, responses should be apparent at intermediate levels of organization. Alternately, if data do not indicate any evidence of exposure, or if biomarker responses indicate only minor effects in the most sensitive and responsive exposure parameters (e.g., genetic damage), but not at any higher levels of biological organization (e.g., histopathological evidence of neoplasia or tumors, or evidence of genetic abnormalities in gametes), it may not be reasonable to attribute any community and population level effects to chemical agents.

Short-Term Predictors of Long-Term Ecological Effects—The ultimate goal of an environmental monitoring program is to prevent deterioration of the environment and to document recovery of affected systems. Biomarkers at the molecular and biochemical level respond quickly to changes in contaminant exposure, while a long latent period may be required before any change is apparent at a population or community level. The rapidly responding biochemical level markers can serve as valuable short-term indicators of the direction of change in the toxic exposure and potential effects at monitoring sites. For example, in a stream in the Oak Ridge area, levels of genotoxic damage have been decreasing significantly in response to improvements in waste treatment at an industrial facility discharging into the stream.[4] Information on the short-term responses to toxic exposure could aid in management decisions regarding the effectiveness of any remedial activities, and the rates of recovery of affected sites. However, it must be recognized that compensatory mechanisms in population responses and the complexity of trophic interactions within ecological communities may limit our capacity to predict ecosystem level effects based on biomarker responses of individual organisms. Nevertheless, on a qualitative basis, it is reasonable to think that changes in biomarker responses in directions indicating decreased exposure and effects presage an improvement in the ecological health of the environment.

APPLICATIONS OF BIOMARKER-BASED MONITORING OF ENVIRONMENTAL CONTAMINATION

Biomarker responses of environmental species can be a valuable component of biomonitoring programs having any of several objectives, including the following:

Surveillance—For example, for routine monitoring of a range of sensitive and important ecosystems, biomarkers could provide a quantitative basis for establishing the current status and long-term trends in the levels of exposure or adverse effects in critical species. An example of the incorporation of biomarker measurements in the Status and Trends Program of the National Oceanic and Atmospheric Administration is described in this volume. Similarly, biomarkers may prove to be an informative component of an Environmental Monitoring and Assessment Program being formulated by the Environmental Protection Agency and a consortium of federal agencies.

Hazard Assessment—Biomarker measurements in organisms living on or near hazardous waste sites or industrial facilities can help provide data needed to assess the potential exposure and effects from contaminants at the sites. Different types of biomarkers may be more appropriate at different tiers in the hazard assessment process (Cairns et al., 1978). For example, for general screening at initial tiers of the assessment process, nonspecific biomarkers that respond to a broad array of insults may be appropriate (e.g., measures of damage to DNA integrity). More specific biomarkers (e.g., identification and quantitation of specific DNA or protein adducts) could provide additional levels of information if more tiers of testing are required to establish whether the site presents an ecological hazard.

Regulatory Compliance—Environmental regulations are often based on meeting criteria levels of chemical concentrations in discharges or on demonstrating no observed effects in standardized bioassays used to test effluents being discharged. The standards for chemical concentrations usually neglect the potential for toxicological interactions of chemical mixtures and the sensitivity of the results can be influenced by sampling design (e.g., composite vs grab samples).

Bioassays, as discussed earlier, are not always sufficient to protect the ecological integrity of an environment. Biomarker measurements are proving to be a useful and informative component of the National Pollution Discharge Elimination System permit for government facilities in Oak Ridge, Tennessee.[1,2]

Remediation—Biomarkers can be sensitive and cost-effective measures demonstrating the effectiveness of remedial actions. Over the last three years, the government facilities in Oak Ridge have instituted major improvements in waste treatment and waste minimization. Biomarker measurements have demonstrated a clear improvement in water quality.[1,4] Extensive studies of community structure and function demonstrated a similar trend, but were much more labor-intensive and far more costly than the biomarker measurements.

In any of these applications, biomarker-based biomonitoring needs to include not only sites of known or suspected contamination but also ecologically comparable reference sites with no known sources of pollutant input. The biomarker responses of species at the pristine sites provide a reference level for comparing responses of species from suspect sites. Statistically significant differences *between sites* (qualitative differences in patterns of response for a suite of biomarkers, as well as quantitative differences in the magnitude of responses) demonstrate differences in the extent or type of pollutant exposure. Differences in biomarker response *over time* at the same site provide information on environmental trends and could prove to be a sensitive and rapid responding tool for assessing whether intervention is required or remediation has been successful. Because many of the biomarkers are short-term indicators of long-term adverse effects, these data may permit intervention before irreversible adverse effects become inevitable.

In addition to monitoring of wild species, specific circumstances may justify the introduction of caged animals on a site to confirm and extend results observed in the trapped animals. Animals collected from pristine environments, or obtained from animal suppliers, can be confined in large cages or fenced areas, and analyzed for biomarkers after weeks or months of exposure. Responses in these introduced individuals would: (1) confirm the relationship between specific sites and the pattern of biomarkers observed in wild animals; (2) provide a finer level of geographic resolution around portions of the site; and/or (3) test specific hypotheses about the contribution of different routes of exposure. Data from confined animals could be required either because not enough animals were trapped in some regions, because the statistical analyses indicated a need for more resolution to improve the analysis of patterns at the sites, or because indigenous animal populations lacked a suitable range of feeding habits or potential exposure pathways.

SUMMARY

This volume of symposium proceedings is intended to make the interested nonexpert aware of the current state of science in biomarker research, including an awareness of the potential value and existing limitations of this approach. Each of the biomarkers discussed in this book is a technical specialty unto itself; consequently, the broad development, application, and validation of biomarker-based environmental monitoring will require coordination and integration of teams of researchers whose expertise encompasses a range of biomarker responses. Hopefully, symposia such as this will make agencies that are responsible for environmental protection in the United States and abroad aware of the need for coordinating and focusing research activities in this area. Without such interagency and international cooperation, valuable resources will be wasted in redundant and incomplete studies that provide only limited and anecdotal validation of the potential of biomarkers for assessing the ecological consequences of environmental contamination.

ACKNOWLEDGMENTS

This work was supported by the Exploratory Studies Program, Oak Ridge National Laboratory. The Oak Ridge National Laboratory is operated by Marietta Energy Systems, Inc., under contract DE-AC05-84OR21400 with the U.S. Department of Energy. Publication No. 3518, Environmental Sciences Division, Oak Ridge National Laboratory.

REFERENCES

1. Adams, S. M., K. L. Shepard, M. S. Greeley, Jr., M. G. Ryon, B. D. Jimenez, L. R. Shugart, J. F. McCarthy, and D. E. Hinton. "The Use of Bioindicators for Assessing the Effects of Pollutant Stress in Fish," *Mar. Environ. Res.* 28:459–464 (1989).
2. Loar, J. M., S. M. Adams, H. L. Boston, B. D. Jimenez, J. F. McCarthy, J. G. Smith, G. R. Southworth, and A. J. Stewart. First Annual Report on the Y-12 Plant Biological Monitoring and Abatement Program, ORNL/TM, Oak Ridge National Laboratory, Oak Ridge, TN (1988).
3. National Research Council. "Committee on Biological Markers," *Environ. Health Perspect.* 74:3–9 (1987).
4. Shugart, L. R. "DNA Damage as an Indicator of Pollutant-Induced Genotoxicity," *Thirteenth Symposium on Aquatic Toxicity and Hazard Assessment,* W.G. Landis and W.H. van der Shalie, Eds., American Society for Testing and Materials, Philadelphia, PA, in press (1990).
5. Cairns, J., Jr., K. L. Dickson, and A. W. Maki, Eds. *Estimating the Hazard of Chemical Substances to Aquatic Life,* STP 657. American Society for Testing and Materials, Philadelphia (1978).

SECTION TWO

Anatomical and Cytological Endpoints

CHAPTER 2

Liver Structural Alterations Accompanying Chronic Toxicity in Fishes: Potential Biomarkers of Exposure

David E. Hinton and Darrel J. Laurén

ABSTRACT

Fish liver microscopic structure is an integrator of physiological and biochemical function which, when altered, may produce biomarkers of prior exposure to toxicants. The liver has a key role in xenobiotic metabolism and excretion, digestion and storage, and production of yolk protein. Thus, alterations in structure are expected under certain toxic conditions. A series of lesions associated with chronic liver toxicity are described, illustrated, and evaluated for their potential usefulness as biomarkers of toxicant exposure. Where possible, the potentially confounding morphological changes associated with infectious disease, season, sex, and nutritional state are differentiated from toxicant-induced lesions. Hepatocyte coagulative necrosis, hepatocyte regenerative foci, spongiosis hepatis, and neoplasia appear to be the most promising liver structural biomarkers of prior exposure to toxicants.

INTRODUCTION

Worldwide, fish are an important source of human nutrition (high in protein and low in saturated lipids) and feral fishes are the major component. Larval forms of many feral species spend their infancy in estuaries and coastal zones.

Associated with urban population densities and industrial complexes, sediments of many estuaries are contaminated. Thus, pollution may threaten human nutrition and health directly, through loss of the food source (viz., through acute toxicity to early life stages of fishes; or, through chronic toxicity by impairment of reproductive potential or by cancer and premature death of older, most productive fish), or indirectly, by human contaminant loading through consumption of tainted fish flesh. Biomarkers of contaminant exposure in indicator fish species are important if we are to maintain a viable fishery and a safe product for human consumption.

Bioindicators and biomarkers are two recently applied terms in environmental assessment of toxicity. With the former, the emphasis is generally at the species level of organization. The sensitivity with which a species responds to a range of contaminants determines its usefulness as a bioindicator. Within individuals of a bioindicator species, careful analysis may detect alterations (biomarkers) in function or structure of specific organs, tissues, and cells brought about by prior exposure to a contaminant. Both terms imply sensitivity (i.e., a response must be shown) yet, if the sensitivity is too great, the species will likely not be present at an impacted site or, in terms of the biomarker, acute lethality would have negated analysis of chronic response. Therefore, for the purposes of this report, we define *biomarker* as any contaminant-induced physiological and/or biochemical change in a not-too-sensitive organism which leads to the formation of altered structure (a lesion) in the cells, tissues, or organs of that individual. Given the relatively low level but chronic nature of most environmental pollution, biomarkers of chronic toxicity should receive most attention.

It is important to realize that physiological and biochemical alterations, if severe enough or protracted, will lead to structural alterations. By selecting the appropriate target organ (in this paper, liver), the investigator may use a variety of morphologic manifestations of chronic toxicity as biomarkers of exposure. Detection of specific alterations (biomarker lesions) signal prior and/or ongoing exposure of the individual to hepatotoxic agent(s). We do not imply that morphologic "fingerprints" exist for each toxicant to which a fish may be exposed. Rather, we emphasize changes in liver morphology which our laboratory experiences tell us are relevant to chemical toxicity. In some instances, field verification of relevancy exists for specific liver lesions in teleost fishes.

Biomarker responses may involve all levels of biological organization (Table 1). For example, changes may be seen in: distribution of molecules (i.e., glycoproteins) on the cell surface; the organelle number, volume, shape or distribution; cell volume, morphology, distribution or number; and organ volume or relative weight. Therefore, morphologic expression of prior exposure is an integrator of biochemical and physiological change.

There are several important reasons for selecting the liver for detection of biomarkers of exposure. First, the liver of teleosts is the major site of the cytochrome P-450, mediated mixed function oxygenase system.[88] This system inactivates some xenobiotics, while activating others to their toxic forms. Secondly, nutrients derived from gut absorption are stored in the hepatocytes and

Table 1. Selected Examples of Biostructural Stratification of
          Liver Toxic Responses.

Level of Organization	Response
Organ	Hepatomegaly
Tissue	
Parenchyma	Proliferation of biliary ductules
Stroma	Stroma
	Macrophage aggregates
Cell	
Hepatocyte	Necrosis
Biliary epithelial	Hyperplasia
Stellate, Ito	Enhanced production of glycoconjugates
Endothelial	Lysis
Macrophage	Enhanced phagocytosis
Organelle	
Mitochondrion	Swollen
Golgi apparatus	Flattened
Vesicle	Enhanced secretion
Endoplasmic reticulum	Proliferation
Lysosome	Increased numbers
Peroxisome	Increased numbers
Glycogen	Depletion
Nucleus	Enlarged, pleomorphic

released for further catabolism by other tissues.[61,97] Third, bile synthesized within hepatocytes[9,75] and released into the proximal intestine aids digestion of fatty acids and also carries conjugated metabolites of toxicants[23] to the gut lumen for excretion or enterohepatic recirculation. Fourth, a yolk protein, vitellogenin, destined for incorporation into the ovum, is synthesized entirely within the liver.[93] Receptors in the liver must bind the hormone, estradiol, for initiation of the signal necessary to begin synthesis of this essential reproductive component. In other conditions, toxic injury to liver may impair synthesis of other proteins with a result being edematous fluid collection within the pleuroperitoneal cavity.[99] Some of the important functions of teleost liver which favor its use as a biomarker locus are given in Table 2.

When exposed to certain toxicants, the liver responds with a variety of morphologic alterations. Certain of the responses of the teleost liver have been

Table 2. Functions of Teleost Liver Favoring Its Use As a Biomarker Locus.

Function	Role
1. Nutrition	Uptake, storage and release of fats and carbohydrates
2. Metabolic activation	Conversion of procarcinogen to ultimate carcinogen
3. Detoxication	Glucuronide formation
4. Endocrine/ reproductive	Synthesis of vitellogenin
5. Digestive	Zymogen from exocrine pancreas of hepatopancreas

validated by exposure of fish to toxicants under controlled laboratory conditions. Good correlation exists for the qualitative similarity and sequential progression of laboratory-induced neoplastic alterations[37] and those seen in some field studies.[63]

Characteristics of Biostructural Alterations Favoring Use as Biomarkers

Before considering specific alterations, general characteristics of the most promising examples need to be discussed. Sensitivity is important since the liver must respond to the contaminant in order for an alteration to be present. The nature of the direct effects as well as host response(s) are important. A direct effect such as acute lethal toxicity may only involve specific cells in the target tissue. However, indirect effects, the responses of the host's cells to the primary insult, are important and may be the more useful biomarker. In this case, a toxicant produces hepatocellular necrosis. The organism survives the cytotoxicity, and the initial loss of hepatocytes stimulates repopulation by surviving cells. The mammalian liver has tremendous regenerative capacity.[50,77,81] Similarly, livers of *Cyprinodon variegatus* and *Oryzias latipes* were shown to repopulate following extensive loss of hepatocytes due to diethylnitrosamine (DENA) toxicity,[38] making this sequel of responses very useful in environmental surveillance. Thus, the presence of islands of small, basophilic hepatocytes, contrasting with the normally pale, glycogen-laden hepatocytes found under control conditions, is a good biomarker of exposure. We have observed karyocytomegaly of unknown etiology in livers of fingerling striped bass (*Morone saxatilis*) taken from aquaculture ponds in California.[26] Moribund fish exhibited pan hepatic megalocytosis (see below) with acute and chronic inflammation. Survivors brought to our laboratory have been followed sequentially over a period of seven months. Although a general return toward normal morphology has been seen, a few megalocytes remain and from time to time extensive regeneration of parenchyma has been observed. Latest samplings indicate normal, glycogen-laden hepatocytes fill most of the parenchyma, but occasional fish have shown extensive areas now occupied by exocrine pancreatic acini. The latter possibly represents a metaplasia of hepatocytes (Figure 1), since both arise from the same stem populations.[1,19] In the above instance, initial intoxicated hepatocytes would be the acute phase biomarker, yet regeneration, patchy karyocytomegaly, and putative pancreatic metaplasia would represent the chronic biomarker.

Another factor for consideration is the ease of recognition of various biomarker alterations. For example, size of lesion must be sufficient to lend itself to resolution by light microscopy. Identification by histological stains is also important. In addition, the theory of what the change means is important. For example, the presence of abundant fat droplets within cells of the liver may signify an alteration in lipid metabolism.[53] Finally, an absence of confusion with other etiological agents is important. For example, coagulative necrosis may well be due to toxic injury by one or more contaminants. However, liquefactive or cytolytic necrosis is much more often associated with microorganisms including bacterial infections.[72]

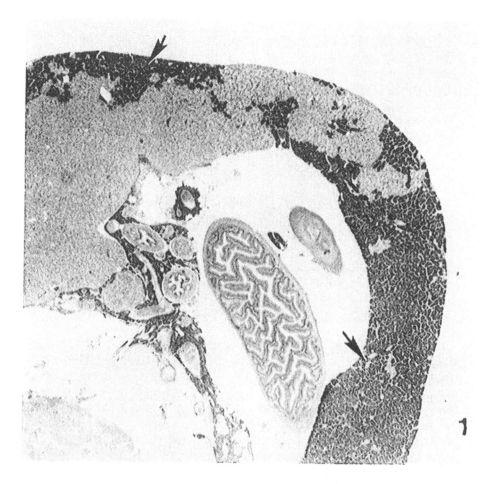

Figure 1. Section of liver from striped bass (*Morone saxatilis*). This fish was removed from
fingerling culture pond in which numerous fish were dying due to liver toxicity.
After 7 months in culture at our aquatics center, the liver has repaired and the
hepatic parenchyma appears normal except for the excessive amount of exocrine
pancreatic tissue (dark areas; arrows) present. Hematoxylin and eosin stain.
Magnification X35.

Review of Relevant Microscopic Anatomy of Teleost Liver

We have described the architectural arrangement, histochemical localization
of key enzymes, high resolution light microscopic and ultrastructural anatomy
of hepatocytes and other cell types, scanning electron microscopy of the sinusoidal wall, and the intrahepatic biliary system of the trout liver.[27,28,37,56] Using
this species as an example, we know that the hepatocytes occupy roughly 80%
of the total liver volume and are the most numerous cell type, being roughly
ten times more prevalent than any other cell type within the liver.[29] There-

fore, it is not surprising to find a multitude of potential biomarkers within hepatocytes.

The arrangement of hepatocytes is tubular.[28,29,82] In this pattern, 5 to 7 hepatocytes have their bases directed toward the sinusoids and their apices directed toward a biliary structure (either a canaliculus—completely bounded by hepatocytes, or a bile preductule which resembles a canaliculus, but is bounded by hepatocytes and one or more bile preductular epithelial cells) (Figure 2). Bile produced in hepatocytes passes through a hierarchy of biliary passageways, many of which are in the parenchyma of the liver, and finally exits the liver by a hepatic duct. Trout liver has numerous biliary epithelial cells and many of these share junctional complexes with hepatocytes, a feature unique to tubular livers.

Although in vivo microscopic observations[37,41] suggest a repeating pattern of lobule-like structures in trout liver, it is difficult if not impossible to recognize portal (afferent) and hepatic (efferent) regions of the lobule. Whereas the mammalian liver has structural associations of hepatic arteriole, lymphatic, bile ductule and portal venule, facilitating recognition of afferent vs efferent portions of the lobule, the divergence of the biliary and portal venous distributions in trout make this differentiation problematic (Figure 3). Therefore we cannot classify fish hepatotoxicants by the zones of the liver which they affect.

In addition to hepatocytes, other cell types play important roles and are involved in toxicity. These include the fenestrated endothelial cells lining sinusoids (Figures 3 and 4) and through which nutrients and other materials must pass to reach the hepatocytes. Between endothelial cells and hepatocytes is the perisinusoidal space of Disse containing the stellate, fat-storing cell of Ito (Figure 4). The ultrastructure of this cell[22,56] is identical to that of the mammalian Ito cell which contains vitamin A in fat droplets,[96] and sends long cytoplasmic processes which course between individual hepatocytes. With special stains these can be seen to completely encircle the sinusoidal endothelium. Under appropriate conditions, these cells become fibroblasts and produce connective tissue in fibrotic lesions of the liver.[105] The basal plasma membrane of hepatocytes borders the space of Disse, while the apical membrane contacts the biliary epithelial cells of bile preductules and ductules. The latter are composed of cuboidal epithelial cells (Figure 2) which eventually drain into larger ducts lined by tall-columnar epithelial cells.[28]

We will show that variations in normal anatomy of the various cells accompany hepatic toxicity.

Chronic Responses of Hepatocytes to Toxicants

Hepatocellular Necrosis

First, differentiation between cytolytic and coagulative necrosis must be made. With the former, disintegration of hepatocytes is rapid and focal cytolytic necrosis

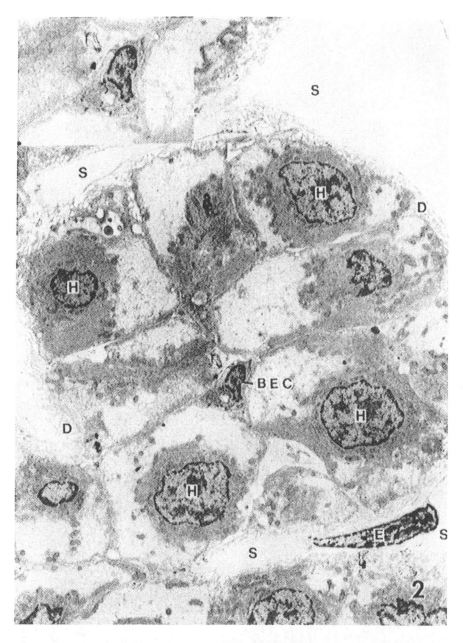

Figure 2.   Electron micrograph showing cross section of hepatic tubule from rainbow trout (*Oncorhynchus mykiss*). Hepatocytes (H) have their bases directed toward sinusoids (S) and their apices toward the biliary passageway which in this instance is surrounded by hepatocytes and a single biliary epithelial cell (BEC; and inset). D, space of Disse; E, endothelial cell. Modified from Hampton et al.,[29] and used with permission. Uranyl acetate and lead citrate. Magnification X4000.

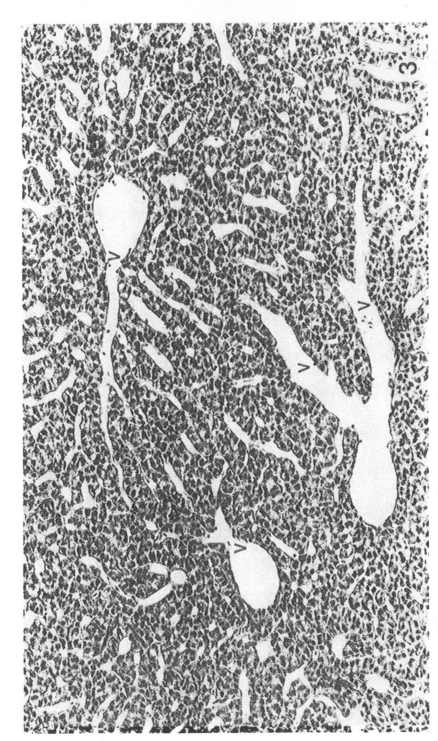

Figure 3. Low magnification view of liver section from rainbow trout (*Oncorhynchus mykiss*) showing four adjacent venular profiles (V). Note absence of bile duct or ductule at perimeter of venules. Hematoxylin and eosin stain. Magnification X175.

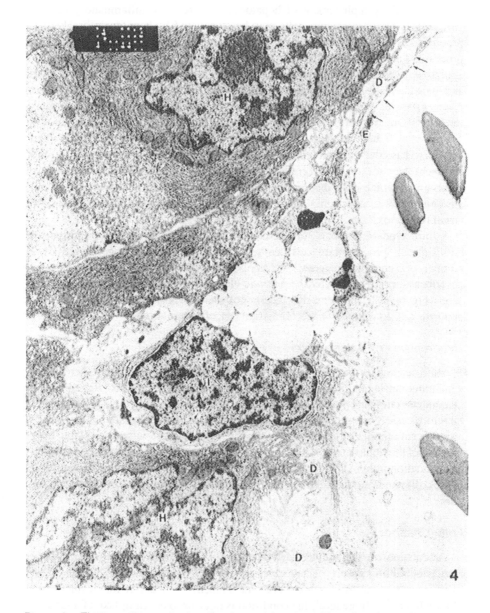

Figure 4. Electron micrograph of section from rainbow trout (*Oncorhynchus mykiss*) liver. Fat storing cell of Ito (I) is shown in space of Disse (D) between hepatocytes (H) and endothelial cell (E) lining sinusoid (S). Arrows point to fenestrae in sinusoidal wall. Picture courtesy of Dr. J. A. Hampton. Uranyl acetate and lead citrate. Magnification X7100.

is recognized by replacement of hepatocytes by reactive inflammatory cells.[70] Chemotaxis of inflammatory cells may be stimulated by hepatocyte breakdown products. Cytolytic or liquefactive necrosis is associated with infectious disease processes and lacks meaning as a direct biomarker of toxicant exposure. Coagulative necrosis is associated with sudden cessation of blood flow to an organ and with damage by toxic agents and represents the meaningful biomarker of prior exposure. With coagulative necrosis, shapes of cells and their tissue arrangement are maintained, facilitating recognition of the organ and tissue. As Trump et al.[92] have stated, it is difficult if not impossible to determine when cell death has occurred. Necrotic changes (Figure 5) occur following death, and represent the sum total of degradative processes. These alterations are useful in determining which cells died and underwent necrosis prior to somatic death and are biomarkers of exposure. The organ in which these are found determines the target of the toxicant.

A third type of necrosis, apoptosis, is now known to be very important in both normal, programmed cell death,[104] and in the cytotoxicity caused by a number of xenobiotic agents.[12.38] Apoptosis, necrosis without inflammation, describes a process whereby the necrotic elements are phagocytized by adjacent cells (Figure 6). Apoptosis has been proposed to play a complementary but opposite role to mitosis in the regulation of cell populations.

Hyperplasia of Regeneration

After necrosis, surviving hepatocytes may undergo hyperplasia, thereby regenerating sufficient hepatocytes to replace those that were lost. Regenerating hepatocytes are basophilic and occur as small islands of irregular shape and therefore easily recognized (Figure 7). This lesion, which signifies prior loss of a significant number of cells, may be quantified using colchicine and establishing the ratio of metaphase to prophase nuclei, or by using tritiated thymidine followed by autoradiography,[108.109] or by immunohistochemistry using monoclonal antibodies directed against the synthetic thymidine analogue, bromodeoxyuridine.[17.58]

Fatty Change

Vacuolation of hepatocytes (fatty change; Figure 8) is a common response associated with exposure of fish to a variety of different agents.[57] Studies in rat liver indicate a multitude of possible mechanisms to account for development of fatty liver.[53.74] In general, the condition is not a problem of uptake of abundant lipid precursors, but a problem of removing the fat from hepatocytes. This could signify one or more of the following biochemical lesions: (1) Inhibition of protein synthesis—the apoprotein is not made in sufficient amount to bind with the lipid so that it can be transported from the cell. (2) Energy depletion—lipid being transported from the cell normally moves through the endoplasmic reticulum (ER) and fuses with the Golgi apparatus. This fusion of ER and Golgi is thought

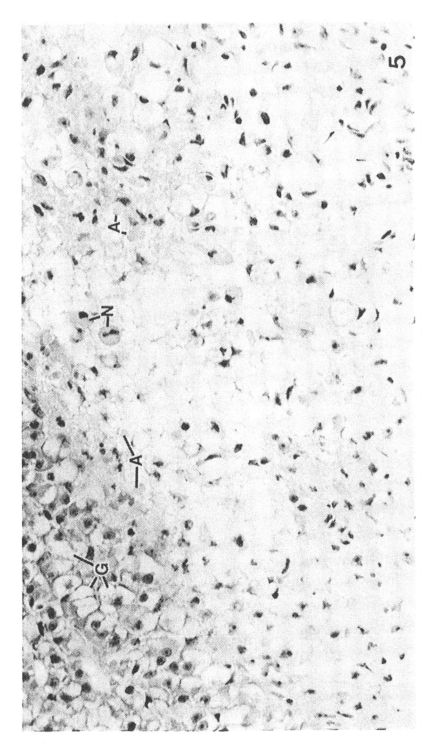

Figure 5. Cytotoxicity in liver of Japanese medaka (*Oryzias latipes*) three weeks after five week bath exposure to 50 ppm DENA. Normal appearing hepatocytes occupy upper left corner of field. Clear areas in hepatocytes are glycogen containing regions (G). A, apoptotic vesicles; N, necrotic hepatocytes with pyknotic nuclei. Necrosis of hepatocytes results in spongiotic areas at center of field. Hematoxylin and eosin stain. Magnification X700.

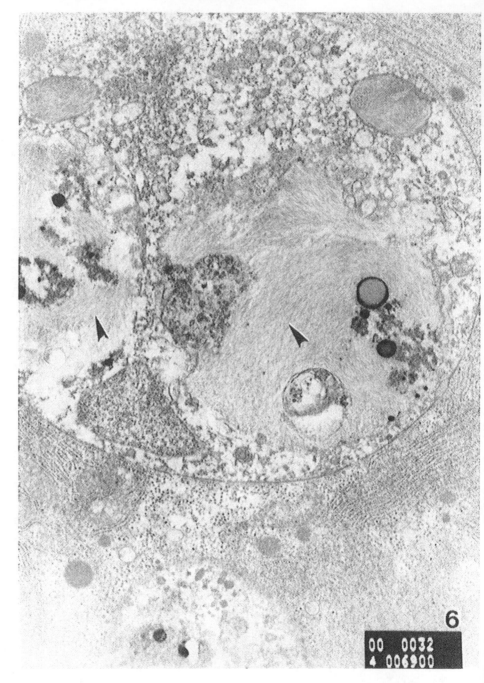

Figure 6. Electron micrograph of liver from Japanese medaka (*Oryzias latipes*). Hepatocyte contains large apoptotic vesicle showing partial lysosomal degradation as evidenced by fibrillar areas (arrows). Uranyl acetate and lead citrate. Magnification X13,800.

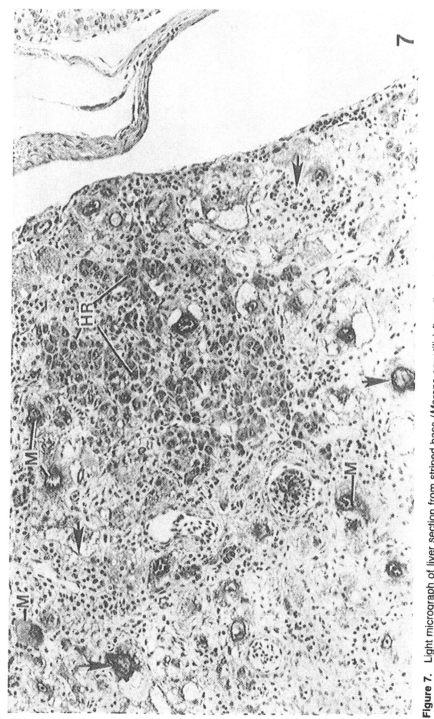

Figure 7. Light micrograph of liver section from striped bass (*Morone saxatilis*) fingerling showing megalocytosis, inflammation (horizontal arrows) and basophilic islands of regenerating hepatocytes (HR). M, megalocytic hepatocytes. Vertical arrows point to nuclear pseudoinclusions. Hematoxylin and eosin stain. Magnification X175.

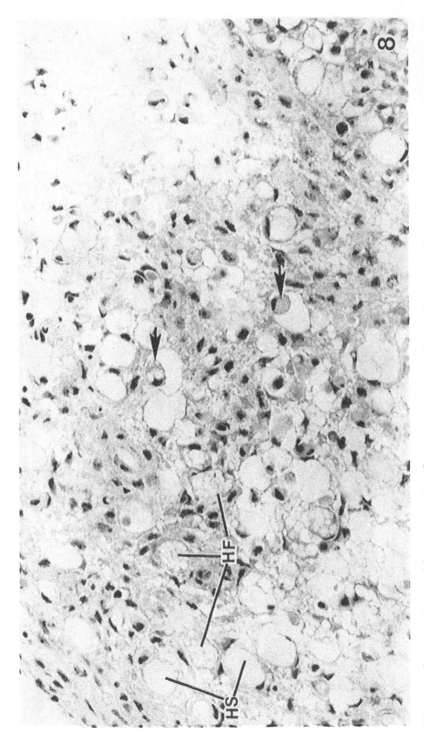

Figure 8. Cytotoxicity in liver of Japanese medaka (*Oryzias latipes*) three weeks after five week bath exposure to 50 ppm DENA. Shrunken, necrotic hepatocytes (arrows) are in round spaces. Other hepatocytes (H-F) show fatty change and others reveal high amplitude swelling (H-S). Hematoxylin and eosin stain. Magnification X700.

to require energy and if the energy levels in the cell are deficient this may cause the build-up of lipid within ER cisternae. (3) Disaggregation of microtubules— once secretory vesicles containing lipoprotein substances have been formed, they must find their way from the Golgi apparatus to the plasma membrane. Microtubules guide movement of vesicles in cells. Thus, disaggregation of microtubules is another mechanism whereby fatty liver can arise. (4) Shifts in substrate utilization—inhibition of metabolic pathways such as the β-oxidation of fatty acids may also lead to the accumulation of lipid stores. Since multiple mechanisms account for fatty liver in the mammal, it is reasonable to assume that fatty change in fish liver may also be useful as a biomarker of exposure. Mechanistic studies of fatty change in teleost liver are needed because fatty change of hepatocytes remains a common toxic response.

High Amplitude Swelling

This form of injury occurs either directly by denaturation of volume regulating ATPases, or indirectly by disruption of the cellular energy transfer processes required for ionic regulation.[86] Extracellular sodium leaks into such injured cells bringing water along its osmotic gradient. Resultant cell swelling (Figure 8) may be transient, or protracted, as has been shown in Winter Flounder in Boston Harbor.[62] These changes are particularly apparent in electron micrographs of swollen hepatocytes (Figure 9). In severely damaged cells, calcium influx and resultant activation of endonuclease is thought to be the final pathway leading to cell death.[55]

Nuclear and Cytoplasmic Inclusions

These lesions are common biomarkers of metal exposure in mammals but have not been commonly reported in fish. Metal containing cytoplasmic inclusions are usually found in lysosomes and may result from metallothionein degradation.[14] Baatrup et al.[3] used autometallography with and without added selenium to detect two different pools of mercury in trout. Both cytoplasmic (lysosomal) and nuclear deposits were found. The mercury found in the nucleus was revealed only after selenium amplification and was suggested to be organic rather than ionic mercury. Leland[51] reported an increase in the number of lysosomes and the presence of electron dense deposits in hepatocytes from trout exposed to both copper and zinc. Sorensen[84] found similar cytoplasmic and nuclear deposits in sunfish (*Lepomis cyanellus*) exposed to high concentrations of arsenic.

Hepatocyte Hypertrophy

This should be distinguished from simple swelling due to ion and water flux. Hypertrophy of cells is due to organelle hyperplasia[1] and is a net gain in the dry mass of a cell, tissue, or organ. An example of hypertrophy of hepatocytes

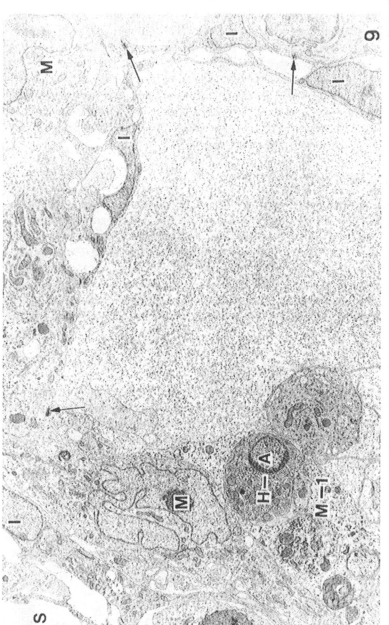

Figure 9. Transmission electron micrograph of liver section from Japanese medaka (*Oryzias latipes*) exposed to aqueous solution of 50 ppm DENA for 6 days. Sinusoid (S) is at upper left and is bordered by endothelial cell. I–Stellate, fat-storing cells of Ito. Arrows point to junctional complexes between adjacent Ito cells. Large spongiotic space delimited by Ito cells is former site of hepatocytes. Apoptotic vesicle within macrophage (M-1) contains hepatocyte (H-A) which shows typical organelles and peripheral heterochromatin clumping in nucleus. Entire cell fragment has been phagocytosed. Uranyl acetate and lead citrate. Magnification ×3,800.

occurs with the exposure to compounds which induce proliferation of endoplasmic reticulum membranes. Exposure to the polycyclic aromatic hydrocarbon (PAH), 3-methylcholanthrene,[76] caused apparent proliferation of endoplasmic reticulum (ER) and cellular hypertrophy. Similarly, proliferation of channel catfish (*Ictalurus punctatus*) hepatocyte ER[52] was found 21 days after exposure to Aroclor 1254.[46] Although the proliferation appears to involve primarily rough ER, chronic exposure caused whorls of smooth ER as well.[46] The resultant changes in endoplasmic reticulum appear to affect numerous hepatocytes and constitute a diffuse change which may be reflected in changes in liver to somatic index. In addition, newer immunohisto- and cytochemical approaches have localized PAH-inducible isozyme of cytochrome P-450 to cell types in livers of trout and scup.[60] The presence of specific isozymes in liver and especially in populations of liver cells will likely serve as sensitive biomarkers of exposure.

Hepatocytomegaly

In addition to hypertrophy, focal and multifocal—but rarely diffuse—hepatocytomegaly was seen in fish from sites in the Kanawha River of West Virginia,[40] as an early change in English sole of Puget Sound, Washington,[63] in sea pen cultures of Atlantic Salmon in Puget Sound,[44] in livers of pond-cultured fingerling striped bass,[26] and have been produced in the laboratory by exposure of trout to pyrrolizidine (senecio) alkaloids,[31] and medaka to DENA.[38] These cells are likely sublethally injured and are able to survive for months.[26,39] Often the enlarged hepatocytes contain enlarged nuclei (Figures 10 and 11), and occasionally, multinucleated forms are seen (Figures 10, 11 and 12).

Neoplastic Hepatocyte Change

In addition to the above, there are a group of changes which comprise the spectrum of lesions associated with the progressive development of neoplasia.[83] Here we are specifically referring to the appearance of populations of hepatocytes. Enzyme-altered foci (EAF) are seen first (Figures 13 and 14). With histochemical procedures to localize selected enzymes, altered phenotypes of "carcinogen-initiated" cells are demonstrated. First described in rodent liver, EAF have been shown to increase in a "dose-dependent" fashion with application of compounds which promote liver tumors.[69] With medaka and the sheepshead minnow (*Cyprinodon variegatus*), Hinton et al.[38] found γ-glutamyltranspeptidase (GGTase) and glucose-6-phosphate dehydrogenase (G6PDH) mark foci of DENA-initiated hepatocytes. These "positive" markers are not found in the surrounding normal parenchyma. Other enzymes are "negative markers" [i.e., glucose-6-phosphatase (G6Pase) and magnesium dependent adenosine triphosphatase (ATPase)] and are deficient in foci but common in the remainder of the liver.[38] Acid and alkaline phosphatase mark foci in a variable fashion (positively or negatively) but in a way different from the noninitiated portions of the liver. To our knowledge, the study by Hinton et al.[38] is the only study to date addressing EAF in fish carci-

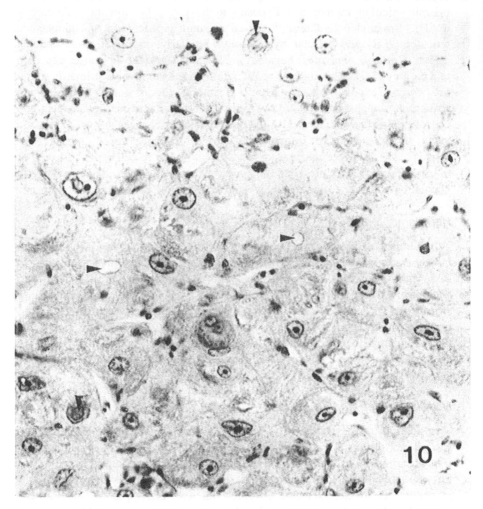

Figure 10.   Electron micrograph of liver section from striped bass (*Morone saxatilis*) fingerling exhibiting chronic liver toxicity. Some enlarged hepatocytes are multinucleated while others show pseudoinclusions (*vertical arrows*). Note enlarged
canaliculi (*horizontal arrows*). Hematoxylin and eosin stain. Magnification X700.

nogenesis bioassay. Under proper conditions, EAF may develop into nodules of
transformed cells. Hendricks et al.[32] reported nodular lesions in *Salmo gairdneri*
liver which were deficient in G6Pase, and Nakazawa et al.[64] used G6Pase and
ATPase to characterize nodular lesions in medaka. The latter worker showed
apparent growth (partial areal analysis) of phenotypically altered hepatocyte
nodules in fish sampled at later intervals after exposure to DENA. From recent
reviews in hepatic carcinogenesis,[20,69] major steps involve initiation, promotion,
and progression. EAF are indicators of initiation and related to promoting agents

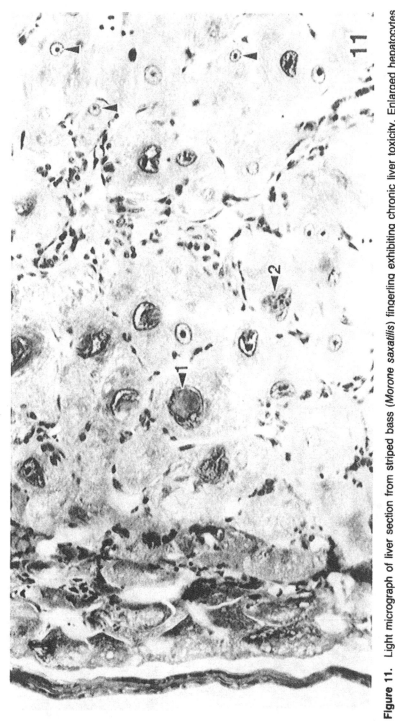

Figure 11. Light micrograph of liver section from striped bass (*Morone saxatilis*) fingerling exhibiting chronic liver toxicity. Enlarged hepatocytes show equally enlarged nuclei, many of which have nuclear pseudoinclusions (horizontal arrows - 1). Vertical arrows point to nuclei of normal size in normal appearing hepatocytes. Horizontal arrow (2) points to multinucleated megalocyte. Hematoxylin and eosin stain. Magnification X700.

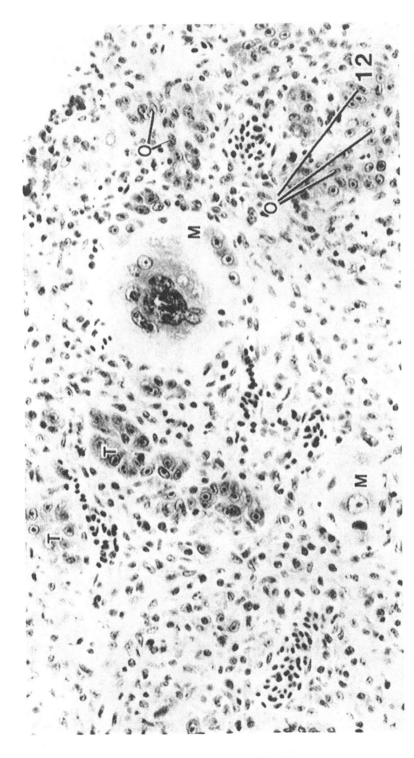

Figure 12. Light micrograph of liver section from striped bass (*Morone saxatilis*) fingerling exhibiting chronic liver toxicity. M, megalocytic hepatocytes. Upper one shows multinucleated form. T, hepatic tubules; O, foci of oval cell proliferation some of which are centrotubular in location. Field shows abundant mononuclear cell inflammation and proliferation of nonhepatocytic cell types. Hematoxylin and eosin stain. Magnification X350.

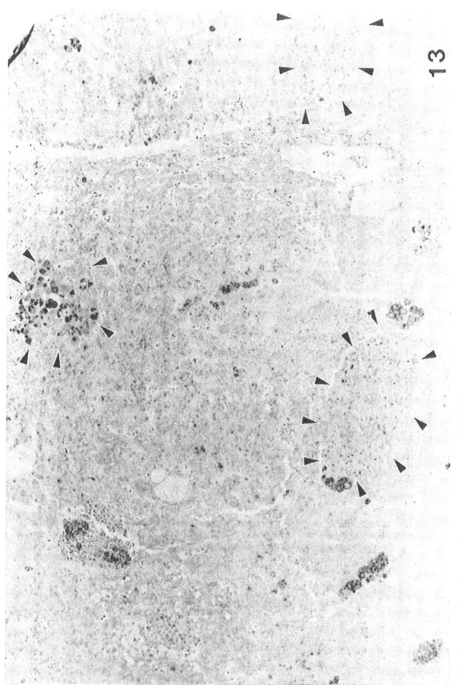

Figure 13. Freeze-dried, glycolmethacrylate-embedded (FDGE) liver section from Japanese medaka (*Oryzias latipes*) exposed to carcinogenic concentration of DENA. Enzyme-altered foci (arrows) are demonstrated by enhanced reaction product of acid phosphatase in hepatocytes. Substrate was naphthol AS BI phosphate. Magnification X140.

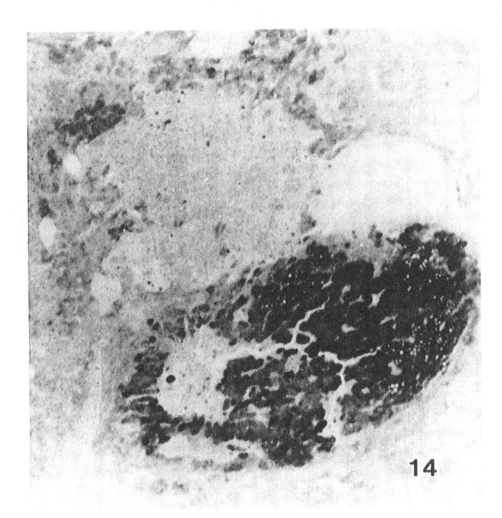

Figure 14. Freeze-dried, glycolmethacrylate-embedded (FDGE) liver section from Japanese medaka (*Oryzias latipes*) exposed to carcinogenic concentration of DENA. Enzyme-altered focus is demonstrated by enhanced reaction product (reduced nitro blue tetrazolium) of glucose-6-phosphate dehydrogenase in hepatocytes. Magnification X140.

in a dose-dependent manner. Therefore, EAF are promising candidates for environmental screens since they would be biomarkers of initiation and promotion. Since one known group of potential carcinogens, the PAH, are widespread environmental contaminants and ubiquitous in sediments, they are likely candidates as initiators of carcinogenesis in livers of certain bottom-dwelling feral fishes. However, whether a tumor actually results from this initiation may be due to the presence or absence of additional xenobiotic substances in sediment, food, and/or water. Mitogenic capacity[8,69] or toxicity with resultant hyperplasia[83] may directly or indirectly stimulate growth of initiated cells. The plasticizer and common mammalian promotor, bis-(2-ethylhexyl) phthalate, is ubiquitous in the aquatic environment and found in high levels in the liver of *Platichthys stellatus*, the starry flounder.[85] It follows that the incorporation of EAF as a biomarker may allow us to detect exposure of feral fish to both initiating and promoting chemicals.

Focal alterations in tinctorial (staining) properties [using the conventional hematoxylin and eosin (H & E) procedure] are frequently encountered after exposure of fish to hepatotoxicants. Glycogen-rich areas are not stained by H & E (Figure 5) and impart a pale homogeneous color to large portions of the hepatocyte. Glycogen loss, a generic and early change in toxicity, may be related to an apparent increased basophilia of the cytoplasm due to the relative increase in ribonucleic acid-rich organelle. Under other conditions, hepatocytes may appear somewhat rounded and show intense basophilic staining. If this occurs in a group of contiguous hepatocytes, the change would be referred to as a focus of basophilic staining (Figure 15). Increased eosinophilia is sometimes encountered over hepatocyte cytoplasm. With this tinctorial change, there has probably been a shearing of the ribosomes and their component nucleic acids from the endoplasmic reticulum membranes. Because substances such as nucleic acids normally react with the basic dye, hematoxylin, and these are now dispersed, less basophilia and more eosinophilia is seen. Also, abundant proliferation of smooth endoplasmic reticulum can lead to a hyalinized appearance over the hepatocyte cytoplasm.[52] When present in contiguous cells, this lesion constitutes the eosinophilic focus (Figure 16). Likewise, "clear" cell foci are probably due to glycogen retention. The last type of focal alteration associated with fish exposed to chemical carcinogens is the fatty focus (Figure 17) in which contiguous hepatocytes show an accumulation of round fat droplets. We have encountered all of these foci in medaka during DENA-induced carcinogenesis (Figures 15 and 17).

Rather than review tumor morphology in detail herein, interested readers should consult the review and atlas of fish liver neoplastic and proliferative disorders.[98] We will focus our attention on hepatocyte structure of these lesions. Fairly normal looking hepatocytes occupy basophilic nodules (Figure 15), hepatoma, and well differentiated hepatocellular carcinomas. However, hepatocytes of less well differentiated tumors have appreciably altered shapes (Figure 18),

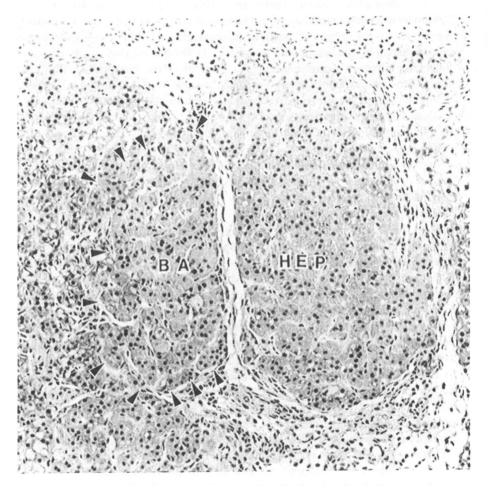

Figure 15.   Basophilic area (BA) to right of arrows and nodule (HEP) in liver section from Japanese medaka (*Oryzias latipes*) exposed to carcinogenic concentration of DENA. Except for staining characteristics, hepatocytes comprising these lesions resemble normal cells. Note the increased thickness of tubules in the hepatoma and its obvious capsule. Hematoxylin and eosin stain. Magnification X350.

and, sometimes, organelle contents. Hepatocytes of poorly differentiated tumors of medaka exposed to DENA vary in shape. Some are extremely attenuated and resemble fibroblasts or spindle cells in light micrographs (Figure 18). However, electron microscopy of the organelle components show that these cells are hepatocytes. Cells of hepatocellular carcinomas from feral English sole (*Parophrys vetulus*)[91] have shown large, branching arrays of swollen granular endoplasmic reticulum cisternae containing densely packed microfilamentous-like material, apparent increases in profiles of Golgi apparatus, and dumbell-shaped mitochondria. In recent ultrastructural studies of DENA-induced tumors in medaka

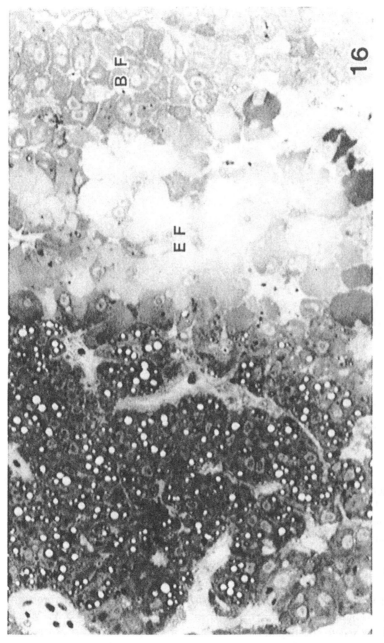

Figure 16. Eosinophilic focus (EF) at right center of field shows rounded hepatocytes with ground glass appearance to cytoplasm. Cells at far right comprised a basophilic focus (BF). The latter cells are triangular to trapezoidal in contour and reveal extensive regions of cytoplasm which stain darkly. Cells at left of field show moderate fat vacuolation but exhibit normal staining characteristics. Toluidine blue stain of liver sections from Japanese medaka (*Oryzias latipes*) given carcinogenic concentration of DENA. Magnification X600.

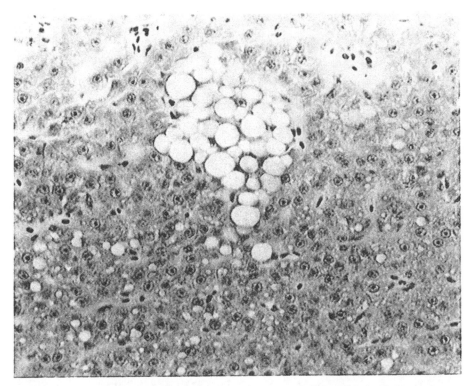

Figure 17. Fatty focus in liver section from Japanese medaka (*Oryzias latipes*) given carcinogenic concentration of DENA. Although not stained to demonstrate fat, smooth margins of circular areas differentiate fat from glycogen which would show a diffuse border. Nuclei are eccentric in location. Hematoxylin and eosin stain. Magnification X558.

liver, one neoplasm was associated with large numbers of apparent peroxisomes (Hinton et al., unpublished). Other tumor hepatocytes contained numerous mitochondria and their plasma membranes were thrown into complicated plicae (Hinton et al., unpublished).

Toxic Changes in Sinusoidal Endothelium

The sinusoidal endothelial cell surrounds the vascular space and separates this from the interstitial space and from the adjacent hepatic parenchymal cells. When trout were exposed to the hepatotoxicant, allylformate, extensive necrosis was induced.[16,18] At early samplings following a single exposure, initial morphologic alteration was destruction of sinusoidal endothelial cells. With loss of sinusoidal endothelium, hemorrhage into space of Disse and interhepatocytic space was associated with hepatocyte shrinkage and a hemorrhagic necrosis.[18]

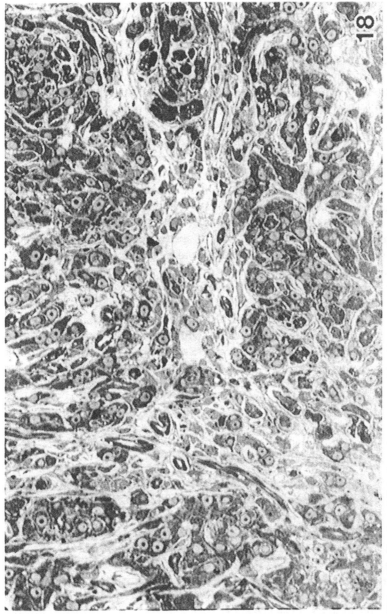

Figure 18. Section of liver from Japanese medaka (*Oryzias latipes*) given carcinogenic concentration of DENA shows hepatocellular carcinoma (poorly differentiated). Field is occupied by pleomorphic hepatocytes. Attenuated forms, originally thought to be fibroblasts, were shown by ultrastructural examination to be hepatocytes. Nuclei vary in shape and have variously sized nucleoli. No capsule is apparent. Hematoxylin and eosin stain. Magnification X320.

Toxic Changes in Ito Cells

The extensive network of cellular extensions from Ito cells in teleost liver is not appreciated in conventional light microscopy and TEM. However, when freeze fracture is followed by SEM,[22] they can be readily seen. These cells possess potential for fibroblastic transformation and are active in hepatic fibrosis.[103] Following the "piecemeal" necrosis (apoptosis) associated with DENA exposure in liver of medaka, the long processes of Ito cells are readily apparent as a framework surrounding altered hepatocytes, spaces (Figure 9), and occasionally other cells. The remaining thin cellular extensions of Ito cells create a sponge-like appearance. Electron microscopy reveals numerous bundles of intermediate filaments within these Ito cell processes.[35] In addition, numerous secretory granules may be responsible for the glycoconjugates inhabiting the appreciable spaces. Bannasch et al.[4] first described this condition in rats exposed to n-nitrosomorpholine as *spongiosis hepatis* (SH). Hinton et al.[35] first reported the occurrence of SH in fish liver. Livers of medaka exposed to methylazoxymethanol acetate, or DENA,[38] showed a similar condition. The similarity of appearance of the rat and fish lesion produced by related nitroso compounds, the identical appearance of secretory vesicles within Ito cells, and the positive reaction with pH 2.5 alcian blue of the glycoconjugates inhabiting cystic spaces of SH is compelling evidence that the two lesions are identical. We have seen proliferation of small basophilic hepatocytes within cystic spaces of SH in medaka liver.[38] Electron microscopic examination of resultant hepatocellular tumors often reveals Ito cell processes in close association with neoplastic hepatocytes (Figure 19). The abundance of secretory vesicles within the former suggests a role in mesenchymal/epithelial interactions which may prove important in neoplastic progression. Interestingly, we have recently found this lesion in feral fish populations in pollutant impacted areas of the Kanawha River in West Virginia.[40]

Toxic Reactions of Biliary Epithelial Cells

Lumina associated with the intrahepatic biliary system may reflect toxic injury states by showing swollen, distended cellular features (Figure 10). In addition, biliary epithelial cells may lose plasma membranes, creating membranous whorls within the ductule lumens. However, under chronic conditions of toxicity, the bile ductules may undergo proliferation with the appearance of foci of numerous, small, oval-shaped nuclei usually in centrotubular location (Figure 12). Proliferation of bile ductules and ducts often elicits a strong connective tissue response leading to the condition termed "adenofibrosis".[31,44] Most of the liver tumors described in the brown bullhead (*Ictalurus nebulosus*) of the Black River involve biliary epithelial cells as well as hepatocytes[5,6] indicating the active nature of these cells in chronic toxic responses.

Toxic Changes in Macrophages

Macrophages appear as aggregates following prior toxicity.[103] Often referred to as melanomacrophage aggregates, these cells are not associated with melanin

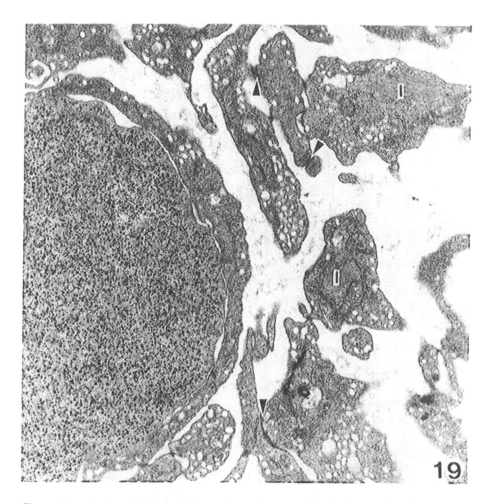

Figure 19. Electron micrograph of section of hepatocellular carcinoma from Japanese medaka (*Oryzias latipes*) exposed to DENA. Glycogen-laden cytoplasmic region in cell at left is from a hepatocyte. All other cellular components are from Ito cells and show numerous circular electron lucent areas (secretory vesicles), junctional complexes (arrows), and abundant intermediate filaments (I). Uranyl acetate and lead citrate stain. Magnification X16,000.

in some species (Figure 20). The area and the number of individual aggregates has been used as a biomarker lesion for prior stress and there is an appreciable literature on this lesion.[10,73,103]

Evaluation of Potential Biomarkers

Because organelles, cells, tissues, and organs of fish are in a state of dynamic equilibrium, their morphology at any one time will likely reflect a spectrum of

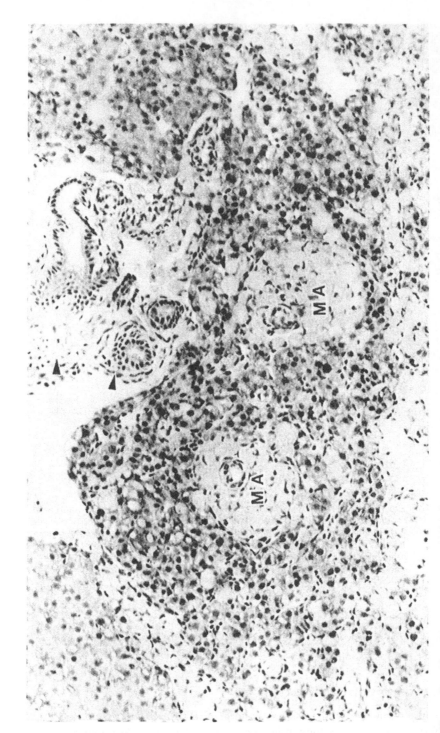

Figure 20. Macrophage aggregates (MA) in liver of adult Japanese medaka (*Oryzias latipes*) exposed to DENA. Arrows point to thickened regions of connective tissue surrounding proliferating bile ducts. Hematoxylin and eosin stain. Magnification X300.

structural possibilities. Given age, diet, environmental factors, seasonal variation, and reproductive cycle, a number of structural states may represent normality and could be potentially confounding issues in our attempt to use morphologic criteria as biomarkers of effect.

With certain of the potential biomarkers described above, overlap between
anticipated toxic state and some aspects of the range of normal morphology
may exist. For this reason, we present strengths and weaknesses for each
below.

Hepatocellular Necrosis

A substantial database exists associating coagulative hepatocellular necrosis
with exposure to anthropogenic environmental toxicants in both mammals and
fish.[70,106,107] However, whereas many toxicants can be classified by the zone
(acinar location) of the liver they affect in mammals,[43] few studies with fish
liver[15,18,23,101] and none with isolated hepatocytes have addressed the question
of hepatocyte heterogeneity. Furthermore, while it is often assumed that the
mechanism of toxicity in fish hepatocytes is the same as in mammalian hepatocytes, mechanistic studies with fish are few.[15,18,23] Nevertheless, as our information from correlated biochemical and morphological studies increases, re-
finement of this important biomarker may provide additional ways to identify
general classes of causative agents. Hepatocellular necrosis certainly signifies
exposure to toxic chemicals, but natural, plant-derived toxins,[30] or even enteric,
bacterial-derived toxins,[2,21] or viruses[102] may cause similar responses. To the
best of our knowledge, comparative research on anthropogenic versus naturally
derived toxicants simply has not been conducted. These reservations apply for
all of the morphological biomarkers for fish and suggest the depth of the need
for these kinds of studies. With these caveats, let us conclude that, in our opinion,
and in the absence of reasonable data suggesting phenomena such as outbreaks
of algal blooms, hepatocellular necrosis is a good biomarker of anthropogenic
toxicant exposure, and may be useful for identifying point sources of toxicant
emissions.

Hyperplasia of Regeneration

In adult fish, hyperplastic regeneration is indicative of extensive prior necrosis.
Either toxicant exposure or infectious disease processes, if severe but not lethal,
could result in this lesion. However, in the absence of evidence for parasitic
infestation (i.e., fibrotic tracks or cystic spaces) or prior bacterial infection, this
is a good biomarker of exposure to toxicants. Differentiation of normal growth
(i.e., somatic growth of larvae) from regeneration is aided by the irregular shape
of islands of basophilic hepatocytes in the latter.[8] The presence of specific types
of inflammatory cells may also be useful in differentiating infectious from tox-
icant-derived etiologies.

Fatty Change

Although fatty change is shown by fish hepatocytes after exposure to a variety of toxicants, other causative agents are known, including nutritional state,[71,79,80] gender and reproductive cycle (i.e., vitellogenesis),[7] and time of year. Furthermore, some species of fish store large amounts of lipid in hepatocytes.[11] Before this "lesion" can be reliably used as a biomarker of toxicant exposure, the researcher must be very familiar with the appearance of the liver in the same species of fish, at the same time of year, under normal conditions.

High Amplitude Swelling

Although swelling is an integral part of adaptation to cell injury, the finding of hepatocyte and biliary epithelial swelling, as the major indication of toxic injury, is rare. The notable exception involved Winter flounder from Boston Harbor where certain of the affected fish also had hepatocellular tumors.[62] In this instance, the magnitude of the change and the percentage of the fish affected constitute a highly specific biomarker. Furthermore, since the related English sole from Puget Sound do not show this lesion, despite the presence of hepatocellular tumors,[63,91] this suggests that this high amplitude swelling may be a biomarker of exposure to a specific toxicant or class of toxicants.

Nuclear and Cytoplasmic Inclusions

Exposure of mammals to lead or bismuth can result in nuclear inclusions. With lead, the inclusions are detected by a positive acid fast stain with light microscopy and by their characteristic ultrastructure.[24,25] We could find no reference to fish studies in which such inclusions were induced. Mercury, silver, cadmium, and selenium may be observed at the light microscopic level using autometallography,[3] and where chemical data support contamination with these metals, nuclear and cytoplasmic inclusions may be a good biomarker. Although both copper[51] and arsenic[84] produce nuclear and cytoplasmic inclusions, these were only observed with electron microscopy. Electron microscopy is also required to rule out a viral etiology of intranuclear inclusions.

Hepatocyte Hypertrophy

Organelle hyperplasia is the usual underlying cause of hepatocyte hypertrophy. Exposure of mammals to phenobarbital results in proliferation of rough endoplasmic reticulum, and, subsequently, of smooth endoplasmic reticulum.[87] Concomitant with this increase in surface area of endoplasmic reticulum is a rise in oxidative demethylase activity.[87] Several studies have shown that fish cytochrome P-450s are refractive to induction by phenobarbital,[47] and other non-coplanar heterocyclic compounds such as the majority of isomers of polychlorinated biphenyls (PCBs) and DDT.[49] However, induction does occur with molecules such as coplanar isomers of PCBs, benzo(a)pyrene, and 3-methylcholanthracene,[49] so that hepatocyte hypertrophy due to proliferation of endoplasmic reticulum may be a valuable biomarker of exposure to polynuclear aromatic hydrocarbons (PAHs). To date only one morphometric study has attempted to quantify endoplasmic reticulum changes in PAH-treated fish.[48] Despite a peak in cytochrome P-450 at 66 hr post exposure, the morphometric parameters in control and treated fish did not differ. Perhaps longer duration of exposure would have produced change. Several authors have reported apparent hepatocyte hypertrophy with exposure to PAHs[76] or PCBs.[46] Furthermore, fingerprint whorls of ER, perhaps not of sufficient magnitude to cause cellular hypertrophy, are nevertheless indicative of cytochrome P-450 induction.[46] These lesions usually require electron microscopy but would constitute a meaningful biomarker of exposure.

Similarly, induction of specific isozymes such as the LM4B form of cytochrome P-450 of trout[100] of P-450E of scup[89] will likely prove important biomarkers of PAH exposure. It is possible to detect these isozymes using immunohistochemical procedures.[60]

Hepatocytomegaly

Hepatocytomegaly has been reported in English sole from Puget Sound[63] (where it is associated with a high incidence of hepatic tumors), in a number of feral fish from the Kanawha River of West Virginia,[40] in cultured hatchling striped bass in California,[26] pen-reared salmon in Hood's Canal,[44] and in rainbow trout treated with pyrrolizidine alkaloids.[31] Electron microscopic examination of nuclear inclusions and pseudoinclusions may be warranted to rule out viral etiologies; this is also true in the case of multinucleated forms which may represent syncytia due to herpes- or reoviruses.[102] However, the lesions seems to be related to toxic injury. The toxicant may be anthropogenic or naturally occurring (i.e., plant alkaloids).

Neoplastic Hepatocyte Change

Fish possess the bioactivation enzymes necessary to convert a number of potential carcinogens to the active, ultimate carcinogenic forms.[90] Furthermore, oncogenes have been isolated and sequenced in some fish species.[65,94] The absence of data to suggest a viral etiology[13,91] coupled with the above, and the findings of epizootics of hepatic neoplasia in fishes inhabiting polluted waters, are strong indicators that liver tumors are good biomarkers of prior chemical exposure.[57] Enzyme altered foci, foci of tinctorial change, neoplastic nodules, and overt tumors are meaningful endpoints in field work. Since transformed cells exhibit altered phenotypes, as in enzyme altered foci, histochemical assays should be incorporated in field studies. Furthermore, since carcinogenesis is a multi-stepped process, chemical analyses should seek to identify noncarcinogens which may nevertheless act as promotors of initiated cells.

Toxic Changes in Sinusoidal Endothelium

Sinusoidal endothelium, like other nonhepatocytic cell types, has received little attention in toxicity studies. However, the involvement of this cell type in the hemorrhagic necrosis of allyl formate[18] suggests that changes may be highly specific to certain chemical agents.

Toxic Changes in Ito Cells

The lesion spongiosis hepatis may prove to be a useful biomarker of exposure. After having examined liver sections from a large number of feral and laboratory fishes, we regard this lesion as due to toxic factors which primarily affect hepatocytes. We have also observed perivascular and peribiliary spongiotic lesions which have an inflammatory component present in nearly all instances. Our observations with medaka in which spongiosis appears to arise from apoptosis (i.e., without involvement of an inflammatory response) suggest that this lesion differs from perivascular and peribiliary spongiotic inflammatory lesions.

Toxic Reactions of Biliary Epithelial Cells

Biliary epithelial cells are responsive to various types of insults. In chronic liver injury, proliferation of ductular and larger passageways may occur. During progression of hepatocellular neoplasia, changes are often seen in biliary epithelium as well. When peribiliary fibrosis and bile ductular proliferation are seen, the possibility of parasitic involvement must be ruled out. We have encountered parasites within biliary passageways with resultant epithelial hyperplasia, duct proliferation, and granuloma formation. The latter may help to distinguish parasitic from toxic etiology. When infectious disease can be reasonably ruled out, biliary change constitutes a meaningful biomarker of exposure.

Toxic Changes in Macrophages

Although there is an abundant literature which suggests that these may be useful as biomarkers of stress, they are also seen following the injury of the liver caused by infectious agents, including parasites. Therefore, their use as a biomarker of xenobiotic exposure may be somewhat limited and must be carefully analyzed for the presence of other existing causative factors.

SUMMARY AND CONCLUSIONS

Naturally occurring and industrial compounds including genotoxic carcinogens, metals, solvents, and pesticides produce a finite number of easily recognizable liver lesions. Where a battery of analytical chemical methods has been applied to sediment,[95] bile,[66] and/or diet[79,80] and correlated with liver morphology, lesions have reflected chemical perturbations in field and laboratory situations. Confounding issues of seasonal variation, age—including life stages, sex, nutrition, and reproductive phase need to be experimentally addressed and used to further laboratory to field correlation. Archives of fixed tissue preparations or, preferably, histopathologic sections of key organs from indicator species need to be established to provide site-specific historical databases. These should be coupled with seasonal or annual population and health estimates of bioindicator or endangered fish species.[41] Since liver morphology is an integration of biochemical and physiological lesions, liver pathology is a realistic biomarker.

ACKNOWLEDGMENT

Supported in part by U.S.P.H.S. grants CA45131 and ESO4699 from the National Cancer Institute and from the National Institute of Environmental Health Sciences (Superfund Basic Research Program).

REFERENCES

1. Arey, L.B. *Developmental Anatomy*, 7th ed. (Philadelphia: W.B. Saunders Co., 1966), p. 255–259.
2. Austin, B., and D.A. Austin. *Bacterial Fish Pathogens: Disease in Farmed and Wild Fish*. (New York: John Wiley & Sons, 1988), p. 364.
3. Baatrup, E., M.G. Nielsen, and G. Danscher. "Histochemical Demonstration of Two Mercury Pools in Trout Tissues: Mercury in Kidney and Liver after Mercuric Chloride Exposure," *Ecotox. Environ. Safety* 12:267–282 (1986).
4. Bannasch, P., M. Bloch, and H. Zerban. "Spongiosis Hepatis. Specific Changes of the Perisinusoidal Liver Cells Induced in Rats by N-nitrosomorpholine," *Lab. Invest.* 44:252–264 (1981).
5. Baumann P., and T.N. Mac. "Polynuclear Aromatic Hydrocarbons and Tumors in Brown Bullhead from the Black and Cuyahoga Rivers—Cause and Effect?" *Transact. Am. Fish. Soc.* in press (1989).
6. Baumann P.C., W.D. Smith, and M. Ribick. "Hepatic Tumor Levels and Polynuclear Aromatic Hydrocarbon Levels in Two Populations of Brown Bullheads (*Ictalurus nebulosus*)," in *Polynuclear Aromatic Hydrocarbons: Physical and Biological Chemistry*, M. Cooke, A.J. Dennis, and G.L. Fisher, Eds. (Columbus, OH: Battelle Press, 1982), p. 93–102.
7. Bohemen C.H., G. Van, J.G.D. Lambert, et al. "Annual Changes in Plasma and Liver in Relation to Vitellogenesis in the Female Rainbow Trout, *Salmo gairdneri*," *Gen. Comp. Endocrinol.* 44:94–107 (1981).
8. Boutwell, R.K. "Some Biological Aspects of Skin Carcinogenesis," *Prog. Exp. Tumor Res.* 4:207–250 (1964).
9. Boyer, J.L., J. Swartz, and N. Smith. "Biliary Secretion in Elasmobranchs. II. Hepatic Uptake and Biliary Excretion of Organic Anions," *Am. J. Physiol.* 230:974–981 (1976).
10. Brown, C.L., and C.J. George. "Age-Dependent Accumulation of Macrophage Aggregates in the Yellow Perch, *Perca flavescens* (Mitchill)," *J. Fish Dis.* 8:135–138 (1985).

11. Byczkowska-Smyk, W. "Observation of the Ultrastructure of the Hepatic Cells of the Burbot (*Lota lota*, L.)," *Zool. Poloniae*. 17:105–128 (1968).

12. Columbano, A., G.M. Ledda-Columbano, P.P. Coni, et al. "Occurrence of Cell Death (Apoptosis) During the Involution of Liver Hyperplasia," *Lab. Invest*. 52:670–675 (1985).

13. Couch, J.A., and J.C. Harshbarger. "Effects of Carcinogenic Agents on Aquatic Animals: An Environmental and Experimental Overview," *Environ. Carcin. Rev.* 3(1):63–105 (1985).

14. Cousins, R.J. "Absorption, Transport and Hepatic Metabolism of Copper and Zinc: Special Reference to Metallothionein and Ceruloplasmin," *Physiol. Rev*. 65:238–309 (1985).

15. Dalich, G.M., and R.E. Larson. "A Comparative Study of the Hepatotoxicity of Monochlorobenzene in the Rainbow Trout (*Salmo gairdneri*) and the Sprague-Dawley Rat," *Comp. Biochem. Physiol*. 80C:115–122 (1985).

16. Droy, B.F., and D.E. Hinton. "Allyl Formate-Induced Hepatotoxicity in Rainbow Trout," *Mar. Environ. Res*. 24:259–264 (1988a).

17. Droy, B.F., M.R. Miller, T. Freeland, et al. "Immunohistochemical Detection of CCL_4-induced, Mitosis—Related DNA Synthesis in Livers of Trout and Rat," *Aquat. Toxicol*. 13:155–166 (1988b).

18. Droy, B.F., M.E. Davis, and D.E. Hinton. "Mechanism of Allyl Formate-Induced Hepatotoxicity in Rainbow Trout," *Toxicol. Appl. Pharmacol*., 98:313–324 (1989).

19. Evarts, R.P., P. Nagy, and E. Marsden, et al. "A Precursor-Product Relationship Exists Between Oval Cells and Hepatocytes in Rat Liver," *Carcinogenesis* 8:1737–1740 (1987).

20. Farber, E., and D.S.R. Sarma. "Hepatocarcinogenesis: A Dynamic Cellular Perspective," *Lab. Invest*. 56:4–22 (1987).

21. Fiume, L. "Pathogenesis of the Cellular Lesions Induced by α-Amanitin," in *Pathology of Transcription and Translation*, E. Farber, Ed. (New York: Marcel Dekker, 1972), p. 105–122.

22. Fujita, H., H. Tatsumi, and T. Ban, et al. "Fine-Structural Characteristics of the Liver of the Cod (*Gadus macrocephalus*), with Special Regard to the Concept of a Hepatoskeletal System Formed by Ito Cells," *Cell Tissue Res*. 244:63–67 (1986).

23. Gingerich, W.H. "Hepatic Toxicology of Fishes," in *Aquatic Toxicology*, L. Weber, Ed. (New York: Raven Press, 1982), p. 55–105.

24. Goyer, R.A., P. May, and M.M. Cates, et al. "Lead Dosage and the Role of the Intranuclear Inclusion Body," *Arch. Environ. Health* 20:705–711 (1970).

25. Goyer, R.A., and B.C., Rhyne. "Pathological Effects of Lead," *Int. Rev. Exp. Pathol*. 12:1–77 (1973).

26. Groff, J. et al. Unpublished data, this laboratory.

27. Hampton, J.A., P.A. McCuskey, R.S. McCuskey, et al. "Functional Units in Rainbow Trout (*Salmo gairdneri*, Richardson) Liver. I. Arrangement and Histochemical Properties of Hepatocytes," *Anat. Rec*. 213:166–175, 1985.

28. Hampton, J.A., R.C. Lantz, and P.J. Goldblatt, et al. "Functional Units in Rainbow Trout (*Salmo gairdneri*, Richardson) Liver. II. The Biliary System," *Anat. Rec*. 221:619–634 (1988).

29. Hampton, J.A., R.C. Lantz, and D.E. Hinton. "Functional Units in Rainbow Trout (*Salmo gairdneri*, Richardson) Liver. III. Morphometric Analysis of Parenchyma, Stroma and Component Cell Types," *Am. J. Anat*. 185:58–73 (1989).

30. Hendricks, J.D. "Chemical Carcinogenesis in Fish," in *Aquatic Toxicology*, Vol. 1, L.J. Weber, Ed. (New York: Raven Press, 1982), p. 149–211.

31. Hendricks, J.D., R.O. Sinnhuber, M.C. Henderson, et al. "Liver and Kidney Pathology in Rainbow Trout (*Salmo gairdneri*) Exposed to Dietary Pyrrolizidine (*Senecio*) Alkaloids," *Exp. Mol. Pathol.* 35:170–183 (1981).

32. Hendricks, J.D., T.R. Meyers, and D.W. Skelton. "Histological Progression of Hepatic Neoplasia in Rainbow Trout (*Salmo gairdneri*)," *Natl. Cancer Inst. Monog.* 65:321–336 (1984).

33. Hinton, D.E., H. Glaumann, and B.F. Trump. "Studies on the Cellular Toxicity of Polychlorinated Biphenyls (PCB's). I. Effects of PCB's on Microsomal Enzymes and on Synthesis and Turnover of Microsomal and Cytoplasmic Lipids of Rat Liver. A Morphological and Biochemical Study," *Virchows Arch. B Cell Pathol.* 27:279–306 (1978).

34. Hinton, D.E., E.R. Walker, C.A. Pinkstaff, et al. "Morphological Survey of Teleost Organs Important in Carcinogenesis with Attention to Fixation," *Natl. Cancer Inst. Monogr.* 65:291–320 (1984).

35. Hinton, D.E., R.C. Lantz, and J.A. Hampton. "Effect of Age and Exposure to a Carcinogen on the Structure of the Medaka Liver: A Morphometric Study," *Natl. Cancer Inst. Monogr.* 65:239–249 (1984b).

36. Hinton, D.E., and J.A. Couch. "Pathobiological Measures of Marine Pollution Effects," in *Concepts in Marine Pollution Measurements*, H. White, Ed. (College Park, MD: Maryland Sea Grant College, University of Maryland, 1984), p. 7–32.

37. Hinton, D.E., R.C. Lantz, J.A. Hampton, et al. "Normal versus Abnormal Structure: Considerations in Morphologic Responses of Teleosts to Pollutants," *Environ. Health Pers.* 71:139–146 (1987).

38. Hinton, D.E., J.A. Couch, S.J. Teh, et al. "Cytological Changes During Progression of Neoplasia in Selected Fish Species," *Aquat. Toxicol.* 11:77–112 (1988).

39. Hinton, D.E., D.J. Lauren, S.J. Teh, et al. "Cellular Composition and Ultrastructure of Hepatic Neoplasms Induced by Diethylnitrosamine in *Oryzias latipes*," *Mar. Environ. Res.* 24:307–310 (1989).

40. Hinton, D.E., D.J. Laurén and E. McCoy. Unpublished observations, this laboratory (1989).

41. Hinton, D.E., D.J. Laurén, P.A. McCuskey, et al. "*In vivo* Microscopy of Hepatic Microvasculature in Rainbow Trout (*Oncorhynchus mykiss*)," *Mar. Environ. Res.* 28:407–410 (1989).

42. Jung, M., and J., Whipple. "Cooperative Striped Bass Study. Appendix 1: Procedures for Histopathological Examinations, Autopsies, and Subsampling of Striped Bass," *Calif. State Water Resources Control Board*, p. 46 (1981).

43. Jungermann, K. "Functional Heterogeneity of Periportal and Perivenous Hepatocytes," *Enzyme* 35:161–180 (1986).

44. Kent, M.L., M.S. Myers, and D.E. Hinton, et al. "Suspected Toxicopathic Hepatic Necrosis and Megalocytosis in Pre-reared Atlantic Salmon *salmo salar* in Puget Sound, Washington, USA," *Dis. Aquat. Org.* 49:91–100 (1988).

45. Kimbrough, R.D., R.E. Linder, and T.B. Gaines. "Morphological Changes in Livers of Rats Fed Polychlorinated Biphenyls. Light Microscopy and Ultrastructure," *Arch. Environ. Health.* 25:354–364 (1972).

46. Klaunig, J.E., M.M. Lipsky, B.F. Trump, et al. "Biochemical and Ultrastructural

Changes in Teleost Liver Following Subacute Exposure to PCB," *J. Environ. Pathol. Toxicol.* 2:953–963 (1979).

47. Kleinow, K.M., M.J. Melancon, and J.J. Lech. "Biotransformation and Induction: Implications for Toxicity, Bioaccumulation, and Monitoring of Environmental Xenobiotics in Fish," *Environ. Health Pers.* 71:105–119 (1987).

48. Kontir, D.M. R.C. Lantz, and D.E. Hinton. "Ultrastructural Morphometry of Rainbow Trout (*Salmo gairdneri*) Hepatocytes During 3-Methylcholanthrene (3-MC). Induction of MFO's," *Mar. Environ. Res.* 14:420–421 (1984).

49. Lech, J.J., M.J. Vodicnik, and C.R. Elcombe. "Induction of Monooxygenase Activity in Fish," in *Aquatic Toxicology*, Vol. 1, L.J. Weber, Ed. (New York: Raven Press, 1982), p. 107–148.

50. Leffert, H.L., K.S. Koch, P.J. Lad, et al. "Hepatocyte Regeneration, Replication, and Differentiation," in *The Liver: Biology and Pathobiology*, 2nd ed., I.M. Arias, W.B. Jakoby, H. Popper, et al., Eds. (New York: Raven Press, 1988), p. 833–850.

51. Leland, H.V. "Ultrastructural Changes in Hepatocytes of Juvenile Rainbow Trout and Mature Brown Trout Exposed to Copper and Zinc," *Environ. Toxicol. Chem.* 2:353–368 (1983).

52. Lipsky, M.M., J.E. Klaunig, and D.E. Hinton. "Comparison of Acute Response to PCB in Liver of Rat and Channel Catfish: A Biochemical and Morphological Study," *J. Toxicol. Environ. Health* 4:107–121 (1978).

53. Lombardi, B. "Considerations on the Pathogenesis of Fatty Liver," *Lab Invest.* 15:1–20 (1966).

54. Masahito, P., T., Ishikawa, and H. Sugano. "Fish Tumors and Their Importance in Cancer Research," *Jpn. J. Cancer Res. (Gann)* 79:545–555 (1988).

55. McConkley, D.J., P. Hartzell, and P. Nicotera, et al. "Stimulation of Endogenous Endonuclease Activity in Hepatocytes Exposed to Oxidative Stress," *Toxicol. Lett.* 42:123–130 (1988).

56. McCuskey, P.A., R.S. McCuskey, and D.E. Hinton. "Electron Microscopy of Cells of the Hepatic Sinusoids in Rainbow Trout (*Salmo gairdneri*)," in *Cells of the Hepatic Sinusoid*, Vol. 1, A. Kirn, D.L. Knook, Eds. (Leiden, The Netherlands: Kupffer Cell Foundation, 1986), p. 489–494.   ·

57. Meyers, T.R., and J.D. Hendricks. "Histopathology," in *Fundamentals of Aquatic Toxicology*, G.M. Rand, and S.R. Petrocelli, Eds. (Washington, DC: Hemisphere Publishing Corp., 1985), p. 283–331.

58. Miller, M.R., C. Heyneman, S. Walker, et al. "Interaction of Monoclonal Antibodies Directed Against Bromodeoxyuridine with Pyrimidine Bases, Nucleosides and DNA," *J. Immunol.* 136:1791–1795 (1986).

59. Miller, M.R., D.E. Hinton, and J.J. Stegeman. "Cytochrome P-450E Induction and Localization in Gill Pillar (Endothelial) Cells of Scup and Rainbow Trout," *Aquat. Toxicol.* 14:307–332 (1989).

60. Miller, M.R., D.E. Hinton, J.B. Blair, et al. "Immunohistochemical Localization of Cytochrome P-450E in Liver, Gill and Heart of Scup (*Stenotomus chrysops*) and Rainbow Trout (*Salmo gairdneri*)," *Mar. Environ. Res.* 24:37–39.

61. Moon, T.W., P.J. Walsh, and T.P. Mommsen. "Fish Hepatocytes: A Model Metabolic System," *Can. J. Fish. Aquat. Sci.* 42:1772–1782 (1985).

62. Murchelano, R.A., and R.E. Wolke. "Epizootic Carcinoma in the Winter Founder (*Pseudopleuronectes americanus*)," *Science* 228:587–589 (1985).

63. Myers, M.S., L.D. Rhodes, and B.B. McCain. "Pathologic Anatomy and Patterns

of Occurrence of Hepatic Neoplasms, Putative Preneoplastic Lesions, and Other Idiopathic Hepatic Conditions in English Sole (*Parophrys vetulus*) from Puget Sound, Washington," *J. Natl. Cancer Inst.* 78(2):333–363 (1987).

64. Nakazawa, T., S. Hamaguchi, and Y. Kyono-Hamaguchi. "Histochemistry of Liver Tumors Induced by Diethylnitrosamine and Differential Sex Susceptibility to Carcinogenesis in *Oryzias latipes*," *J. Natl. Cancer Inst.* 75:567–573 (1985).

65. Nemoto, N., K. Kodama, A. Tazawa, et al. "Extensive Sequence Homology of the Goldfish ras Gene to Mammalian ras Genes," *Differentiation* 32:17–23 (1986).

66. Oikari, A., and T. Kunnamoojala. "Tracing of Xenobiotic Contamination in Water with the Aid of Fish Bile Metabolites: A Field Study with Caged Rainbow Trout (*Salmo gairdneri*)," *Aquat. Toxicol.* 9:327–341 (1987).

67. Okihiro, M., D.E. Hinton, and S.J. Teh. Unpublished observations, this laboratory (1988).

68. Olsen, P.S., S. Boesby, P. Kirkegaard, et al. "Influence of Epidermal Growth Factor on Liver Regeneration after Partial Hepatectomy in Rats," *Hepatology* 8:992–996 (1988).

69. Pitot, H.C. "Hepatic Neoplasia: Chemical Induction," in *The Liver: Biology and Pathology*, I.M. Arias, W.B. Jakoby, H. Popper, et al., Eds. (New York: Raven Press, 1988), p. 1125–1146.

70. Popper, H. "Hepatocellular Degeneration and Death," in *The Liver: Biology and Pathobiology*, 2nd ed., I.M. Arias, W.B. Jakoby, H. Poppe, et al., Eds. (New York: Raven Press, 1988), p. 1087–1103.

71. Rafael, J., and T. Braunbeck. "Interacting Effects of Diet and Environmental Temperature on Biochemical Parameters in the Liver of *Leuciscus idus melanotus* (Cyprinidae:Teleostei)," *Fish Physiol. Biochem.* 5(1):9–19 (1988).

72. Robbins, S.L. "Cell Injury and Cell Death," in *Pathologic Basis of Diseases*, (Philadelphia: W.B. Saunders Co., 1974), p. 40.

73. Roberts, R.J. "Melanin-Containing Cells of Teleost Fish and Their Relation to Disease," in *The Pathology of Fishes*, W. Ribelin, and G. Migaki, Eds. (Madison, WI: University of Wisconsin Press, 1975), p. 399–428.

74. Scarpelli, D.G., and B.F. Trump. *Cell Injury*, (Kalamazoo, MI: The Upjohn Co., 1971), p. 67.

75. Schmidt, D.C., and L.J. Weber. "Metabolism and Biliary Excretion of Sulfobromophthalein by Rainbow Trout (*Salmo gairdneri*)," *J. Fish. Res. Bd. Can.* 30:1301–1308 (1973).

76. Schoor, W.P., and J.A. Couch. "Correlation of Mixed-Function Oxidase Activity with Ultrastructural Changes in the Liver of a Marine Fish," *Cancer. Biochem. Biophys.* 4:95–103 (1979).

77. Schulte-Hermann, R. "Reactions of the Liver to Injury: Adaptation," in *Toxic Injury of the Liver*, E. Farber and M.M. Fisher, Eds. (New York: Marcel Dekker Inc., 1979), p. 385.

78. Segner, H., and H. Möller. "Electron Microscopical Investigations on Starvation-Induced Liver Pathology in Flounders (*Platichthys flesus*)," *Mar Ecol. Progr. Ser.* 19:193–196 (1984).

79. Segner, H., and T. Braunbeck. "Hepatocellular Adaptation to Extreme Nutritional Conditions in Ide, (*Leuciscus idus melanotus*) L. (Cyprinidae). A Morphofunctional Analysis," *Fish Physiol. Biochem.* 5(2):79–97 (1988).

80. Segner, H., and J.V. Juario. "Histological Observations on the Rearing of Milkfish, (*Chanos chanos*), by Using Different Diets," *J. Appl. Ichthyol.* 2:162–173 (1986).
81. Shafritz, D.A., and A. Panduro. "Protein Synthesis and Gene Control in Pathophysiologic States," in *The Liver: Biology and Pathobiology*, 2nd ed., I.M. Arias, W.B. Jakoby, H. Popper, et al., Eds. (New York: Raven Press, 1988), p. 83–101.
82. Shore, T.W., and H.L. Jones. "On the Structure of the Vertebrate Liver," *J. Physiol. (London)* 10:408–428 (1889).
83. Solt, D.B., and E. Farber. "New Principle for the Analysis of Chemical Carcinogenesis," *Nature (London)* 263:702–703 (1976).
84. Sorensen, E.M.B. "Ultrastructural Changes in the Hepatocytes of Green Sunfish (*Lepomis cyanelus*) Rafinesque Exposed to Solutions of Sodium Arsenate," *J. Fish Biol.* 8:229–240 (1976).
85. Spies, R.B., D.W. Rice, and J. Felton. "Effects of Organic Contaminants on Reproduction of the Starry Flounder (*Platichthys stellatus*) in San Francisco Bay. I. Hepatic Contamination and Mixed-Function Oxidase (MFO) Activity During the Reproductive Season," *Mar. Biol.* 98:181–189 (1988).
86. Spring, K.R., and A.C. Erickson. "Epithelial Cell Volume Regulation, *J. Membr. Biol.* 69:167–176 (1982).
87. Staubli, W., R. Hess, and E.R. Weibel. "Correlated Morphometric and Biochemical Studies on the Liver Cell. II. Effect of Phenobarbital on Rat Hepatocytes," *J. Cell. Biol.* 42:92–112 (1969).
88. Stegeman, J.J., R.L. Binder, and A. Orren. "Hepatic and Extrahepatic Microsomal Electron Transport Components and Mixed-Function Oxygenases in the Marine Fish (*Stentomus versicolor*)," *Biochem. Pharmacol.* 28:3431–3439 (1979).
89. Stegeman, J.J., and P.J. Kloepper-Sams. "Cytochrome P-450 Isozymes and Monooxygenase Activity in Aquatic Animals," *Environ. Health. Persp.* 71:87–95 (1987).
90. Stegeman, J.J., B.R. Woodin, and R.L. Binder. "Patterns of Benzo(a)pyrene Metabolism by Varied Species, Organs and Developmental Stages of Fish," *Natl. Cancer Inst. Monogr.* 65:371–377 (1984).
91. Stehr, C.M., L.D. Rhodes, and M.S. Myers. "The Ultrastructure and Histology of Hepatocellular Carcinomas of English Sole (*Parophrys vetulus*) from Puget Sound, Washington," *Toxicol. Pathol.* 16:418-431 (1988).
92. Trump, B.J., E.M. McDowell, and A.U. Arstila. "Cellular Reaction to Injury," in *Principles of Pathobiology*, 3rd ed., R.B. Hill, and M.F. LaVia, Eds. (New York: Oxford University Press, 1980), p. 20–111.
93. Vaillant, C., C. Le Guellec, F. Padkel, et al. "Vitellogenin Gene Expression in Primary Culture of Male Rainbow Trout Hepatocytes," *Gen. Comp. Endocrin.* 70:284–290 (1988).
94. Van Beneden, R.J., D.K. Watson, T.T. Chen, et al. "Cellular myc (C-myc) in Fish (Rainbow Trout): Its Relationship to Other Vertebrate myc Genes and to the Transforming Genes of MC29 Family Viruses," *Proc. Natl. Acad. Sci. USA* 83:3698–3702 (1986).
95. Varanasi, U., J.E. Stein, and M. Nishimoto, et al. "Chemical Carcinogenesis in Feral Fish: Uptake, Activation, and Detoxification of Organic Xenobiotics," *Environ. Health. Pers.* 71:155–170 (1987).
96. Wake, K. "Perisinusoidal Stellate Cells (Fat-Storing Cells, Interstitial Cells, Lipocytes), Their Related Structure in and Around the Liver Sinusoids, and Vitamin A-Storing Cells in Extrahepatic Organs," *Int. Rev. Cytol.* 66:303–353 (1980).

97. Walton, M.J., and C.B. Cowey. "Aspects of Intermediary Metabolism in Salmonid Fish," *Comp. Biochem. Physiol.* 73B:59–79 (1982).
98. Ward, J., J. Hendricks, and D.E. Hinton. "Neoplasms and Related Disorders of Liver," in: *Atlas on Neoplasms and Related Disorders in Fishes*, C. Dawe, Ed. (in press, 1989).
99. Wester, P.W., and J.R. Canton. "Histopathological Study of Medaka (*Oryzias latipes*) after Long-Term B-Hexachlorocyclohexane Exposure." *Aquat. Toxicol.* 9:21–45 (1986).
100. Williams, D.E., R.C. Bender, and M.T. Morrissey. "Cytochrome P-450 Isozymes in Salmonids Determined with Antibodies to Purified Forms of P-450 from Rainbow Trout," *Mar. Environ. Res.* 14:13–21 (1984).
101. Wofford, H.W., and P. Thomas. "Effect of Xenobiotics on Peroxidation of Hepatic Microsomal Lipids from Striped Mullet (*Mugil cephalus*) and Atlantic Croaker (*Micropogonias undulatus*)," *Mar. Environ. Res.* 24:285–289 (1988).
102. Wolf, K. *Fish Viruses and Fish Viral Diseases*, (Ithaca, NY: Cornell University Press, 1988), p. 476.
103. Wolke, R.E., R.A. Murchelano, C.D. Dickstein, et al. "Preliminary Evaluation of the Use of Macrophage Aggregates (MA) as Fish Health Monitors," *Bull. Environ. Contam. Toxicol.* 35:222–227 (1985).
104. Wyllie, A.H., J.F.G. Kerr, and A.R. Cumi. "Cell Death: The Significance of Apoptosis," *Int. Rev. Cytol.* 68:251–306 (1980).
105. Yamamoto, K., P.A. Sargent, M.M. Fisher, et al. "Periductal Fibrosis and Lipocytes (Fat-Storing Cells or Ito Cells) During Biliary Atresia in the Lamprey," *Hepatology* 6:54–59 (1986).
106. Zieve, L., W.R. Anderson, C. Lyftogt, et al. "Hepatic Regenerative Enzyme Activity after Pericentral and Periportal Lobular Toxic Injury," *Toxicol. Appl. Pharmacol.* 86:147–158 (1986).
107. Zimmerman, H.J. "Chemical Hepatic Injury and Its Detection," in *Toxicology of the Liver*, G. Plaa, and W.R. Hewitt, Eds. (New York: Raven Press, 1982), p. 1–45.
108. Zuchelkowski, E.M., R.C. Lantz, and D.E. Hinton. "Effects of Acid-Stress on Epidermal Mucous Cells of the Brown Bullhead (*Ictalurus nebulosus*, Le Seur): A Morphometric Study," *Anat. Rec.* 200:33–39 (1981).
109. Zuchelkowski, E.M., R.C. Lantz, and D.E. Hinton. "Skin Mucous Cell Response to Acid-Stress in Male and Female Brown Bullhead Catfish (*Ictalurus nebulosus*, Le Seur)," *Aquat. Toxicol.* 8:139–148 (1986).

CHAPTER 3

Histopathology of Atlantic Tomcod: A Possible Monitor of Xenobiotics in Northeast Tidal Rivers and Estuaries

Susan M. Cormier and Richard N. Racine

ABSTRACT

Atlantic tomcod, *Microgadus tomcod*, from the Hudson River, New York exhibited a very high incidence of hepatocellular carcinoma while less industrially impacted rivers such as the Pawcatuck River, Rhode Island and the Saco River, Maine were much lower. A comparative pathological study revealed that hepatic function was compromised by fatty infiltration, basophilic foci, areas of cellular alteration, and hepatocellular carcinoma in samples from Hudson River fish. Age and size may also be affected in Hudson River tomcod. An etiology involving xenobiotics is highly suspect for these effects since it is known that elevated detoxification of some organics results in increased hepatic lipid levels. Juvenile tomcod removed from the Hudson after six months and reared in clean water showed no incidence of liver disease six months later.

BIOMARKERS AND BIOMONITORS

Biomarkers at the cellular and biochemical level are useful indicators of chemical presence and impact to organisms when they can be correlated with chemical analyses and biologically meaningful effects. Table 1 relates the different levels of testing with some of the processes they evaluate using a number of different measurable endpoints. In this general scheme there are three levels of evaluation: chemical load, biomarker, and biomonitor. Additionally the entire ecosystem could be evaluated but this greatly increases the complexity. At each level new strategies are necessary to evaluate the relationship between the level above and below that being tested. Some of the processes that can be evaluated at the biomarker level include the availability of chemical species and their subsequent accumulation, transformation and excretion by organisms. These processes are measured using biomarkers as endpoints. Since these processes may occur much earlier than the overt biomonitor effects, they may be more sensitive in predicting environmental degradation which may allow us to act before biomonitor or even ecosystem effects become irreversible.

There are several experimental approaches that can be taken to select good biomarkers and biomonitors for environmental evaluation. We have elected to study a problem in the Hudson River ecosystem because at least one species of fish, the Atlantic tomcod (*Microgadus tomcod*), has exhibited overt pathologies that were potentially linked to xenobiotic inputs. We have begun to document several changes at the biomonitor or organismic level and plan to begin correlating these changes with appropriate biomarkers. It is hoped that these biomarkers can then be used to evaluate the health of other tidal rivers or even totally different habitats at the earlier and more sensitive endpoint. Table 2 outlines the different endpoints that have been evaluated thus far in the tomcod and which will be discussed further.

THE ATLANTIC TOMCOD

Liver cancer in the Hudson River population of Atlantic tomcod was first identified and reported by Smith et al.[1] Further confirmation of hepatocellular carcinoma and ultrastructural analyses were reported by Cormier.[2] All of these researchers postulated a chemical etiology for the very high incidence of cancer in tomcod since portions of the Hudson River are known to contain high levels of polyaromatic hydrocarbons (PAH), polychlorinated biphenyls (PCB), pesticides and heavy metals.[3] In fact, high levels of PCBs in striped bass have kept that fishery closed to commercial exploitation in the Hudson for several years. A chemical survey of priority pollutants showed very high levels of PCBs, chlordane, DDT, and heavy metals in tomcod liver.[4] PAHs were not detected; however, other workers have shown that PAHs have a rapid clearance rate in fish but the effects of metabolites can be very significant.[5]

More than 560 tomcod have been evaluated during this study for gross and histological evidence of hepatocellular carcinoma. Incidences from 24 to 100%, depending on age/size class, were reported for the Hudson River population as compared to 3 to 10% for another less impacted population from the Pawcatuck River, Connecticut.[4] However, a clear link with xenobiotics could not be made. One portion of the present study expands histopathological monitoring to include

Table 1.   Differing Levels of Analysis for Measuring the Effects of Chemicals in the Environment.

Processes	Level	Endpoints
Loadings	Direct chemical	Air, water, soil concentrations
Bioavailability, bioaccumulation, biotransformation, and excretion	Biomarker	Chemical analyses of tissue, hemoglobin/DNA adducts, DNA breakage, induced enzymes, and metabolites
Biologically meaningful effects	Biomonitor	Tumors, tissue damage, shortened life span, reduced fecundity, and mortality

evaluation of cancer incidence in tomcod from rivers in Maine, as well as more detailed analysis of the Hudson River histological data.

Collection of Fish

During the winter months (Nov.–Jan.) tomcod spawn in many of the coastal rivers of northeastern United States and Canada, and many remain in the rivers throughout the year. Tomcod were collected during the spawning seasons (1985-89) using both large and small box traps set along the shore of selected sites of the Hudson River, New York (n = 64), Pawcatuck River, Connecticut (n = 35), and the Saco (n = 68) and Royal (n = 4) Rivers, Maine (Figures 1a and 1b).

Tomcod spawning occurs earlier in Maine, then further south and finally in the Hudson River allowing a single research team to follow the spawning runs and collect at all sites. Theoretically, earlier collections could also be made further north in Maine and Canada. Summer collections (1987) for juveniles using an otter trawl in the Sheepscot and Penobscot Rivers, Maine also confirmed the presence of adults (n = 2 and 19, respectively) in the rivers during the summer.

Collected fish were measured, sexed, and aged by the number of rings of their otolith. The livers were excised and examined for gross pathology. Liver samples

Table 2.   Endpoints Evaluated for the Atlantic Tomcod.

Direct chemical	Water and sediment analyses
Biomarker	Priority pollutants in liver hepatic lipid levels
Biomonitor	Hepatocellular carcinoma Other histopathologies Reduced adult size Shortened life span

of fish collected during the winter months were fixed in Bouin's solution, buffered formaldehyde or glutaraldehyde. Glutaraldehyde-fixed tissue was subsequently fixed with OsO_4. All tissue samples were dehydrated in ethanol and then embedded in either paraffin or epoxy resin. Paraffin and epoxy resin sections were cut and stained with hematoxylin and eosin (H & E) or toluidine blue, respectively. H & E stained sections were evaluated for the presence of basophilic foci, areas of cellular alteration and hepatocellular carcinoma. Toluidine blue stained sections were quantified for lipid content using standard morphometric analyses.[6] Parasites were also apparent, but were not critically evaluated.

Incidence of Hepatocellular Carcinoma

The liver of the Atlantic tomcod possesses two large lobes and one smaller central lobe that is suspended over the digestive tract. Normal appearing liver is brown to tan (Class 1) while hepatocellular carcinomas appear as translucent areas, cream colored nodules or multinodular, hemorrhagic masses (Class 2). Very large tumors can encompass the entire lobe of a liver and often are cystic with mixtures of blood and bile (Class 3).

Gross evidence of hepatocellular carcinoma was only detected in tomcod from the Hudson River. No gross tumors or small lesions were seen in any of the other rivers. However, histological examination showed two small neoplasms in one fish from the Saco, a 2% incidence and 1 neoplasm in one fish from the Pawcatuck, a 4% incidence. None of the other rivers sampled had detectable carcinoma (Table 3). Dey et al.[4] also detected two fish with neoplastic nodules in the Pawcatuck, an incidence of 5%. The low, but nevertheless detectable, incidence of cellular transformation in some of these other rivers indicates that disease in the Hudson River population is not an isolated case or due solely to some genetic predisposition. Since the Hudson is on the southern extreme of the species range, the synergistic effect of temperature with xenobiotics is possible, but thermal stress alone is not the likely cause of these tumors since fish in colder waters also developed carcinomas. Both the Pawcatuck and the Saco have had past industrial inputs from textile mills and a tannery, respectively, so xenobiotics are still the most likely cause.

Normal livers in paraffin section were typically composed of cuboidal epithelial cells having an oval or round nucleus with a vacuolated cytoplasm. Tubules of cells were separated by blood sinuses. Pancreatic tissue was normally present within the liver, but was generally located near the margins of the liver and near ducts. Pancreatic tissue was more darkly stained but was easily distinguished from hepatocellular carcinoma cells by their smaller nuclei, uniform cell shape, and the presence of pink staining zymogen granules.

Several pathological abnormalities in hepatic tissue were identified in tomcod livers. These included fatty degeneration, basophilic foci, areas of cellular alteration, and hepatocellular carcinoma. Fatty degeneration was characterized by a highly vacuolated cytoplasm in paraffin sections that were filled with lipid in

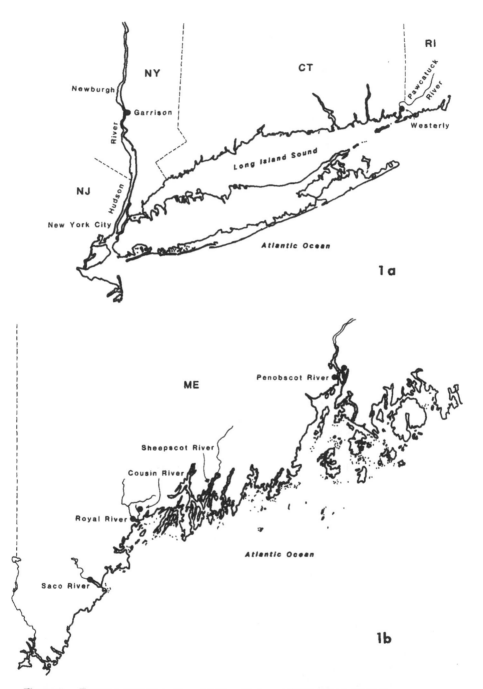

Figure 1.   Tomcod collecting sites. (a) New York and Connecticut; (b) Maine.

Table 3.   Incidence of Liver Tumors in Atlantic Tomcod Based on
Gross Observations and Histological Analyses.

River	Gross (Class 2, 3)	Histology (Class 1)	Total
		All Fish	
Field collection			
Hudson	27/58 (47%)	6/38[a] (16%)	33/64 (52%)
Saco	0/68	1/56 (2%)	1/56 (2%)
Royal	0/4	0/4	0/4
Penobscot	0/19	0/17	0/17
Pawcatuck	0/35	1/25 (4%)	1/25 (4%)
		Fish <175 mm	
Field collection			
Hudson	0/21	2/21 (10%)	2/21 (10%)
Laboratory reared			
Hudson	0/18	0/18	0/18
Sheepscot	0/3	0/3	0/3

[a]Includes 8 fish used for breeding stock at SP, Inc., MA that were not grossly classified—2/8 were positive.

plastic sections. Basophilic foci were also vacuolated but the cytoplasm was more darkly stained than the surrounding tissue. Basophilic areas of cellular alteration and hepatocellular carcinoma were very similar in appearance and differed only in the amount of area affected and the degree of invasion. The cells of these two pathologies were larger and possessed larger nuclei. They were less vacuolated and were more basophilic than the surrounding cells (Figure 2). The cells often were not cuboidal but were elongated in the same plane of section and the tissue architecture was greatly altered. Large areas of these cells compressed and displaced the normal hepatic parenchyma. Necrosis, hemorrhage, and cysts containing eosinophilic material were characteristic of large tumors.

Other Histopathologies

Plastic sections of normal tissue revealed three cell types forming the hepatic tissue proper (Figure 3). The most common cell type corresponded to the cells previously described in paraffin sections. They were cuboidal, stained light blue with variable amounts of green lipid droplets. However, a second kind of cell not distinguishable in the paraffin sections had an irregular shape and stained more darkly. The function of these darkly stained cells is unclear, but they were not found lining the sinusoids and therefore were not Kupffer cells. They may be pre-ductule or fat-storing cells. The third cell type that was easily seen lining the Space of Disse were endothelial cells.

Plastic sections of hepatic tissue from the Hudson River sample were clearly different from those of the other rivers (Figure 4). First, the darkly stained cells were completely absent and second, the lipid within the hepatocytes was sig-

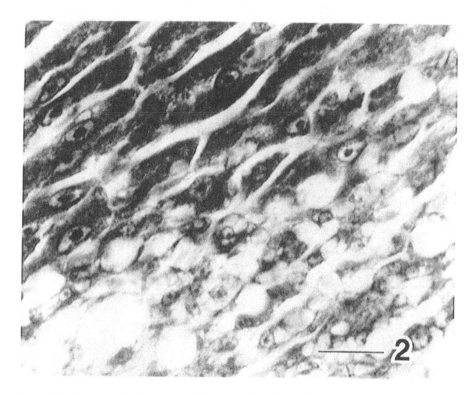

Figure 2. Liver sample from tomcod from Hudson River. Hepatocellular carcinoma cells to the left of micrograph are elongated and larger than vacuolated hepatocytes, H & E. Bar is 20 μm.

nificantly elevated. Morphometric analysis showed a 36.3% vs 11.0% lipid level for the Hudson and Pawcatuck samples, respectively.[7] By comparison, other fish species (n = 4) had only 4.0% lipid. There was also a significant difference between males and females in the Hudson sample. The darkly stained cells were not present in the hepatocellular carcinoma nor the surrounding tissue. Hepatocellular carcinoma cells contained less lipid accounting for both the less opaque appearance in gross observations of the liver and the less vacuolated appearance in paraffin section.

The high incidence of hepatocellular carcinoma and other histopathologies indicates that hepatic function is probably compromised. Of particular noteworthiness was the absence of darkly staining cells from the Hudson River sample. Although the function of these cells is not known, it seems reasonable to assume that the loss or alteration of these elements is not beneficial. Fatty infiltration of hepatocytes was found to be pathological and generally elevated in the Hudson sample. This metabolic change may be due to detoxification as

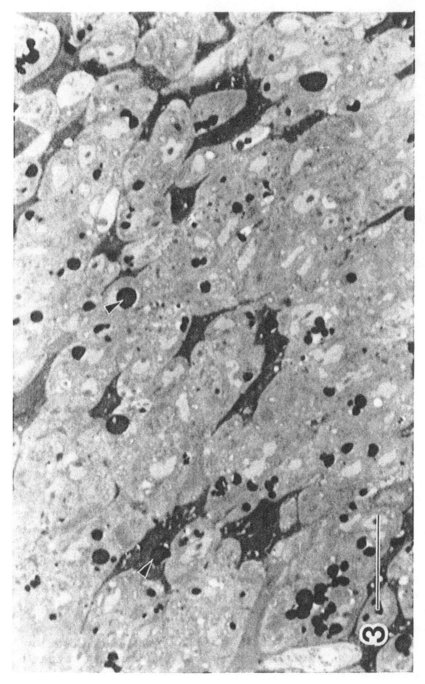

Figure 3. Liver sample from tomcod from Pawcatuck River. Hepatocytes and darkly stained cells both contain moderate amount of lipid (arrows) droplets. Plastic section, toluidine blue. Bar is 20 μm.

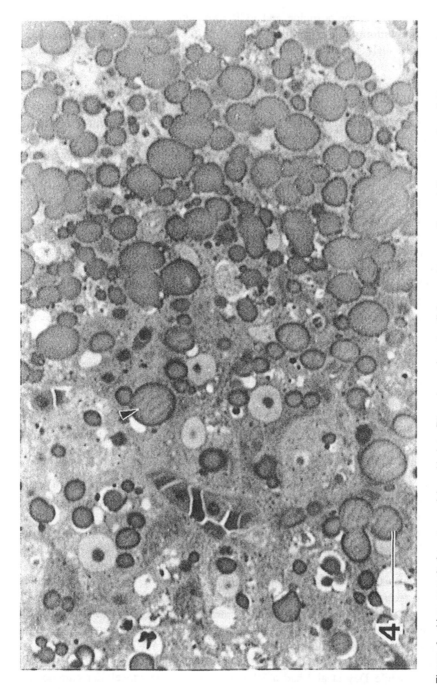

Figure 4. Liver sample from tomcod from Hudson River. Large hepatocellular carcinoma cells on left possess less lipid (arrows) than area of fatty degeneration on right. Plastic section, toluidine blue. Bar is 20 μm.

has been reported for pesticide and PCB exposures.[8] Furthermore, the larger lipid depot within the liver probably augments bioaccumulation of lipid soluble xenobiotics such as PCBs and pesticides. In fact, Dey et al.[4] showed very high levels of these compounds and heavy metals in tomcod livers.

Environmental Correlates

Although the histopathologies and chemical analyses of sediments and tissues suggest a correlation between pollutants and tumor incidence, this comparison is currently limited since chemical analyses for sediment are only available from the Hudson River[9,10] and chemical analyses of livers have only been performed on the tomcod from the Pawcatuck and Hudson Rivers.[4] Also, the exacerbating role of higher temperatures in the Hudson, which is at the southern extreme of the tomcod's range, cannot be ignored as a possible synergistic factor.

Nevertheless, levels of contaminants are believed to be much lower in the other rivers studied than in the Hudson River and the minimal and lack of hepatocellular carcinoma would support this contention. However, we do know that although input into the Saco River is limited, an upstream dam retains sediments that may contain potential carcinogens from a tannery plant. These sediments being upstream from the spawning area, however, are not directly accessible to the native tomcod population. The other rivers in Maine that were sampled do not have appreciable industrial input, but may be impacted by agricultural runoff; however, the Maine Department of Natural Resources considers these to be among the cleanest tidal rivers in the state.[11] Pristine, chemical-free rivers probably no longer exist in the United States; however, this study would suggest that some levels of chemical mixtures can be tolerated, but not levels as high as have been found in the Hudson River.

Rearing Experiment

Although the survey data from several rivers seem to link river condition with the incidence of hepatocellular carcinoma, this evidence is strictly circumstantial. A different approach was used to test this argument. Juvenile tomcod were collected from the Hudson River in June of 1987 and reared in the laboratory in dechlorinated water obtained from the municipal water supply for Louisville, Kentucky. Juvenile tomcod were also collected from the Sheepscot River, Maine to serve as a negative control. Both groups of fish were reared in circular tanks using recirculating water systems and fed an artificial diet. In December of 1987, surviving fish were sacrificed, autopsied, and their livers were fixed in buffered formalin. Surprisingly, no hepatocellular carcinoma was found at the gross and histological levels for either group. Our samples from the Hudson for fish of a similar size, less than 170 mm, had a 10% incidence of hepatocellular carcinoma (Table 3) while Dey et al.[4] had a 39% incidence of hepatocellular carcinoma for fish less than 175 mm. Fish that died prior to the final autopsy also lacked any incidence of hepatocellular carcinoma or areas of cellular alteration.

Although caution must be exercised in interpreting these findings due to our low sample size, the lack of tumor formation in the laboratory-reared Hudson River population suggests that maternal loading of yolk may not be sufficient to induce hepatocyte transformation, and that exposure during the first six months of life is not sufficient to induce high levels of tumor expression. This is interesting since laboratory experiments have shown that yolk sac larvae are more sensitive to carcinogens than older fish.[12] The insufficient exposure period may simply be too short or it may be that tomcod are exposed to the causative agents closer to the mouth of the Hudson and the juveniles that we had collected had not yet migrated down to contaminated areas. Alternatively, data linking size of fish with a higher tumor incidence[4] would suggest that carcinogen dosage through food is even more likely. Another possible factor that may be responsible for the lack of tumor expression in the laboratory-reared fish is that light cycle and temperature were not matched with those of the Hudson and none of the fish were in reproductive condition when sacrificed, as were most of the field collected fish. Hormonal stimulation of some cancers has been well documented and may be required for the promotion of tumor growth. Although titillating, these findings should be reproduced with a larger sample size and rearing conditions must be more closely matched with seasonal conditions normally found in the river.

Population Effects

Age—The age distribution of the Hudson River population of tomcod may be shifted toward younger individuals due to the reduced fitness of fish compromised by hepatocellular carcinoma. Of the more than 560 tomcod collected from the Hudson, only one was found to be 3 years old.[4] We found one 3-year-old in the Saco out of 68 collected. Two of the 19 fish collected from the Penobscot were 2+ in the summer and would be considered age 3 in December. Three-year-olds and even older tomcod have been more commonly reported from rivers north of the Hudson and in Canada.[13-15] Since the incidence of hepatocellular carcinoma is almost 100% in two-year-old Hudson tomcod[4] this may now be the upper age limit for this population.

Size—The tomcod from Maine also appear to be larger for a given age (Table 4). This size difference was reflected in both our field collections and our laboratory-reared fish (field data: $t = 2.60$, $df = 67$, $p < 0.02$; laboratory data: $t = 2.26$, $df = 19$, $p < 0.05$). Since spawning occurs earlier in Maine, the difference may just be due to age differences but we also collected earlier in Maine. Further effort on these population parameters is certainly needed.

CONCLUSIONS

Atlantic tomcod from the Hudson River clearly show pathologies that are almost certainly related to the presence of xenobiotics. No other species in the Hudson have shown similar problems which suggests that the tomcod is highly

Table 4.   Lengths of One-Year-Old Adult Atlantic Tomcod Collected from Rivers in Maine and New York During the Annual Spawning Run and Final Lengths of Laboratory Reared Juveniles.

Site	Date	n	Mean (mm)	SD	Range (mm)
Adults					
Saco, Royal	12/86	38	194.5	22.38	151–253
Saco	12/87	16	190.1	14.84	164–218
Hudson	1/87	4	153.5	35.62	117–193
Hudson	1/88	11	176.7	28.42	130–216
Hudson[a]	12/83–3/84	458	155.2	—[b]	109–235
Laboratory					
Sheepscot	12/87	3	151.7	10.62	139–165
Hudson	12/87	18	135.3	16.28	107–170

[a]From Dey et al.[4]
[b]Not provided.

sensitive. This sensitivity may be specific for a particular class of compound or to a wide range of compounds. In either case, the tomcod would be an ideal biomonitor in the rivers in which it is found. We would like to be able to detect the presence of xenobiotics well before hepatocellular carcinoma becomes commonplace in the population. Efforts are now underway to analyze biomarkers such as DNA adducts and DNA breakage which may allow much more sensitivity in determining environmental impacts. Whether the tomcod will prove suitable for biomarker analysis is yet to be determined but it clearly is a useful biomonitor of serious degradation.

Questions raised by this study can be further answered by additional sampling of both highly and slightly impacted sites for liver pathologies in the tomcod and then correlating these findings with pollutant levels. We believe that this study has demonstrated that the tomcod should be considered as a valuable biomonitor of tidal rivers in the northeast. A close relative, *Microgadus proximus*, inhabits rivers on the west coast of North America and could prove to be an equally useful biomonitor.

ACKNOWLEDGMENTS

We would like to thank Charlie Smith, Dennis Suszkowski, and John Waldman for their helpful reviews of the manuscript. This research was supported by grants from the Hudson River Foundation and the University of Louisville Graduate School.

REFERENCES

1. Smith, C.E., T.H. Peck, R.J. Klauda, and J.B. McClaren. "Hepatomas in Atlantic Tomcod *Microgadus tomcod* (Walbaum) Collected from the Hudson River Estuary in New York," *J. Fish Dis.* 2:313–319 (1979).

2. Cormier, S.M. "Fine Structure of Hepatocytes and Hepatocellular Carcinoma of the Atlantic Tomcod, *Microgadus tomcod* (Walbaum)," *J. Fish Dis.* 9:179–194 (1986).
3. Rohmann, S. *Tracing a River's Toxic Pollution: A Case Study of the Hudson*; (New York: Inform, Inc., 1985), 154 pp.
4. Dey, W., T. Peck, C. Smith, S. Cormier, and G. Kraemer. "A Study of the Occurrence of Liver Cancer in Atlantic Tomcod (*Microgadus tomcod*) from the Hudson River Estuary," Final Report to Hudson River Foundation, (New York: Hudson River Foundation, 1986), 58 pp.
5. Neff, S. "Polycyclic Aromatic Hydrocarbons," in *Fundamentals of Aquatic Toxicology,* G. Rand and S. Petrocelli, Eds. (New York: Hemisphere Publ., 1985), pp. 416–454.
6. Weibel, E.R. *Stereological Methods, Vol. 1, Practical Methods for Biological Morphometry* (New York: Academic Press, 1979), pp. 101–161.
7. Cormier, S.M., R.N. Racine, C.E. Smith, W.P. Dey, and T.H. Peck. "Hepatocellular Carcinoma and Fatty Infiltration in the Atlantic Tomcod, *Microgadus tomcod* (Walbaum)," *J. Fish Dis.* 12:105–116 (1989).
8. Couch, J.A. "Histological Effects of Pesticides and Related Chemicals on the Live of Fishes," in *The Pathology of Fishes*, W.E. Ribelin and G. Migaki, Eds. (Madison: University of Wisconsin Press, 1975), pp. 559–584.
9. Bopp, R.F., H.J. Simpson, C.R. Olsen, and N. Kostyk. "Polychlorinated Biphenyls in Sediments of the Tidal Hudson River, New York," *Environ. Sci. Technol.* 15:210–216 (1981).
10. Bopp, R.F., H.J. Simpson, C.R. Olsen, R.M. Trier, and N. Kostyk. "Chlorinated Hydrocarbons and Radionuclide Chronologies in Sediments of the Hudson River and Estuary, New York," *Environ. Sci. Technol.* 16:666–676 (1982).
11. Flagg, L. Personal communication (1986).
12. Hendricks, J.D. "Chemical Carcinogenesis in Fish," in *Aquatic Toxicology*, L.J. Weber, Ed. (New York: Raven Press, 1982), pp. 149–211.
13. Howe, A.B. "Biological Investigations of the Atlantic Tomcod, *Microgadus tomcod* walbaum, in the Weweantic River Estuary, Massachusetts, 1967," M.S. Thesis, University of Massachusetts, Amherst, (1971), 89 pp.
14. Legendre, V. and R. Lagueux. "The Tomcod (*microgadus tomcod*) as a Permanent Fresh-Water Resident of Lake St. John, Province of Quebec," *Can. Field Nat.* 65:157 (1948).
15. Roy, J.M., G. Beaulieu, and G. Labrecque. "Observations sur le Poulamon, *Microgadus tomcod* (Walbaum), de l'Estuaire du Saint-Laurent et de la Baie de Chaleurs," Ministere de Industrie et du Commerce, Direction des peches maritimes, Direction de la recherche. Cahiers d'information 70:1–56 (1975).

CHAPTER 4

The Croaker (*Nibea mitsukurii*) and the Sea Catfish (*Plotosus anguillaris*): Useful Biomarkers of Coastal Pollution

Mitsuko Yamashita, Naohide Kinae, Ikuo Kimura, Hirojji Ishida,
Hidemi Kumai, and Genji Nakamura

During the course of our epizootiological surveys on neoplasms in coastal fishes, an estuarine area (station A) was unique in the high incidences of chromatophoroma in the croaker, *Nibea mitsukurii*, and of skin pigment cell hyperplasia in the sea catfish, *Plotosus anguillaris*, respectively. The following results of our studies suggest that the pigment cell disorders in these species may be useful biomakers to monitor the coastal water pollution by carcinogens:

1. The skin lesions in both species were induced by administration of certain kinds of known carcinogens.
2. The effluents from a kraft pulp mill located near station A contained mutagenic substances.
3. The effluents induced skin pigment cell hyperplasia on 70 to 100% of tank-reared sea catfish.
4. Four chloroacetones and three α-dicarbonyl compounds were detected from the effluents as mutagens by GC-MS analysis.

INTRODUCTION

More than 70,000 kinds of natural and authentic chemicals are used for industrial purpose in the world. After use, some of them are directly and/or indirectly discharged into aquatic environments such as lake, river, and seashore.

As a result of these phenomena, several aquatic organisms are suffering from the attack of chemical contaminants.

Since the first observation of naturally occurring fish tumors was reported at the end of 18th century, investigators have diagnosed neoplastic diseases in 250 kinds of fish species so far. Lucké and Schlumberger[1] have described a very impressive report in which environmental contaminants would be potent candidates for induction of fish tumors. In recent years, Pierce et al.[2] found hepatic tumors in English sole at Puget Sound, Washington. Then Malins et al.[3] studied what kinds of chemicals exist in the sediment of the coastal area. They assumed that certain polycyclic aromatic hydrocarbons (PAHs) are related to the induction of marine fish tumors. Actually, there was a good correlation between the incidences of hepatocellular carcinoma or papilloma of these fish species and the concentration of PAHs in the sediment samples. Black[4] reported that a high prevalence of epidermal papilloma, epidermal carcinoma, and hepatocellular carcinoma was observed in brown bullhead inhabiting the Buffalo river in which the sediment was contaminated with PAHs including benzo(a)pyrene (BaP). They also reported that the same skin and liver tumors were induced by painting or by oral administration of the sediment extracts to the brown bullhead.

In Japan, Kimura et al.[8] first observed the chromatophoromas in the croaker (*Nibea mitsukurii*, Sciaenidae, Percina, Percida, Teleostei) in 1973. In 1974, a project team financially supported by the Agency of Science and Technology of Japan was started to investigate the geographic distribution of the fish tumors and to explore the causal factors of the tumors.

The epizootiological study showed that the geographic distribution of chromatophoroma in the croaker was greatest at the Pacific coastal area of Japan islands, station A. At the same station, the skin pigment cell hyperplasia in the sea catfish (*Plotosus anguillaris*, Plotosidae, Sclurina, Cyprinida, Teleostei) was also found. Both species were bottom-feeding fish inhabiting the northeastern coast to southwestern coast of Japan. Through these experiments, the hypothesis that the fish tumors were caused by parasitic and genetic factors was denied. But an effect of environmental chemicals on induction of these fish tumors was obscure. We continued the epizootiological study for six years. Our focus was to research whether the occurrence of the skin tumors in the feral fish was related to environmental chemicals, especially xenobiotics in the effluents discharged from a kraft pulp mill, which is located near station A.

In this paper, we discuss the relationship between skin tumors of these fishes and the chemical constituents of the kraft pulp mill effluents, and the possibility of both fish species as biomarkers in the aquatic environment which is contaminated with chemcial substances.

MATERIALS AND METHODS

Epidemiological Studies of Croaker and Sea Catfish

An epidemiological survey of chromatophoroma in the croaker was done at 70 stations in Honshu, Shikoku, and Kyushu islands from 1973 to 1981.[5] Fish samples were caught by trawl net or hook-and-line fishing. The sampling location, body size, body weight, and color of the skin tumors were recorded on a standardized data sheet. The survey at station A which showed the highest incidence of the tumor had continued up until this year.

In our experiment, the skin pigment cell hyperplasia in the sea catfish was also surveyed from 1979 to 1980 at 11 stations.

Sampling of Kraft Pulp Mill Effluents

Effluents A and B taken from a kraft pulp mill located by station A were sewages before and after final treatment, respectively. B was mainly discharged from the kraft pulp mill into the coastal seashore.

Induction of Skin Diseases to Tank-Reared Fish

In the exposure experiment of carcinogens to fish, we used 550 tank-reared 5-month-old croakers and divided them into five groups. Group 1 received a single subcutaneous injection of 1–2 mg of dimethylbenz(a)anthracene (DMBA, Sigma Chemical Co., St. Louis, Missouri) dissolved in olive oil. Group 2 received a single gastric intubation of 2–3 mg of N-methyl-N′-nitro-N-nitrosoguanidine (MNNG, Aldrich Chemical Co., Milwaukee, Wisconsin) dissolved in saline. Group 3 received a single subcutaneous injection of 0.1–0.2 mL of olive oil and a single intubation of 0.2–0.3 mL of saline. Group 4 was held as an untreated control. All fish used in the experiment had been exposed to a solution of 6-hydroxymethyl-2-[2-(5-nitro-2-furyl)-vinyl]pyridine (NP: nifurpironol, Dainihon Pharmaceutical Co., Japan) when they were moved from one tank to another; i.e., the fish of groups 1, 2, 3, and 4 were exposed to 1 ppm of NP solution each for 1 hr. NP has been used as an antibacterial agent in fish culture. Group 5 was the control group without NP treatment. The details of this experiment have been described elsewhere.[5]

The sea catfish (6 cm total body length) were also exposed to sea water containing 1 ppm of NP for 2 hr. The exposure was done seven times (once every 2 weeks).

Exposure experiments of paper mill effluents to both fish species were also performed in the hatchery. The croaker and the sea catfish were exposed to sea water containing 10% of effluent A or B. One hundred 50-day-old croakers were exposed to each effluent, once a month for 8 hr-exposure. Fifty sea catfish (6 cm total body length) were exposed to the same test waters once per 2 weeks for 12 hr-exposures (7 times). In the exposure experiment to effluents, NP was not used in all groups. After the exposure, experimental and control groups were reared in the unpolluted sea water.

Isolation of Mutagens from the Kraft Pulp Mill Effluents

After determination of physicochemical data (pH, COD), each water sample was passed through an Amberlite XAD-2 column. An adsorbate was eluted by diethylether and then methanol by using a Soxhlet extracting apparatus. Both extracts obtained were subjected to the mutagenicity test using *S. typhimurium* TA98 and TA100 according to the method of Ames et al.[6]

Determination of Chloroacetones

1,3-Dichloroacetone and hexachloroacetone were purchased from Tokyo Kasei Kogyo Co. Ltd., Japan) and 1,1,3-trichloroacetone and pentachloroacetone were purchased form Aldrich Chemical Co. Ltd. 1,1,3,3-Tetrachloroacetone (bp718 180–182°C) was obtained by reduction of pentachloroacetone dissolved in methanol with Pd-C.[7]

One liter of the effluent A or B was extracted with three 200-mL portions of diethylether. Five chloroacetones in the ether extract were determined by GC-MS-SIM method (gas chromatography-mass spectrometry-selected ion monitoring; Hitachi M-80B, Hitachi Ltd., Japan).

Determination of α-Dicarbonyl Compounds

Glyoxal and diacetyl were purchased from Tokyo Kasei Kogyo Co. Ltd., Japan and methylglyoxal was purchased from Sigma Chemical Co. Each 0.2 g of o-phenylenediamine (OPD) was added to 1 L of effluent A or 2 L of effluent B. The mixed solution was stirred for 2 hr at room temperature. The reaction solution was treated with three 100-mL portions of chloroform. The solvent of the chloroform extract was evaporated and the residue was dissolved in 5 mL of chloroform. A 1-mL aliquot of the chloroform solution was subjected to a Sep-Pak Silica-cartridge C18 (Waters Associates, U.S.A.) and eluted with 9 mL of chloroform. Three standard quinoxaline derivatives (quinoxaine, 2-methylquinoxaline, and 2,3-dimethylquinoxaline which were derivatives corresponding to glyoxal, methyglyoxal, and diacetyl, respectively) were prepared as follows: A mixture of 0.1 mole of α-dicarbonyl compound and 0.2 moles of OPD was incubated and treated in the same manner as the cases of the effluent. Three α-dicarbonyl compounds were determined by GC-MS-SIM method. Each sample was injected onto Ultra-1 column (crosslinked methyl silicone gum, 0.32 mm × 25 m, Hewlett Packard, U.S.A.) using cooled-on-column injector (Gaskuro Kogyo Inc., Japan). The column temperature for chloroacetones was programmed at 5° C/min up to 280°C after keeping at 50°C for 2 min. For quinoxaline derivatives, the temperature was programmed at 20° C/min up to 280° C after keeping at 125° for 7 min. In both experiments, the temperatures of injection and interface ports were held at 150° C and 280° C, respectively. Helium was used for carrier gas. In GC-MS-SIM analysis two characteristic ions (m/Z) were used for monitoring each compound: 77 and 126 for 1,3-dichloroacetone, 77 and 83 for 1,1,3-trichloroacetone, 83 and 111 for 1,1,3,3-tetrachloroacetone, 83 and 117 for pentachloroacetone, 117 and 119 for hexachloroacetone, 103 and 130 for quinoxaline, 117 and 144 for 2-methylquinoxaline, and 117 and 158 for

Figure 1. *Nibea mitsukurii* bearing chromatophoroma caught at station A.

2,3-dimethylquinoxaline. SIM areas of each sample and external reference were used for quantitative analysis.

RESULTS

The croaker, *N. mitsukurii*, is a common bottom feeder inhabiting shallow waters along the Pacific coast from northeastern to southwestern portions of Japan islands.

A photograph of the croaker, which was caught at station A, bearing skin tumor, chromatophoroma, is shown in Figure 1. The survey stations and the incidence of the chromatophoroma in the croaker are shown in Figure 2. The geographic distribution of the tumor was greatest at station A. The average incidence of 47% at the station from 1973 to 1981 was significantly higher than that (0–8.5%) of the other 23 stations. The annual changes of tumor incidence of the croaker at station A from 1973 to 1987 is shown in Figure 3. The high average incidence (47%) between 1973 and 1983 decreased to about 20% between 1984 and 1987. In all cases, no tumor was observed in fish having total body length less than 209 mm. Namely, all tumor fish were older than 1.5 years. This fact shows that the fish tumor incidence increases with an increase of total body length.

The sea catfish having a skin lesion was also caught at the same area that the croaker inhabited (Figure 4). The lesion was diagnosed as a skin pigment cell hyperplasia by histological examination. The average incidence of this fish at

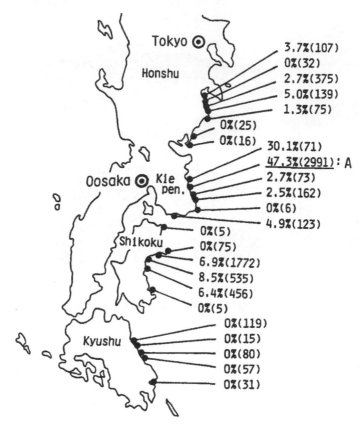

Figure 2. Geographic distribution of chromatophoroma in *Nibea mitsukurii* in Japan (1973 to 1981). The figures in parentheses showed the number of fish observed.

station A was 13.5% in 1979 to 1980 and was significantly higher than that of the other 10 survey stations (Table 1).

The croakers which were tank-reared 5 months old were exposed to well known carcinogens, DMBA and MNNG. Small melanotic lesions were developed within 3 months. After 8 months, the incidences of the chromatophoroma were 100% for DMBA and 70% for MNNG (Table 2). But another chemical, NP, which had been used as an antibacterial agent in fish culture, unexpectedly induced chromatophoroma in the croaker; even for a single exposure of NP, the incidence was 40% after 8 months. These results show that the chromatophoroma can be induced by exposure of carcinogens in the croaker as well as in mammals.

When another species of fish, the sea catfish, was exposed to NP, black pigmented lesions were developed after 10 months of exposure. The lesions were diagnosed as skin pigment cell hyperplasia, melanosis, from the histological examination, and were very similar to those found in feral sea catfish caught at

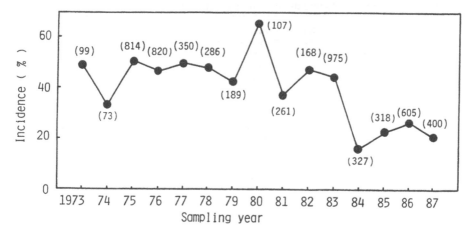

Figure 3. Incidence of chromatophoroma in *Nibea mitsukurii* (station A). The figures in parentheses showed the number of fish observed.

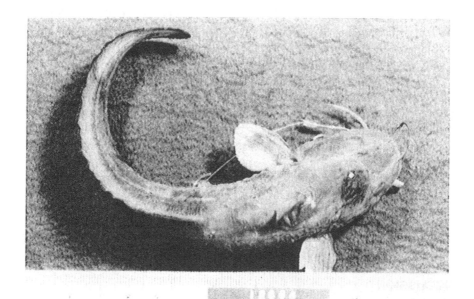

Figure 4. *Plotosus anguillaris* bearing melanosis caught at station A.

Table 1.   Incidence of Melanosis in *Plotosus anguillaris*.

Station	No. of Fish Melanosis/ No. of Fish Examined	Incidence (%)
A	15/111	13.5
B	10/210	4.8
C	1/40	2.5
D	0/19	0
E	0/39	0
F	0/10	0
G	0/43	0
H	0/44	0
I	0/32	0
J	0/28	0
K	0/123	0

station A. The incidence of melanosis in the sea catfish by exposure of NP was 91% after 18 months (Table 3).

The sea catfish was exposed to effluents A and B which were taken from a kraft pulp mill located by station A. Skin melanosis appeared on the surface of the body. After 18 months, the incidences of melanosis reached 100% (7/7) for effluent A and 70% (14/20) for effluent B, respectively (Table 3).

An exposure experiment of effluents A and B to the croaker was also carried out. When the croaker was exposed to sea water containing 10% of effluent B, one fish developed chromatophoroma on the dorsal fin after 13 months of the exposure.

The physicochemical data of the effluents submitted to tumor induction test and the mutagenic activity of their ether extracts after pasing through an Amberlite XAD-2 column are described in Table 4. They showed mutagenic activity toward *S. typhimurium* TA100 without metabolic activation.

We purified the mutagens in the ether extracts of both effluents by gel filtration using a Sephadex LH-20 column and a high performance liquid chromatography (HPLC) with LiChrosorb Si100 and Si60 columns. 1,1,3,3-Tetrachloroacetone was detected from a mutagenic fraction of effluent A. Three α-dicarbonyl compounds such as glyoxal, methylglyoxal, and diacetyl were identified as quinoxaline derivatives from effluents A and B.

These compounds were determined by GC-MS-SIM and the results are described in Table 5. In effluent A, 1,3-dichloroacetone (0.688 μg/L), 1,1,3-trichloroacetone (44.2 μg/L), and pentachloroacetone (12.2 μg/L) as well as 1,1,3,3-tetrachloroacetone (35.4 μg/L) were detected. No chloroacetones were detected in effluent B. The concentrations of glyoxal, methylglyoxal, and diacetyl were 64.5, 140, and 10.0 μg/L in effluent A, and 26.3, 52.1, and 14.2 μg/L in effluent B.

DISCUSSION

The geographic distributions of chromatophoroma in the croaker and skin pigment cell hyperplasia in the sea catfish were greatest at station A. Several

Table 2. Induction of Chromatophoroma in *Nibea mitsukurii* by DMBA or MNNG with Exposure to NP.

Group	Initial No. of Fish	Compound Given and Dosage	Exposure Route	Exposure to NP	No. with Lesion/No. Examined (%)	
					After 7 months	After 8 months
I	50	DMBA (1–2 mg)	sc injection	Twice	11/15 (73)	4/4 (100)
II	50	MNNG (2–3 mg)	Gastric intubation	Twice	26/44 (59)	7/10 (70)
III	50	Olive oil (0.1–0.2 mL) and saline (0.2–0.3 mL)	Olive oil sc + saline by gastric intubation	Twice	9/40 (23)	6/10 (60)
IV	300	300	Untreated	Once	32/246 (13)	4/10 (40)
V	100	Untreated	None			1/38 (2.6)

Table 3.   Incidence of Melanosis in *Plotosus anguillaris* Exposed to the Kraft Pulp Mill Effluents and NP.

Group	Exposure conditions Effluent	Exposure conditions hr × times	No. of Fish Examined	No. of Fish with Melanosis	Incidence (%)	No. of Melanosis/ Fish
I	A	12 × 7	7	7	100	3.00
II	B	12 × 7	20	14	70.0	1.55
III	NP	2 × 7	23	21	91.3	3.09
IV	Control	—	10	1	10.0	0.10

other fish including sole (*Areliscus joynerii*), black sea bream (*Mylis latus*), and perch (*Lateolablax japonicus*) had adenomatous polyps in their stomachs at the same station, but these tumors were found in too few numbers to do an epizootiological study.

In the tumor induction test, using well-known carcinogens, chromatophoroma and melanosis were easily induced in the croaker and the sea catfish, respectively. These lesions were very similar to those observed in feral fish. The specific difference of sensitivity of carcinogens against each organ was observed in fish as shown in mammals. The metabolic activation of xenobiotics in fish may change depending on the species. As these lesions appear on the skin surface, it is very easy to find the development of the tumor.

We have been researching the causes of lesions in fish. To analyze the causes of this geographic distribution of the tumor fish, epizootiological studies have been done using several environmental factors. Viral, genetic, and parasitological factors have been negative so far. In consideration of the above results, we supposed that environmental contaminants, which are present in the coastal area, might be related to the development of fish tumors. From this point of view, we studied the relationship between fish tumor and postulated environmental contaminants, especially the mutagenic constituents in a kraft pulp mill effluents A and B, induced skin melanosis in the sea catfish at high prevalence. Chromatophoroma was induced on the dorsal fin of one croaker by exposure to effluent B. These results suggest that there is a certain correlation between the chemical constituents of effluent and the development of the skin lesion.

Table 4.   Water Qualities of the Kraft Pulp Mill Effluents Used to the Exposure Experiment.

Effluent	pH	COD (mg/L)	Yield of Ether Ext. (mg/L)	Mutagenic Activity (TA100, −S9) His$^+$ revs/100 μg of Ether Ext.
A	4.3	150	12.6	152
B	6.7	56	4.3	70

Table 5.   Quantification of Identified Compounds in the Kraft Pulp Mill Effluents.

Compounds	Mutagenic Activity TA100, revs/μg	Conc. in Effluent (μg/L)	
		A	B
1,3-Dichloroacetone	292	0.688	0
1,1,3-Trichloroacetone	52	44.2	0
1,1,3,3-Tetrachloroacetone	19	35.4	0
Pentachloroacetone	4.0	12.2	0
Hexachloroacetone	1.5	0.1	0
Glyoxal	6.5	64.5	26.3
Methylglyoxal	60	140	52.1
Diacetyl	0.08	10.0	1.42

By GC-MS-SIM analysis, four mutagenic chloroacetones were detected in effluent A (0.688–44.2 μg/L). And three α-dicarbonyl compounds were found in effluents A and B at a concentration of 10.0–140 μg/L and 1.42–52.1 μg/L, respectively.

Recently, other investigators reported the presence of some mutagens, neoabietic acid[8] and chloropropenes,[9] in pulp and paper mill effluents. Furthermore, chloroacetones,[10] 2-chloropropenal,[11] and chlorofuranone[12] were also detected as mutagens in the effluents on the chlorine-treatment process. But there are few reports which examined the carcinogenicity of the mutagenic compounds identified from the effluent of pulp and paper mills. In our preliminary tumor-induction test to fish, when 1,3-dichloroacetone (0.1 ppm) or pentachloroacetone (10 ppm) was exposed to red medaka (*Oryzias latipes*), hyperplasia was observed on gill and GI tract in the former group. Recently, Nagao et al.[13] reported that methylglyoxal induced sarcoma on rats by subcutaneous injection. Furihata et al.[14] reported that glyoxal and diacetyl were promoters as well as initiators.

The mutagen content identified from kraft pulp mill effluent is not great, but it is a big problem because of the enormous volume of the effluent discharged into the aquatic environment. It is very important to determine whether these mutagens in effluents were carcinogenic toward fish and mammals. On final discussion, it will be a very useful tool to use the croaker and the sea catfish as biomarkers.

REFERENCES

1. Lucké, B., and H.G. Schlumberger. "Transplantable Epitheliomas of the Lip and Mouth of Catfish. I. Pathology. Transplantation to Anterior Chamber of Eye and Into Cornea," *J. Exp. Med.* 74:397–408 (1941).
2. Pierce, K.V., B.B. McCain, and S.R. Willings. "Pathology of Hepatomas and Other Liver Abnormalities in English Sole (*Parophrys vetulis*) from the Duwamish River Estuary, Seattle, Washington," *J. Natl. Cancer Inst.* 50:1445–1449 (1978).

3. Malins, D.C., B.B. McCain, D.W. Brown, S.-L. Chan, M.S. Myers, J.T. Landahl, P.G. Prohaska, A.J. Friedman, L.D. Rhodes, D.G. Burrows, W.D. Gronlund, and H.O. Hodgins. "Chemical Pollutants in Sediments and Disease of Bottom-Dwelling Fish in Puget Sound, Washington," *Environ. Sci. Technol.* 18(9):705–713 (1984).
4. Black, J. "Field and Laboratory Studies of Environmental Carcinogenesis in Niagara River Fish," *J. Great Lake Res.* 9:326–334 (1983).
5. Kimura, I., N. Taniguchi, H. Kumai, I. Tomita, N. Kinae, K. Yoshizaki, M. Ito, and T. Ishikawa. "Correlation of Epizootiological Observations with Experimental Data: Chemical Induction of Chromatophoromas in the Croaker, *Nibea mitsukurii*," *Natl. Cancer Inst. Monogr.* 65:139–154 (1984).
6. Ames, B.N., J. McCann, and E. Yamasaki. "Methods for Detecting Carcinogens and Mutagens with the Salmonella/Mammalian-Microsome Mutagenicity Test," *Mutat. Res.* 31:347–364 (1975).
7. Wyandotle Chemicals Corp. Brit. 1, 1140, 434, 22 Jan 1969, U.S. Appl. 25 Apr. 1966, 3 pp.
8. Nestmann, E.R., E.G.-H. Lee, J.C. Mueller, and G.R. Douglas. "Mutagenicity of Resin Acids Identified in Pulp and Paper Mill Effluents Using the Salmonella/Mammalian-Microsome Assay," *Environ. Mutagenesis* 1:361–369 (1979).
9. Nestmann, E.R., E.G.-H. Lee, T.I. Matula, G.R. Douglas, and J.C. Mueller. "Mutagenicity of Constituents Identified in Pulp and Paper Mill Effluents Using the Salmonella/Mammalian-Microsome Assay," *Mutat. Res.* 79:203–212 (1980).
10. MaKague, A.B., E.G.-H. Lee, and G.R. Douglas. "Chloroacetones: Mutagenic Constituents of Bleached Kraft Chlorination Effluent," *Mutat. Res.* 91:301–306 (1981).
11. Kringstad, K.P., P.O. Ljungquist, F. de Sousa, and L.M. Strömberg. "Identification and Mutagenic Properties of Some Chlorinated Aliphatic Compounds in the Spent Liquor from Kraft Pulp Chlorination," *Environ. Sci. Technol.* 15(5):562–566 (1981).
12. Holmbom, R.B., R.H. Voss, R.D. Mortimer, and A. Wong. "Isolation and Identification of an Ames-Mutagenic Compound Present in Kraft Chlorination Effluents," *Tappi* 64(3):172–174 (1981).
13. Nagao, M., Y. Fujita, and T. Sugimura. "Methylglyoxal in Beverages and Foods: Its Mutagenicity and Carcinogenicity," in *The Role of Cyclic Nucleic Acid Adducts in Carcinogenesis and Mutagenesis*, B. Singer and H. Bartsch, Eds. (IARC Scientific Publication No. 70, 1986), pp. 283–291.
14. Furihata, C., S. Yoshida, and T. Matsushima. "Potential Initiating and Promoting Activities of Diacetyl and Glyoxal in Rat Stomach Mucosa," *Jpn. J. Cancer Res. (Gann)* 76(9):809–814 (1985).

SECTION THREE

Detoxication, Adaptive, and Immunological Responses

CHAPTER 5

Sublethal Responses of *Platichthys stellatus* to Organic Contamination in San Francisco Bay with Emphasis on Reproduction

Robert B. Spies, John J. Stegeman, David W. Rice, Jr., Bruce Woodin,
Peter Thomas, Jo Ellen Hose, Jeffrey N. Cross, and Mari Prieto

ABSTRACT

In this field study of the effects of organic contaminants in San Francisco Bay on starry flounder *Platichthys stellatus,* we have focused on indicators of contaminant exposure and their relationship to changes in reproductive physiology and biochemistry. We also measured micronucleus occurrence as an indication of the exposure of these fish to environmental mutagens. Two large collecting efforts were made—one during the middle part of the reproductive cycle (November–early December 1986) and one at the time of spawning (February–March 1987). For the former period *P. stellatus* were collected from five localities (four in San Francisco Bay and one near the mouth of the Russian River). A variety of measures were made on these fish that might indicate the nature and extent of contaminant exposure, the alteration of normal reproductive processes, and genetic effects: (1) trace organic contaminants in the liver, mainly chlorinated hydrocarbons, (2) constituents and in vitro enzymatic activities of the hepatic P-450 microsomal system, (3) oocyte development in maturing females, (4) plasma concentrations of vitellogenin and titers of sex steroids, and (5) the incidence of micronuclei in circulating erythrocytes.

The midbay sites of Oakland and Berkeley receive a variety of wastes from the surrounding urban areas. The San Pablo and Vallejo sites are up-estuary and are somewhat less contaminated. A site off the mouth of the Russian River, and later off Santa Cruz, served as less contaminated oceanic comparison sites.

In general hepatic concentrations of chlorinated pesticides did not differ significantly between sites but p,p'-DDT, o,p'-DDT, transnanochlor and mirex were highest at Oakland. Also, the less chlorinated (2 to 4 chlorine atoms per molecule) polychlorinated biphenyls (PCBs) did not differ between sites, except for (1) IUPAC congener 8, with those from the Russian River having the highest concentrations, and (2) IUPAC congener 101, with Oakland having the highest concentrations. Many of the moderately to highly substituted PCB congeners (5 to 10 chlorine atoms) had their highest concentrations in fish collected at the Oakland and Berkeley sites. Ten of the thirteen PCB congeners with five or more chlorine atoms had their highest concentrations in Oakland fish.

For the hepatic P-450 proteins and their catalytic activities the most obvious overall pattern was that Oakland fish had the greatest indications of contaminant induction and those from the Russian River the least. While most mixed-function oxidase (MFO) parameters indicated little difference between sites other than Oakland, P-450E was clearly elevated in Berkeley fish (except in mature females) relative to the other sites. The differences in abundance of P-450E between stations was most consistent with our notion of contaminant distribution prior to the study (based mainly on contaminant concentrations in sediments).

Site differences were generally not observed in plasma steroids. However, there were significant differences between sites in some sex-maturity groups of fish. For example, plasma testosterone concentrations were elevated in San Pablo and Vallejo males and estradiol plasma concentrations were significantly elevated in immature Berkeley females over other sites.

Micronuclei in circulating erythrocytes were significantly elevated in fish from all Bay sites relative to those from the Russian River.

Two of the eight large females from the most contaminated site, Oakland, did not have vitellogenic oocytes, indicating complete inhibition of vitellogenesis in these fish during the 1986–1987 reproductive season. Since their plasma had concentrations of vitellogenin and titers of estradiol typical of immature females it appears probable that the gonadotropin signal was either inhibited or ineffective in triggering estrogenesis.

Overall, ΣPCB concentrations were correlated with EROD activity and EDOD activity, in turn, with hepatic microsomal P-450E concentrations. These relationships are consistent with the known induction of P-450 isozymes by PCBs and the postulated catalyzation of EROD by P-450E.

In immature females plasma estradiol was correlated with P-450E and P-450 content of hepatic microsomes, suggesting that induced fish may either have higher rates of estradiol production or lower rates of estradiol hydroxylation, or both.

A total of 11 females were captured and spawned during January–February 1987. Except for one female captured in Oakland, all of these females were either from Berkeley, or the Santa Cruz coastal site. Nearly all measured parameters were quite variable and not different between these two sites. However, embryological success was significantly greater in Santa Cruz than in Berkeley fish.

INTRODUCTION

The general objectives of our research in San Francisco Bay have been to determine if the reproductive success of an important flatfish species on the Pacific Coast of North America and specifically in its largest estuary, San Francisco Bay, is being impaired by organic contaminants and to further develop methods for assessing reproductive effects in fish from contaminated environments. The efforts during 1986–1987 were done in conjunction with a series of related studies in San Francisco Bay attempting to evaluate the sensitivity of various organisms to contaminated sediments through the use of sediment bioassays.[1]

The results of our earlier research indicated relationships between certain measures of organic contaminant exposure and reproductive success.[2,3] For example, the maternal hepatic aryl hydrocarbon hydroxylase (AHH) activity at the time of spawning was found to be inversely related to three measures—egg viability, fertilization success and successful development from fertilization through hatching. The objectives of this present study were to investigate additional measurements, or indicators, of contaminant exposure that might provide further insight into the mode of action of contaminants on the reproductive system, to assess potential genetic effects of contaminant exposure, and to determine what measures might be useful in monitoring programs of estuarine fish health.

BACKGROUND

One possible explanation for the relationship of AHH activity, which is well known to be induced in fish by certain organic contaminants, and reproductive success is that contaminant responses of this system may interfere with the ability of this general class of P-450 proteins to regulate sex steroids. There is evidence that induction of cytochrome P-450 isozymes from contaminant exposure results in altered patterns of sex steroids metabolism. A cytochrome P-450 isozyme (P-450E) has been isolated from the hepatic microsomes of induced scup, *Stenotomus chrysops*, that hydroxylates both benzo[a]pyrene for which it is the major catalyst, and testosterone,[4,5] the immediate precursor of estradiol. There has also been a similar finding with rats.[6] Because of this surprisingly broad substrate specificity, P-450E, and perhaps other induced isozymes as well, may accelerate testosterone clearance rates in contaminant-induced fish. Evidence in support of this hypothesis was obtained from a series of short-term exposures of salmon and winter flounder to petroleum, resulting in generally lowered total plasma and bile titers of testosterone, 11-ketotestosterone, and 17β-hydroxytestosterone in sexually mature males, but not in males and females exposed during gonadal recrudesence.[7]

Relatedly, treatment of rainbow trout with radiolabeled estradiol and β-naphthoflavone, a polynuclear aromatic hydrocarbon (PAH)-type mixed-function oxidase (MFO) inducer, resulted in increased excretion of the label in the bile.[8]

However, recent in vitro studies with winter flounder and scup hepatic microsomes indicate that P-450E does not catalyze estradiol 2-hydroxylase activity and, in fact, induction may suppress estradiol 2-hydroxylation.[9] Another isozyme, P-450A, may, however, catalyze such activity.[5]

Because all the steps in steroidogenesis from cholesterol to estradiol, as well as steroid metabolism, are mediated by P-450 enzymatic functions, there also exists the potentital for contaminants to interfere in the formation of the proper quantities of steroids at the right times for normal reproductive function. Thus, kidney tissue from cod fed Aroclor 1254 (PCB)-contaminated food exhibited altered patterns of androgen synthesis.[10] Effects of contaminants on steroidogenesis may occur at several different levels in fish. These contaminants may alter production of brain and pituitary hormones that control the secretion of steroid hormones.[11] There can also be direct effects of contaminants on steroidogenesis.[12] The catalyzation of steroid transformations by MFOs may also affect steroidogenesis as the quantities of precursors, for example of testosterone, available for conversion to another steroid could be affected.

During final oocyte maturation, surges of pituitary gonadotropins are the presumed initiating steps in the formation of maturation steroids, e.g., 17-α OH, 20β-dihydroprogesterone, that in turn initiate germinal vesicle breakdown (GVBD) and ovulation. During this period there appear to be multiple opportunities for contaminant-induced cytochrome P-450 activity to interfere in steroid balance and proper reproductive function. One possibility is that the estrogenic effect of chlorinated hydrocarbons[13] could result in the prolonged secretion of estradiol by the ovaries. At the time for final oocyte maturation, high concentrations of estradiol could result in suppression of final oocyte maturation. Evidence of such a phenomenon was obtained from the in vitro incubation of brook trout oocytes where the concentration of estradiol in the culture medium was inversely related to the percentage of ooctyes that underwent germinal vesicle breakdown.[14]

The formation of yolk precursor, vitellogenin, is under the influence of estradiol and fish exposed to PCBs have been reported to be less capable of producing vitellogenin.[15] This may be another manifestation of contaminant effects on reproductive fitness.

Micronuclei are small, secondary nuclei that are formed following chromosomal breakage or spindle damage during mitosis.[16] Micronuclei may arise spontaneously, but the induction of micronuclei above background is a sensitive indicator of genotoxic damage resulting from mutagen exposure.[17] Genotoxic damage has been shown to be higher in fishes from contaminated urban areas compared to fishes from less contaminated reference sites.[18]

METHODS

Sampling Strategy and Site Selection

Collection stations for *P. stellatus* were made as coincident as possible to those used in the sediment bioassay portion of the program (see Long et al., Chapter 20), with some adjustments necessary for the local abundance of fish (Figure 1). For example, the reference site was moved to a location off the mouth of the Russian River. The location of the stations was chosen to represent a possible gradient of contamination within San Francisco Bay, with the Russian River station representing a potentially least-contaminated reference site. Implicit in this design was the assumption that the mid-Bay samples, Berkeley and Oakland, would be the most contaminated, the San Pablo Bay and the Vallejo sites of intermediate contamination and the north coast site, the Russian River, would be the least contaminated. This seemed to be reasonable based on the limited amount of previously reported data (e.g., Spies et al.[19]).

The general strategy was to sample twice during the reproductive season. To assess the potential effects of contaminants part way through gametogenesis we sampled in late Novermber–early December 1986 and to assess effects at the time of spawning we sampled again in January–February 1987. Greater emphasis was placed on the November–December collection because of the movement of flounder out of the estuary at the time of spawning. Although we have managed to successfully capture and spawn females late in the reproductive season over several years,[2,3] their abundance greatly decreases later in the reproductive cycle. We also have less confidence that in an estuary-wide study fish collected later in the reproductive cycle will represent site differences as faithfully as those collected earlier. In other words, site differences in reproductive parameters are best investigated as late in the reproductive cycle as possible but before there is a substantial large-scale movement of adults. Since we do not yet have a precise understanding of the factors controlling movement of adult fish and how these change from year to year, these late-season collections have varying success.

Collection and Dissection

Fish were collected with 5- and 7-m otter trawls towed for 20 min in depths from 2.5 to 7 m at the stations indicated in Figure 1. At the station off the mouth of the Russian River fish were collected to a depth of 50 m. The San Francisco Bay stations are located on extensive subtidal mud flats that slope gently toward the deeper main channels. Flounder less than 20 cm were not retained. When more than 15 flounder were captured at a station we retained more of the larger, sexually mature fish. Although females grow larger than males,[20] each collection included several males. Previous analyses indicated that size and hepatic aryl hydrocarbon hydroxylase (AHH) activity were unrelated in this species.[21] Also, handling stress has been shown not to have a measurable effect on hepatic AHH

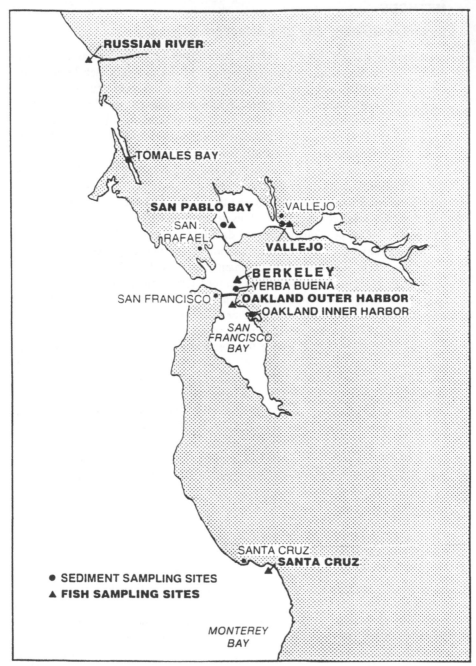

Figure 1. Location of sampling stations in the 1986–1987 studies of contaminant effects in San Francisco Bay.

activity in salmon.[22] Captured fish were maintained on the vessels in flowing bay or sea water until transported to the recirculating marine aquaria at Lawrence Livermore National Laboratory (LLNL). Fish not held for spawning were sacrificed the day following capture. Solvent-rinsed tools were used to remove the livers and the gonads, which were weighed and aliquots of each put aside for subsequent analyses.

Hepatic Microsomal Enzyme Activity and Cytochrome P-450 Determinations

Slices less than 5 mm thick were removed from the posterior liver for the in vitro assay of microsomal enzyme activities and cytochrome P-450 abundance. These were immediately frozen on dry ice and transferred within an hour to a freezer maintained at $-76°C$. One or more liver slices were later used to prepare a microsomal pellet. Analyses were carried out at both Lawrence Livermore National Laboratory (LLNL) and at Woods Hole Oceanographic Institution (WHOI).

An enzyme analyses were done at least in duplicate; some were done in up to ten replicates. Enzyme assays as well as spectral and immunoblot analyses of P-450 were repeated for samples giving unusual results or results near the limits of detection. Some samples were analyzed three or more times. Repeated analysis showed less than 15% variation in the results for any given sample.

Aryl hydrocarbon hydroxylase (AHH) activity. Microsomes were assayed for AHH activity at LLNL using the method of Nebert and Gelboin,[23] as described previously.[24] A portion of the hepatic microsomes from each fish was reassayed with 10^{-4} M 7,8-benzoflavone (7,8-BF), an inhibitor of some contaminant-induced P-450 isozymes.[25] This concentration was determined to cause maximal suppression of AHH activity. Twenty-five degrees (C) had been previously determined to be a more optimal assay temperature than either 18°C or 37°C. At 25°C reaction kinetics were linear for 10 min. An Aminco Bowman spectrofluorometer was used to quantify the 3-OH benzo[a]pyrene metabolite. The spectrofluorometer was calibrated with quinine sulfate standards. In addition a standard of microsomes from 3-methylcholanthrene-induced mice was assayed with each batch of microsomes as a control for assay conditions. Fluoresence values were corrected according to the quinine sulfate standard. Based on duplicate and triplicate assays from seven fish the mean coefficient of variation was 14.6% for AHH activity, and 2% for the mean percent change in AHH activity with the addition of the AHH inhibitor, 7,8-BF.[2] Protein concentrations were determined by the method of Lowry et al.[26] using bovine serum albumin (BSA) as the standard. Hepatic AHH specific activiites are reported as picomoles (pmol) 3-OH-benzo(a)pyrene mg protein^{-1} min^{-1}.

The effects of freezing whole liver tissue on hepatic AHH activity had been previously determined for *P. stellatus* by comparing activities of fresh liver from two males with liver samples from the same individuals frozen for 5, 19, and 40 days and assayed as outlined above. Mean AHH activities of frozen samples were found to be within 15% of mean values of freshly prepared microsomes.[2] This variation was nearly the same as the coefficient of variation for replicate determinations, 14.6%. The effects of freezing whole liver tissue on P-450 monoxygenase activities and UDP glucuronyl transferase were investigated for *Salmo gairdneri*.[27] Little loss activity was found when small slices of tissue were preserved for up to 3 days in liquid nitrogen and subsequently assayed; however, at dry ice temperature ($-75°C$) a loss of 72–85% of monoxygenase activity occurred in 24 hr. This is in contrast to our results, where there was no detectable loss of AHH activity in frozen liver tissue maintained at $-76°C$ for up to 40 days. We attribute this to intraspecific differences in the susceptibility of P-450 enzymes to degradation by freezing at -75 to $-76°C$. It would therefore appear appropriate to test each new species for susceptibility to degradation of enzyme activity under various conditions of freezing when undertaking field studies that necessitate preservation of whole tissue before microsomal fractions can be prepared.

A portion of the frozen microsomes prepared at LLNL were shipped to WHOI for analysis of ethoxyresorufin O-deethylase activity, estradiol 2-hydroxylase activity, cytochrome b_5 content, total cytochrome P-450, and for immunochemical assay for the inducible isozyme P-450E.

Ethoxyresorufin O-deethylase activity. Ethoxyresorufin O-deethylase (EROD) activity was measured at WHOI by the spectrophotometric method described by Klotz et al.[28] This method directly measures product formation, like the fluorometric analysis described by Burke et al.,[29] except that the resorufin is detected by absorbance.

Estradiol 2-hydroxylase. Estradiol 2-hydroxylase (E$_2$OHase) activity was assayed by ^3H_2O release from [2-^{3}H-E$_2$] (Kupfer et al.[30]) with some modifications.[9]

Cytochromes P-450 and b$_5$. Cytochrome P-450 (e $= 91$ mM^{-1} cm^{-1}) was measured by sodium dithionite difference spectra of CO-treated samples and cytochrome b_5 content (e $= 185$ mM^{-1} cm^{-1}) was determined from NADH difference spectra as previously described.[31]

Cytochrome P-450E homologue. Immunoblot ("Western" blot) analysis was accomplished with monoclonal antibody MAb 1-12-3 against P-450E, the major PAH-induced isozyme isolated from the marine fish soup.[25] The characterization of this antibody and its specificity in immunoblotting has been described.[32,33]

Positive Control Samples for P-450 Induction

Individual *P. stellatus* (immature females) that had been obtained in San Francisco Bay, at Berkeley and San Pablo, were used. These fish had been obtained 2 to 3 months prior to use and held at LLNL. Fish were treated with a known MC-type inducer, 7,8-benzoflavone (7,8-BF) (also called naphtholflavone) at 20 mg/kg given intraperitoneally, and were killed 5 days after treatment. Microsomes were prepared from fresh liver according to methods employed at WHOI. These microsomes were prepared to serve as a positive control, to validate the test methods, and to provide a measure of the degree of response which might be possible in *P. stellatus*.

Neutral Organic Contaminants

Residues of chlorinated hydrocarbons in liver were determined using methods similar to those of Ozretich and Schroeder,[35] but had some modifications.[36] Chlorinated hydrocarbons of interest were analyzed based on retention times and response factors of authentic external standards (National Bureau of Standards/NOAA standards). Values of analytical blanks were subtracted and final concentrations were corrected for recovery of internal standards. Analyte identifications were confirmed using a Hewlett Packard Mass Selective Detector (MSD), model 5970, operating in the ion scan mode. The MSD was interfaced with a Hewlett Packard gas chromatograph, model 5880, equipped with a DB-1 fused-silica capillary column. Instrumental conditions included an ionization voltage of 2800 eV and scan conditions of m/z 45-450 at one scan per second. Selected ion searches were used to obtain ion chromatograms for compounds with known retention indexes that were suspected to be present in the samples. If necesary, the mass spectrum and retention time of an identified peak was retrieved and compared with an authentic standard or to a mass spectrum library to aid identification.

The PCBs were identified on the basis of International Union of Pure and Applied Chemists (IUPAC) congener numbers.[37,38] Since there are many more PCB congeners in contaminated marine environments than we could quantify in this study, 18 major congeners were chosen for analysis (Figure 2). These congeners represent the different degrees of chlorination encountered in Aroclor mixtures, are indicative of specific Aroclors, or are known to be biologically active. Congeners 18, 87, and 180 were used to estimate the concentrations of Aroclor 1242, 1254, and 1260, respectively.[39] For example, congener 18 represents 9.38 percent of the total congeners in Aroclor 1242. In order to estimate the concentration of Aroclor 1242, the measured concentration of congener 18 is multiplied by (100/9.38). Congener 87 represents 3.32% of the congeners in Aroclor 1254. Congener 180 represents 6.5% of the congeners in Aroclor 1260. Congener numbers 66, 87, 118, 128, 153, 180, 195, and 206 are included since they are possible inducers of mixed-function oxidase (MFO) activity in animals.[40] Further, cogener numbers 66, 118, 128, 180, 195, and 206 have chlorine atoms in position 4 and 4', and are thought to be preferentially degraded in marine sediments.[41] Congener numbers 87, 101, and 187 lack chlorines in the 4 and 4' positions and are included in order to compare 4,4' congeners and non-4,4' congeners. For example, in an unaltered Aroclor 1254, congener 118 is expected to be about 3.5 times more abundant than congener 87 (11.5%/3.3% = 3.5). For Aroclor 1260, the expected proportion of congener 180 to 187 is 2.6. In the past use of these analytical methods resulted in 70% of recoveries being within 50–90%. Analysis of split samples have produced values within 10% of the

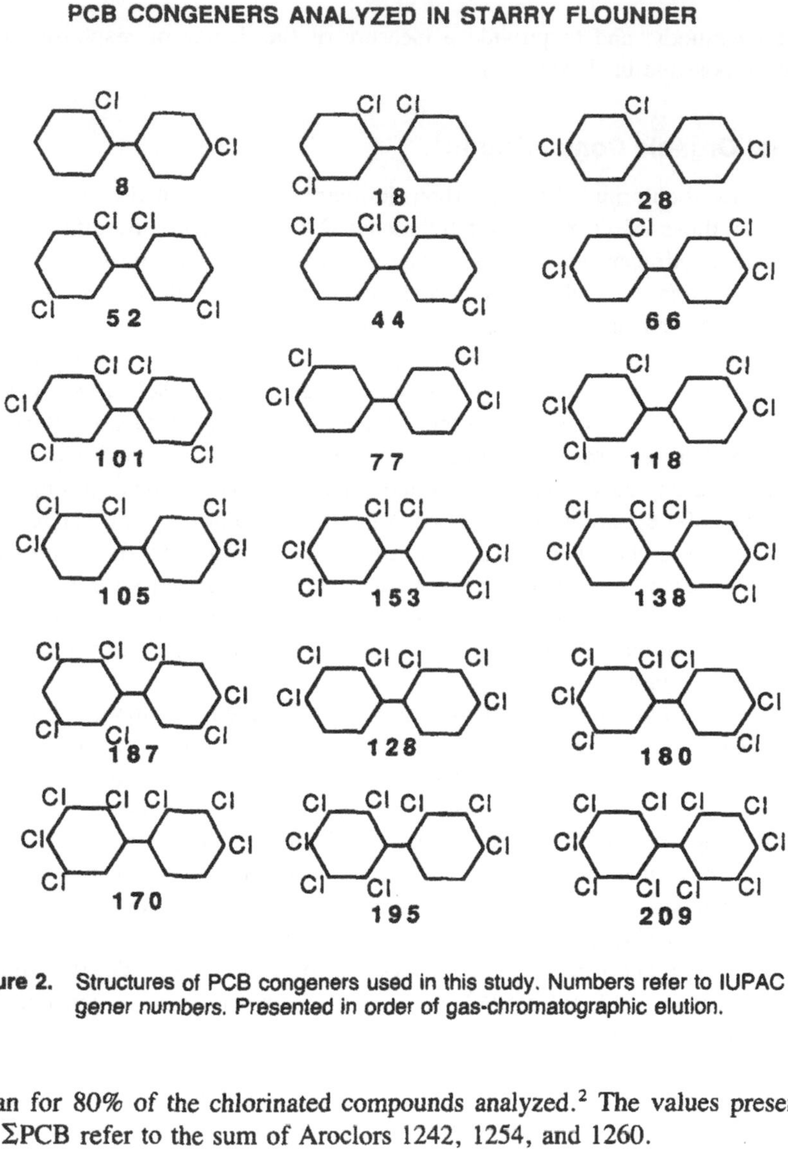

PCB CONGENERS ANALYZED IN STARRY FLOUNDER

Figure 2. Structures of PCB congeners used in this study. Numbers refer to IUPAC congener numbers. *Presented in order of gas-chromatographic elution.*

mean for 80% of the chlorinated compounds analyzed.[2] The values presented for ΣPCB refer to the sum of Aroclors 1242, 1254, and 1260.

Determination of Sex Steroids

Estradiol-17β and testosterone in plasma samples were analyzed by radioimmunoassay (RIA) techniques. The estradiol-17β antiserum generated against estradiol-17β-3-carboxymethyl-ether BSA (Radioassay Systems Laboratories) cross reacted 22.3% with 16-ketoestradiol, 2.46% with estriol, and 1.32% with estrone. The assay could detect 2.5 pg estradiol-17β per assay tube. The testosterone antiserum prepared against testosterone-3-BSA (Cambridge Medical

Diagnostics) was relatively specific and cross reacted 28.2% with dihydrotestosterone, 17.2% with 11-ketotestosterone, and 1.46% with androstenedione. This assay could detect 1.25 pg testosterone per assay tube.

Radioimmunoassay for testosterone and estradiol-17β were performed on the same plasma extract. One hundred microliters of plasma was extracted with 2 mL hexane/ethyl acetate (70:30) in 12 × 25 borosilicate tubes. Prior to extraction tritiated testosterone (1800 cpm) was added to each sample for the determination of extraction efficiency. The extract was dried under a stream of nitrogen and reconstituted in 250 μL of gelatin assay buffer (51.2 mM phosphate buffer, pH 7.6). Two aliquots (50 μL and 25 μL) of the reconstituted sample were measured in each RIA to test for parallelism. In the estradiol-17β assay 50 μL of tritiated steroid tracer (approx. 4000 cpm, Radioassay Systems Laboratories) and 75 μL of antiserum (1 to 23,200 dilution) were added to the samples and the standards. In the testosterone assay 25 μL of tritiated steroid trace (approx. 4000 cpm, Research Products International) and 25 μL of antiserum (1 to 9,000 dilution) were added to the samples and standards. Assay mixtures were incubated overnight at 4°C and bound steroid was separated from free steroid and dextran-coated charcoal.

The intra-assay coefficient of variation of replicate determinations of estradiol-17β and testosterone in a plasma control pool were 16.8% and 8.3%, respectively (estradiol-17β mean, 1.14 ng/mL, s.e. ± 0.15, N = 3; testosterone, 2.12 ng/mL, s.e. ± 0.04, N = 3). Recovery of steroid standards (40250 pg/tube) added to the control plasma ranged from 94.5–119%.

Evaluation of Reproductive Success

For the November–December collections a small section of the gonad of each female was removed during necropsy and preserved in Davidson's fixative. Appropriate subsections were taken from female gonads in these samples and made up into paraffin sections at 6 μm, mounted on microscope slides and stained with hematoxylin and eosin. These were later examined to evaluate the predominant egg stages present[42] and for the occurrence of atreitic oocytes.

To evaluate reproductive success of fish captured in January–February they were transported to LLNL and acclimated for several days to the seawater system, which is maintained at 11–13°C and at a salinity of 29–30‰. The holding aquaria measured 58 × 58 × 47 cm (158.1 L) and two to three females were placed in each aquarium. Males were generally maintained separately.

One to three days after capture gonadally mature females were started on a course of carp pituitary extract injections (1 mg kg^{-1} d^{-1}, i.m.) to induce spawning. During 1985 in San Francisco Bay, a majority of females were in maturation stage VII[42] late in the reproductive season (Spies et al., unpublished). We assume this was also the case for the 1986 season. Stage VII is the tertiary yolk stage and the nuclei at this stage are still close to the center of the developing oocytes. Some movement of nuclei to peripheral positions and germinal vesicle breakdown is usually observed as early as 10 days after the start of pituitary injections. Hydration and ovulation appeared to follow shortly thereafter, but the total time from the start of injections to hydration varied substantially between females. Females generally spawned between two and five times.

In order to allow for the possibility that various effects of contaminant exposure might find expression at specific developmental stages[43] we adopted measures that indicated survival through separate early life history stages. Thus, where

N = total number of eggs spawned,
V = the number of eggs that float (viable eggs),
F = the number of fertilized eggs,
H = the number of eggs that hatched, and
L = the total number of normal larvae

we define the following measures of survival through early life-history stages:

(1) % floating eggs   = $(V/N) * 100$
(2) % fertilization success   = $(F/V) * 100$
(3) % embryological success = $(H/F) * 100$
(4) % normal larvae   = $(L/H) * 100$

Sources of variability for these measures in more than 100 spawnings have been assessed and a standard protocol adopted for spawning and evaluation of developmental success.[3]

Evaluation of Micronucleus Occurrence in Peripheral Erythrocytes

Blood was collected from each fish immediately after capture in a heparinized syringe and blood smears were prepared on board ship. The smears were air dried and fixed in absolute methanol for 15 min.

In the laboratory the smears were stained with May-Grunwald-Giesma and examined with a high power ($1000\times$) microscope according to the method of Hose et al.[18] The number of micronucleated erythrocytes (MN) was determined on coded slides and scored in a blind review. Only micronuclei completely detached from the main nucleus were included in the counts. The number of micronucleated erythrocytes was the average of two counts of 1000 erythrocytes each.

Statistical Analyses

For analyses all data were first log-transformed [$\ln(x)$ or $\ln(x + 1)$]. The significance of site differences was then tested by one-way analysis of variance with site pair differences tested with the Fisher PLSD test.

In the cases where a factor such as sexual maturity was thought to have some bearing on a variable, e.g., concentrations of plasma estradiol, the variable was

Table 1.   Summary of Field Collections of *P. stellatus* during the 1986–1987 Reproductive Season.

Date	Site	No. Collected	No. Retained
11/17–18/86	Berkeley	31	31
11/18/86	Vallejo	4	4
11/18/86	San Pablo Bay	11	11
11/19/86	Oakland	11	11
11/20/86	San Pablo Bay	20	20
11/20/86	Vallejo	20	20
11/21/86	Oakland	9	9
12/4/86	Russian River	15	15
12/12/86	Russian River	9	9
1/12/87	Berkeley	8	1
1/13/87	Berkeley	10	3
1/14/87	Vallejo	3	0
1/15/87	Vallejo	1	0
1/15/87	Oakland	3	0
1/16/87	Oakland	1	1
1/20/87	Russian River	0	0
2/4/87	Santa Cruz	10	5
2/7/87	Santa Cruz	8	8
2/12/87	Berkeley	9	1

analyzed for several sex-size groupings. Relationships between individual parameters were examined by linear regression analysis.

In analyzing data for which there were individuals with values below detection limits, these were assigned values of zero.

RESULTS

Field Collections

The field work is summarized in Table 1. We were able to capture as many fish as planned in the November–December collection; however, the collections in January–February did not provide enough sexually mature females for a balanced design to determine reproductive success measures at our five sites. We were, however, able to compare reproductive success in females captured at the mid-Bay site, Berkeley, and from a new coastal comparison site, Santa Cruz.

November–December Collections

Hepatic Concentrations of Contaminants

The lipid-normalized concentrations of contaminants in liver were not related to size of the fish. Total weight was not significantly related to ΣDDT (F =

Table 2. Mean (± Standard Deviation) Lipid-Normalized Hepatic Concentrations (μg/g lipid) of Chlorinated Hydrocarbons in *Platichthys stellatus* at Five Sites.

Underlined sites are not significantly different by Fisher's PLSD. Statistical tests done on log (x + 1)-transformed data. Sites are as follows: BK, Berkeley; OK, Oakland; SPB, San Pablo Bay; VJ, Vallejo; RR, Russian River.

Chemical	BK	OK	SPB	VJ	RR	Effect of Site (one-way ANOVA) F	P
p,p'-DDE	0.89	1.0	0.54	1.06	0.76	0.86	0.483
	(0.513)	(1.3)	(0.255)	(1.2)	(0.87)		
p,p'-DDD	0.34	0.509	0.34	0.28	0.08	7.02	0.0001
	(0.316)	(0.34)	(0.21)	(0.16)	(0.09)		
o,p'-DDT	0.007	0.151	0.004	0.004	0	2.58	0.0475
	(.0.27)	(0.356)	(0.013)	(0.011)	(0)		
p,p'-DDT	0.049	0.043	0.022	0.011	0.034	1.05	0.39
	(0.076)	(0.064)	(0.031)	(0.016)	(0.068)		
ΣDDT	1.28	1.68	0.92	1.37	0.93	1.51	0.21
	(0.081)	(1.74)	(0.37)	(1.19)	(0.89)		
Lindane	0.036	0.019	0.026	0.025	0.0004	1.78	0.14
	(0.039)	(0.033)	(0.026)	(0.044)	(0.014)		
Heptachlor	0.052	0.032	0.011	0.027	0.031	0.68	0.61
	(0.058)	(0.091)	(0.03)	(0.072)	(0.077)		
Heptachlor epoxide	0.016	0.024	0.024	0.028	0.051	0.86	0.49
	(0.012)	(0.034)	(0.021)	(0.048)	(0.11)		
Aldrin	0.001	0.011	0.003	0.006	0.037	2.57	0.046
	(0.004)	(0.024)	(0.005)	(0.009)	(0.074)		
Chlordane	0.043	0.058	0.051	0.052	0.01	2.76	0.035
	(0.037)	(0.077)	(0.039)	(0.031)	(0.012)		
Transnanochlor	0.155	0.217	0.091	0.065	0.05	4.35	0.0035
	(0.098)	(0.24)	(0.12)	(0.072)	(0.05)		
Dieldrin	0.18	0.22	0.17	0.109	0.214	0.69	0.60
	(0.09)	(0.18)	(0.10)	(0.064)	(0.43)		
Mirex	0.007	0.12	0.01	0.013	0.04	5.30	0.0009
	(0.016)	(0.16)	(0.023)	(0.024)	(0.06)		

0.018, P = 0.67) or ΣPCB (F = 1.84, P = 0.18) in fish from all sites. Similar comparisons with fish just from Oakland indicated that neither ΣDDT (F = 2.12, P = 0.17) nor ΣPCB (F = 0.58, P = 0.81) were significantly correlated with total weight. Likewise Russian River fish showed no relationships between total weight and concentration of ΣDDT (F = 0.12, P = 0.74) or ΣPCB (F = 0.32, P = 0.58). We therefore did not take size into account in the analyses of variance reported below.

Pesticides. Hepatic concentrations of pesticides are given in Table 2. Five separate DDT-type compounds were resolved in sufficient numbers of samples to analyze for site differences. Significant site differences occurred in concentrations of o,p'-DDT and p,p'-DDD. Since o,p'-DDD coeluted with the polychlorinated biphenyl, IUPAC congener 77, concentrations of this single gas-chromatographic peak are considered later. No site differences occurred with p,p'-DDE or p,p'-DDT. The sum of the resolved forms of DDT-type compounds, ΣDDT, was also not significantly different between sites.

Most of the other chlorinated pesticides were resolved in sufficient quantities to test for site differences. Site differences were found for concentrations of aldrin. Lindane was not significantly different between sites, but the Russian River fish did have significantly lower concentrations than those from Bay sites. Concentrations of heptachlor were not significantly different between sites, nor were those of heptachlor epoxide. Site differences were not found for dieldrin. There were significant differences attributable to site for transnanochlor, with the highest concentrations being found in fish from Oakland and Berkeley. There were also significant site differences in mirex, with Oakland having significantly higher concentrations than Berkeley, San Pablo Bay, or Vallejo.

Table 3.   Mean (± SD) Lipid-Normalized Hepatic Concentrations of PCBs (μg/g lipid) in *Platichthys stellatus* at Five Sites.

Underlined sites are not significantly different by Fisher's PLSD. Statistical tests done on log (x + 1)-transformed data.

PCB Congener IUPAC No.	BK[a]	OK	SPB	VJ	RR	Effect of Site (one-way ANOVA) F	P
8	0 (0)	0.027 (0.046)	0.001 (0.005)	0 (0)	0.059 (0.101)	3.77	0.008
18	0 (0)	0.002 (0.008)	0.011 (0.023)	0.01 (0.01)	0.029 (0.085)	1.2	0.32
28	0.029 (0.033)	0.02 (0.029)	0.009 (0.032)	0.029 (0.063)	0.034 (0.105)	0.39	0.81
52	0.15 (0.08)	0.15 (0.17)	0.08 (0.07)	0.16 (0.14)	0.068 (0.12)	2.1	0.09
44	0.038 (0.055)	0.035 (0.062)	0.015 (0.028)	0.64 (0.17)	0.021 (0.047)	0.61	0.65
66	0.05 (0.083)	0.014 (0.025)	0.019 (0.034)	0.029 (0.085)	0.001 (0.004)	1.48	0.21
101	0.186 (0.12)	0.253 (0.15)	0.08 (0.03)	0.111 (0.07)	0.064 (0.084)	10.23	0.0001
77/o,p'-DDD	0.21 (0.22)	0.22 (0.17)	0.09 (0.05)	0.16 (0.09)	0.16 (0.19)	1.48	0.22
118	0.35 (0.27)	0.43 (0.33)	0.10 (0.09)	0.14 (0.11)	0.10 (0.10)	8.5	0.0001

(continued)

Table 3. Continued

PCB Congener IUPAC No.	BK[a]	OK	SPB	VJ	RR	Effect of Site (one-way ANOVA)	
						F	P
105	0.09 (0.05)	0.15 (0.10)	0.07 (0.05)	0.05 (0.04)	0.11 (0.15)	3.04	0.023
153	0.66 (0.50)	0.87 (0.65)	0.17 (0.07)	0.21 (0.14)	0.23 (0.22)	12.5	0.0001
138	0.63 (0.46)	0.83 (0.51)	0.13 (0.10)	0.19 (0.16)	0.15 (0.14)	18.02	0.0001
187	0.28 (0.21)	0.33 (0.23)	0.09 (0.05)	0.095 (0.10)	0.024 (0.06)	4.42	0.0032
128	0.053 (0.07)	0.086 (0.074)	0.075 (0.19)	0.061 (0.08)	0.046 (0.053)	0.40	0.81
180	0.43 (0.35)	0.60 (0.50)	0.05 (0.05)	0.14 (0.19)	0.14 (0.14)	12.1	0.0004
170	0.034 (0.10)	0.023 (0.063)	0.024 (0.085)	0.001 (0.003)	0.001 (0.005)	0.73	0.57
195	0.009 (0.01)	0.009 (0.02)	0.001 (0.004)	0.001 (0.004)	0.001 (0.005)	2.44	0.06
206	0.074 (0.076)	0.03 (0.05)	0.022 (0.026)	0.029 (0.036)	0.059 (0.075)	2.35	0.06
209	0.024 (0.039)	0.065 (0.12)	0.005 (0.01)	0.003 (0.009)	0.0008 (0.018)	3.02	0.02
Aroclors							
1242	0 (0)	0.02 (0.09)	0.12 (0.24)	0.11 (0.19)	0.31 (0.91)	1.3	0.25
1254	3.04 (2.4)	3.7 (2.9)	0.89 (0.79)	1.2 (0.99)	0.91 (0.85)	9.11	0.0001
1260	6.6 (5.4)	9.2 (2.5)	0.71 (0.81)	2.1 (3.0)	2.2 (2.2)	15.3	0.0001
ΣPCB	9.7 (7.7)	12.9 (8.7)	1.7 (1.5)	3.5 (3.6)	3.5 (2.9)	14.8	0.0001

[a]Site abbreviations as in Table 2.

Polychlorinated biphenyls. Fish livers were analyzed for 18 individual PCB congeners. In addition the traditional measures of PCB concentrations, Aroclor 1242-, 1254-, and 1260-equivalents, as well as total Aroclors (the sum of the above three) were also calculated. The chlorine-substitution patterns of 18 of the congeners are illustrated in Figure 2. The mean concentrations for the PCBs by site are presented in Table 3.

In general site differences were more commonly encountered with greater chlorine substitution. Additionally, many of the lesser chlorinated PCB congeners were below detection limits in many of the samples. In the following discussion the individual congeners will be referred to by their IUPAC (International Union of Pure and Applied Chemistry) numbers.

Site did not explain a significant amount of the variance in concentrations of the following congeners with 2 to 4 chlorine atoms per molecule: 18, 28, 52 (although Vallejo had fish with higher concentrations than those of the Russian River), 44, and 66. Only two congeners among the lesser chlorinated PCBs exhibited site differences: 8 (with the Russian River fish having significantly higher concentrations than Berkeley, San Pablo Bay, or Vallejo fish) and 101 (with Oakland and Berkeley fish having higher concentrations than those of Vallejo, San Pablo Bay, and the Russian River). It should be mentioned that congener 18 was only detected in a total of 14 fish. Both o,p'-DDD and congener 77 coeluted; but the variability in concentrations calculated for this peak was not explained by site (although San Pablo Bay values were significantly less than the Oakland values).

With some exceptions the congeners with a greater degree of chlorine substitution, 5 to 10 chlorine atoms per molecule, showed significant variability due to site. The results of statistical analyses indicate significant differences due to site for the following congeners: 118, 105, 153, 138, 187, 180, and 209. Congeners 195 and 206 had marginally significant effects due to site, both with P values of about 0.06. Congener 126 was detected in three fish, and since it has not been reported from the marine environment and we could not confirm it by mass spectrometry we chose not to report it. In each case the horizontal lines under the sites in Table 3 indicate which site pairs differed significantly in their concentration. Those sites appearing on the same line did not differ significantly. In most cases fish from one or both of the mid-Bay stations, Berkeley and Oakland, had greater hepatic concentrations of PCB congeners with 5 to 10 chlorines per molecule than the other sites.

For the Aroclors a pattern emerges that is generally consistent with that seen for specific congeners: Aroclor 1242, composed mainly of congeners with lesser numbers of chlorines and quantified here using congener 8, did not exhibit site differences, but site differences did occur with Aroclor 1254 as well as with Aroclor 1260, that have more chlorine substitution. Since ΣPCB is the sum of these three Aroclors, it is not surprising that there are also significant differences between sites in this parameter.

Hepatic MFO Activity and P-450 Enzymes

Because there are potential effects of sexual maturity on P-450 enzymes,[2,31] particularly with supression expected in spawning females, we have stratified, or analyzed separately, site differences among three categories (mature females, immature females, and mature males), where the effects of sex and maturity might otherwise obscure site differences. The results of analyses for P-450 content of hepatic microsomes and P-450-mediated enzyme activities of microsomal fractions are given in Table 4.

AHH activity. In mature females the highest activities were observed in Oakland fish, which had significantly greater activities than those from the Russian River. No other site differences occurred and, overall, site did not explain a significant amount of the variance. For immature females a significant amount of the variance was accounted for by site. Oakland fish had significantly higher activities than Berkeley, Vallejo, and Russian River fish. In mature males a significant amount of variance was not explained by site. The only significant differences between sites were those between Vallejo and San Pablo Bay and between the Russian River and Vallejo.

A marginally significant amount of the variance in hepatic AHH activity in the presence of the inhibitor, 7,8-BF, in mature females was explained by site. The activity of the Oakland fish was only significantly higher than the activity of the Russian River fish. Site did not explain a significant amount of the variance of this same activity in immature females and none of the site differences were significant. In males the variance in this same activity was not explained by site and no site pairs were significantly different.

Ethoxyresorufin O-deethylase (EROD). In mature females a marginally significant amount of the variance in EROD activity was explained by site: Oakland was significantly higher than Berkeley. In immature females a very significant amount of the variance was due to site: Oakland had activities significantly higher than Berkeley, San Pablo Bay, Vallejo, and the Russian River. In mature males a very significant amount of the variance was explained by site: Berkeley and Oakland fish had the highest activities and were generally significantly higher than fish from the other sites (with the exception of the Oakland-San Pablo Bay and the Oakland-Vallejo comparisons).

P-450E content. In mature females a marginally significant amount of the covariance was explained by site: Oakland fish had significantly more P-450E than those from the rest of the stations. In immature females a very significant amount of the variance was explained by site with Oakland and Berkeley fish having significantly higher values than those from all the other stations (Table 4; Figure 3). In males a significant amount of the variance was due to site, with Berkeley and Oakland fish having the highest values. However, Berkeley fish had values that were not significantly different than those of fish from Vallejo.

P-450. The spectrophotometric spectra revealed that considerable amounts of P-420 were present in microsomal preparations made at LLNL. This appears to

Table 4. Mean (±SD) Hepatic Microsomal P-450 Activities and Content of *Platichthys stellatus* by Sex-maturity Groups.

Parameter Group		BK	OK	SPB	VJ	RR	F	P
Activities								
AHH (pmol-mg^{-1} min^{-1})	MF	73.6 (28)	200 (145)	86 (0)	147 (65)	66 (94)	2.62	0.07
	IF	70.2 (35.5)	437 (268)	169 (65)	193 (173)	74.8 (44.9)	4.37	0.01
	MM	71.7 (35)	75 (0)	81.8 (80)	433 (381)	31.7 (24.5)	2.67	0.09
AHH with 7,8-BF (pmol-mg^{-1} min^{-1})	MF	20.6 (9.2)	48.3 (37.3)	45 (0)	37 (18.2)	17.2 (6.6)	2.82	0.06
	IF	26 (17.5)	60 (33.3)	41.5 (26.1)	47.5 (38)	29.8 (14.9)	1.2	0.32
	MM	22.2 (18.4)	16 (0)	22.7 (30.1)	86.3 (38.3)	22 (9.2)	2.47	0.12
EROD (pmol-mg^{-1} min^{-1})	MF	0.01 (0.02)	0.14 (0.12)	0.050 (0)	0.037 (0.02)	0.06 (0.04)	2.64	0.07
	IF	0.18 (0.035)	0.44 (0.20)	0.07 (0.017)	0.20 (0.21)	0.037 (0.025)	8.2	0.0004
	MM	0.198 (0.05)	0.12 (0)	0.04 (0.02)	0.09 (0.04)	0.02 (0.01)	2.47	0.12
P-450E (pmol-mg^{-1})	MF	0.84 (0.81)	16.2 (19.9)	0.0001 (0.0001)	0.01 (0.02)	1.26 (1.73)	2.85	0.06
	IF	29.5 (28.1)	44.5 (22)	2.76 (0.56)	5.85 (9.87)	2.77 (3.05)	7.87	0.0006
	MM	25.2 (39)	14 (0)	0 (0)	2.6 (2.6)	1 (1.7)	4.65	0.02
P-450 (nmol-mg^{-1})	MF	0.10 (0.07)	0.16 (0.07)	0.27 (0)	0.076 (0.26)	0.14 (0.64)	2.3	0.099
	IF	0.20 (0.04)	0.16 (0.08)	0.10 (0.034)	0.12 (0.056)	0.11 (0.05)	2.56	0.07
	MM	0.19 (0.04)	0.11 (0)	0.07 (0.03)	0.19 (0.03)	0.13 (0.05)	7.07	0.005

Effect of Site (one-way ANOVA) spans the F and P columns.

Abbreviations: MF = mature female; IF = immature female; MM = mature male. Site abbreviations as in Table 2.

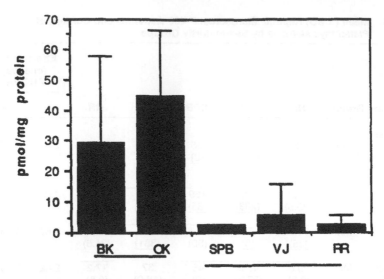

Figure 3. Hepatic microsomal ethoxyresorufin O-deethylase (EROD) activity enzyme activity in immature female *P. stellatus*. BK = Berkeley, OK = Oakland, SPB = San Pablo Bay, VJ = Vallejo, and RR = Russian River. Underlined sites are not significantly different. Error bars are 1 standard deviation.

be a result of the type of buffer and possibly other factors. Use of fresh microsomes, different buffers, and the addition of NADH to microsomal preparations made at WHOI does not give rise to this spectrophotometrically detected alteration. This alteration appears not to make a difference to the P-450-catalyzed enzyme activities.[2] In spite of this alteration of P-450, some site differences in measured P-450 were observed (Table 4).

Site was not a significant source of variance in P-450 content of microsomes of mature females, but San Pablo Bay fish had significantly higher concentrations than Berkeley fish. Site was a marginally significant source of variance in immature females: Berkeley fish had significantly higher concentrations than all sites except for Oakland. Site was also a significant source of variance in P-450 content of mature male hepatic microsomes: Berkeley and Vallejo fish had concentrations that were significantly higher than those of San Pablo Bay.

Estradiol hydroxylase activity. Several mature and several immature females were assayed for estradiol hydroxylase activity from the two sites that were thought to represent the most and least contaminated, Oakland and the Russian River. A two-way analysis of variance indicated that site was not a significant source of variance in this P-450 activity (F = 1.31, P = 0.30), but that sexual maturity was a significant source of variance (F = 15.9, P = 0.0073) (Figure 4), with mature females having lower activities than immature females.

Micronucleii Occurrence in Circulating Erythrocytes

For this parameter there was a significant effect due to site (F = 5.13, P = 0.0024) with none of the sites within the Bay differing significantly from one

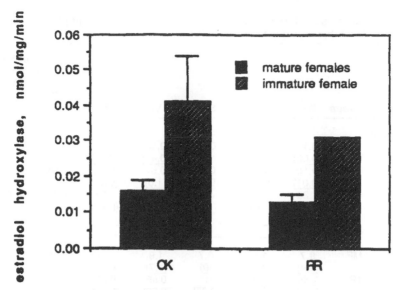

Figure 4. Hepatic microsomal enzyme activities in *P. stellatus*: estradiol 2-hydroxylase activity in females. Symbols as in Figure 3.

another, but all of them being significantly higher than the Russian River value (Figure 5).

Plasma Sex Steroids

Testosterone. Site did not account for a significant proportion of the variance in plasma concentrations in mature females or immature females. In mature

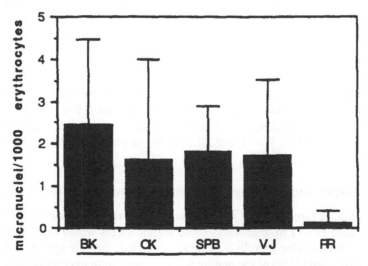

Figure 5. Incidence of micronuclei in circulating erythrocytes of *P. stellatus*. Symbols as in Figure 3.

Table 5. Mean (±SD) Plasma Concentration of Sex Steroids (ng/mL) in
Platichthys stellatus from Five Sites.

Steroid	Group	BK[a]	OK	SPB	VJ	RR	Effect of Site (one-way ANOVA) F	P
Testosterone								
	MF	0.14 (0.06)	0.38 (0.35)	0.14 (0)	0.48 (0.49)	0.21 (0.07)	1.29	0.32
	1F	0.10 (0.01)	0.24 (0.25)	0.22 (0.25)	0.57 (1.0)	0.12 (0.05)	0.54	0.72
	MM	1.20 (0.41)	1.6 (0)	2.4 (1.0)	2.7 (0.68)	0.57 (0.43)	6.7	0.007
Estradiol								
	MF	9.7 (4.4)	10.9 (10.7)	8.7 (0)	1.9 (1.7)	7.8 (4.7)	0.85	0.52
	1F	1.89 (1.3)	0.80 (0.47)	0.63 (0.08)	0.55 (0.22)	0.38 (0.13)	5.7	0.0031
	MM	0.29 (0.13)	0.81 (0)	0.27 (0.16)	0.32 (0.22)	0.34 (0.16)	0.76	0.57

[a]Site abbreviations as in Table 2.

males site did account for a marginally significant proportion of variance (P = 0.045) (Table 5). Both San Pablo Bay and Vallejo males had significantly greater plasma testosterone concentrations than males from off the Russian River.

Estradiol. Site did not account for a significant proportion of variance in plasma estradiol of mature females, but did for immature females, with Berkeley females having significantly higher concentrations than those of all other sites and Oakland having higher concentrations than the Russian River fish.

Vitellogenin in Sexually Mature Female Flounder

Mean concentrations of vitellogenin, as measured by protein phosphorus, in plasma samples of sexually mature females ranged from 8.6 µg/mL in Vallejo fish to 17 µg/mL in Berkeley fish (Table 6). There was no significant effect of site (one-way ANOVA, F = 0.468, P = 0.76) on plasma vitellogenin concentrations.

Oocyte Stages and Occurrence of Atreitic Ooytes in Female Flounder

Microscopic examination of the ovarian sections revealed that most females were in stages IV–VI of oogenesis. These are the stages where rapid growth of the oocytes is occurring as yolk is being accumulated; they are spawned at stage XI. However, two (25%) of the mature females (both were over 40 cm SL) from

Table 6.   Mean (±SD) Plasma Vitellogenin Concentrations as Measured by Protein Phosphorous (μg/mL) in Sexually Mature Starry Flounder from Five Sites.

					One-way ANOVA	
BK[a]	OK	SP	VJ	RR	F	P
17.2	12.5	12.5	8.6	12.7	0.47	0.76
(4.4)	(6.9)	(0)	(0)	(5.7)		

[a]Site abbreviations as in Table 2.

Oakland were completely without any yolky eggs. Fish number 5166 was a 41-cm (SL) female weighing 1428 g and had oocytes in advanced stage II (previtellogenic). The nucleoli were peripherally arranged in the nucleus and there were some vesicles beneath the cortex in the outer ooplasm. Fish number 5218 was a 42.4-cm (SL) female weighing 1431 g and had oocytes that were even less mature, all being strongly basophilic and showing no signs of yolk vesicles. All flounder previously collected by us from San Francisco Bay over 34 cm (SL) in the autumn and winter have had yolky eggs. Additionally, Orcutt[20] found that all female fish over 30 cm (SL) taken from Monterey Bay during this time of year were sexually mature, i.e., containing yolky eggs that would be presumably spawned.

The occurrence of atresia in mature females was much lower than expected based on our previous preliminary investigations. In all the microscopic sections we examined of mature female gonads, six to eight sections of approximately 10–20 cm^2 (6μm thick) for each, we found one atreitic egg in a female fish from Oakland and about 5% of the eggs atreitic from one female from Vallejo. It was therefore obvious that in this collection of females the rate of atresia in oocytes was extremely small (less than 1%) and not different between sites.

Relationships between Chemical Contaminants, Hepatic MFO Activity and Plasma Steroid and Vitellogenin Concentrations

To determine if relationships might exist between contaminants in the fish hepatic MFOs, sex steroid concentrations in the plasma and erythrocyte micronucleus occurrence an extensive series of regressions were carried out. The regressions were carried out on data from all individual fish and did not use site average for any parameters. The following discussion highlights some of the relationships found. Since not all of the regressions between chemical measures and the other parameters have been carried out (there were more than 40 separate chemicals measured) we have dealt mainly with total PCBs and DDTs. We consider these first as correlations between all fish and then between fish in the same sex-maturity categories. It should be remembered that when multiple correlations are carried out the chance of spurious correlations increases with the number of regressions. This is normally accounted for by the application of the

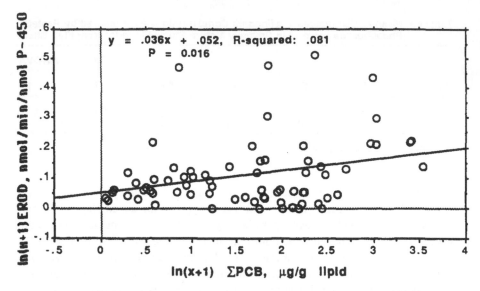

Figure 6. Relationship between hepatic EROD activity and concentrations of ΣPCB in all fish.

Bonferroni technique, where the probability (P) is divided by the number of correlations performed. Our experience tells us, however, that many of the correlations found in these types of data where a plausible cause and effect mechanism exists are still fairly weak and would be eliminated by application of this technique. This presented a dilemma, but we chose not to use this technique and to report some of the correlations that appear to be interesting in terms of the potential mechanisms that may explain them. We are aware that some of these may be spurious.

All fish. For all fish ΣPCB was correlated with EROD activity (F = 6.11, P = 0.016) (Figure 6), and P-450E (F = 6.7, P = 0.018) (Figure 7).

Not surprisingly many of the MFO parameters were interrelated, since induced hepatic microsomal P-450 tends to form a significant proportion of total P-450 in fish chronically exposed to contaminants.[21] Many intercorrelations of this type were also observed in the sex-maturity groups discussed below.

Mature females. As with all fish ΣPCB and EROD activity in liver were marginally correlated (F = 3.4, P = 0.080). In addition ΣDDT was correlated with testosterone (F = 6.6, P = 0.02) and EROD activity was correlated negatively with plasma estradiol concentrations (F = 6.32, P = 0.024). This last correlation seems to be strongly affected by one fish which had exceptionally low estradiol and high EROD activity.

The mean plasma estradiol concentrations of mature females collected from Oakland (10.7 ng/mL) was the highest of any of the sites, as was the P-450E content. Although the relationship between site means of these two parameters was not significant and site differences were not significant for plasma estradiol,

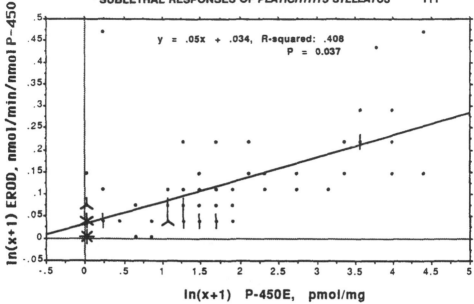

Figure 7. Relationship between EROD activity and P-450E in all fish. Overlapping data points are indicated by radiating lines, with the number of lines corresponding to the number of coincident values.

the data may indicate the potential for a threshold effect of P-450E on plasma estradiol. These data are by no means convincing in this respect, but in light of the correlation of plasma estradiol with increasing P-450E in immature females (see below) it may be worthwhile investigating the possiblity of a threshold with further sampling.

A marginally significant relationship was found between plasma protein phosphorus (vitellogenin) and hepatic concentrations of PCB congener 105 (Figure 8).

Immature females. In immature females plasma estradiol was significantly positively correlated with P-450E content (F = 7.02, P = 0.015) (Figure 9) and total P-450 content (F = 11.8, P = 0.0023) of hepatic microsomes.

Mature males. In mature males there was a significant positive relationship between hepatic EROD activity and ΣPCB (F = 0.0287, P = 0.029).

January–February Collections

Summary of Mature Females Captured and Spawned

As mentioned earlier, the only site within the Bay that had significant numbers of sexually mature fish to collect in January and February of 1987 was Berkeley. In addition, the coastal reference site that we had chosen, off the mouth of the Russian River, did not have *P. stellatus* during this period. The only other location that we could locate with sexually mature fish was off Santa Cruz. The females we obtained from the Santa Cruz population were generally much larger

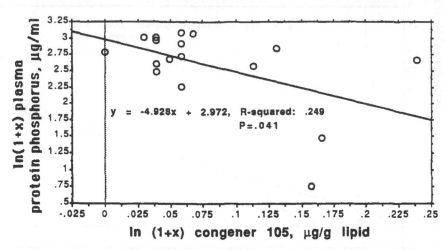

Figure 8. Relationship between plasma vitellogenin and hepatic concentrations of PCB congener 105 in mature female *P. stellatus*.

than fish we captured in San Francisco Bay. They were also generally much closer to spawning than fish we normally capture from San Francisco Bay, as at least two fish spawned with a week of capture and we normally expect at least 2 weeks of pituitary injections to induce spawning. There were, however, some complicating factors. Most of the Santa Cruz collection died eventually in our aquaria. We do not know the cause of this. It was perhaps due to the longer trawling intervals with subsequent damage occurring to the fish in the cod end of the net. Fish may have incurred slight abrasions that later became infected. Also, females that are closer to spawning may not be as resistant to the stress of capture and transport.

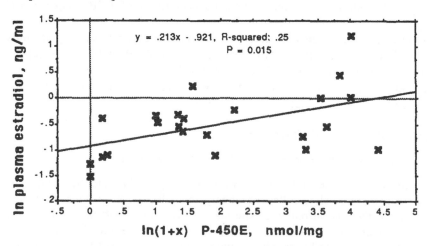

Figure 9. Relationship between plasma estradiol and hepatic P-450E in immature female *P. stellatus*.

Table 7. Reproductive Success, Hepatic AHH Activity and Plasma Steroid Concentrations of Female Starry Flounder Collected in January–February 1987.

Site/ MSB[a]	1st sp[b] (days)	Reproductive Success[c]				Hepatic MFO		Steroids	
		% Float	% Fert	% Embryo	% Normal Larvae	AHH[d]	AHH Corr[e]	Test[f]	E-2[g]
Berkeley									
5253	28	50.5	50.6	3.6	45	27	8	0.47	20
5255	26	64.6	67.8	14.3	92.	BD	BD	1.3	25
5263	26	65.8	56.8	14.3	100	57	20	1.09	19
5319	17	41.8	60	35	93	NA	NA	1.09	17
Oakland									
5268	24	56.2	76.9	28.1	73	10	8	1.42	26
Santa Cruz									
5311	2	23.6	46	60	84.6	NA	NA	2.76	9.1
5314	6	76.1	92.3	86.3	97.9	11	1	4.67	21
5315	6	66	66	47.8	92.9	NA	NA	3.77	18
5316	12	62.2	57	65.3	92.1	NA	NA	1.41	50
5318	10	77	71.6	80.6	98	NA	NA	1.49	15
5276	31	63.4	62.3	1.2	100	63	13	0.21	11

[a]MSB = fish collection number.
[b]1st sp = no. of days in the laboratory until spawning took place.
[c]% Float = percentage of spawned eggs that floated; % Fert = percentage of floating eggs that were fertilized; % Embryo = percentage of fertilized eggs that hatched; % Normal Larvae = percentage of hatched larvae that were normal in appearance.
[d]AHH = pmol 3-OH B[a]P/mg protein/min.
[e]AHH Corr = AHH activity with 10^{-4} M 7,8-benzoflavone.
[f]Test = plasma testosterone at time of capture (ng/mL).
[g]E-2 = plasma estradiol at time of capture (ng/mL).

Reproductive Success, Hepatic MFO Activities, and Plasma Steroids

With the understanding that reproductive physiology and biochemical measures of Santa Cruz fish may have been affected by handling or disease, we present the summary data on all our spawnings in Tables 7 and 8.

In Table 7 site means and results of statistical analyses for differences are presented for reproductive success measures, plasma steroids, and AHH activity at the time of spawning. In none of these comparisons did site explain a significant amount of the variance. However there was a significant difference in percent embryological success between Berkeley-captured fish (17%) and those from Santa Cruz (57%). Since our previous studies of fish captured in San Francisco Bay showed that Berkeley-captured fish had embryological success rates of 41.8%,[3] we consider this latest mean value as low and perhaps a result of either small sample size or year-to-year variability. Although embryological success was the only significant difference noted, all of the other measures of reproductive success were higher in Santa Cruz fish. Other measures of spawning fish, including hepatic AAH activity, in Table 7 are comparable to values obtained

Table 8. Hepatic Concentrations of PCB Congeners (μg/g lipid) in Sexually Mature Females Collected in January and February 1987.

Congener IUPAC No.	Berkeley		Santa Cruz		Site Differences	
	Mean	SD	Mean	SD	F	P
8	0.12	NA	0.05	0	1.4	0.29
18	0.11	0.07	0	0	1.5	0.3
28	0.14	0.08	0.18	0	0.39	0.56
52	0.4	0.13	0	0	1.9	0.22
44	0.08	0.08	0.07	0	0.14	0.22
66	0.1	0.06	0.0001	NA	1.48	0.28
101	0.57	0.01	0.19	0.12	1.67	0.25
77/o,p'-DDD	0.26	0.19	0.6	0.35	0.58	0.48
118	0.96	0.12	0.19	0.09	6.8	0.001
105	0.15	0.06	0.15	0.14	0.1	0.95
153	1.1	0.64	0.17	0.08	5.8	0.06
138	0.9	0.7	0.16	0.06	3.1	0.13
187	0.54	0.35	0.18	0.07	4	0.1
128	0.08	0.01	0.02	0	1.38	0.29
180	0.74	0.65	0.33	0.13	1.69	0.25
170	0	0	0	0	NA	NA
195	0.02	0.01	0	0	10	0.05
206	0.09	0.1	0.01	0.01	1.2	0.31
209	0.05	0.07	0	0	0.71	0.44
Aroclors						
1242	1.18	0.77	0	0	1.5	0.27
1254	8.34	1.06	1.63	0.77	3.26	0.13
1260	11.38	9.99	5.1	1.97	1.68	0.25
ΣPCB	18.23	12.84	5.03	2.14	2.93	0.15

NA = not available or applicable.

previously from studies with San Francisco Bay *P. stellatus*.[3] One exception is that the percent normal larvae from Santa Cruz fish (85–100%) was higher than previous values from San Francisco Bay fish (means, 77–75%).

Hepatic Contaminant Concentrations

In Table 8 the results of chemical analyses of livers of four of the spawning females collected from Berkeley and three spawning females collected from Santa Cruz are presented. Very few of the mean site concentrations were significantly different (only PCB congeners numbers 118 and 195), but this was a small sample and differences would have to be great in order for them to be significant with contaminant concentrations that are so innately variable. There was a tendency for the Berkeley-captured fish to have higher concentrations of some PCB congeners and Aroclors.

Positive Controls—Laboratory Induction of P-450 Enzyme Activity

Treatment of immature *P. stellatus* with 7,8-BF resulted in a strong induction of cytochrome P-450 in liver microsomes (Table 9). The concentrations of cy-

Table 9.   Positive Control: Hepatic Microsomal P-450 Content and Activities of *P. stellatus* Treated with 7,8-BF.

Parameter	Treated		Control	
	Mean	SD	Mean	SD
Cytochrome P-450 (nmol/mg)	0.326[a]	0.046	0.135	0.013
Cytochrome b_5 (nmol/mg)	0.06	0.013	0.044	0.009
Cytochrome P-450E (nmol/mg)	171.5	48.1	11	8.5
Cytochrome P-450E (% of total P-450)	52.2	10.4	8.56	6.5
EROD (nmol/min/mg protein)	2.17	0.79	0.186	0.097
EROD (nmol/min/nmol P-450)	6.8	2.68	1.43	0.83
E_2-OHase (nmol/min/mg protein)	0.051	0.018	0.052	0.011
E_2-OHase (nmol/min/nmol P-450)	0.19	0.07	0.4	0.08

NA = not available or applicable.
[a]Treatment means in bold type are significantly different than controls.

tochrome P-450/mg protein in the treated fish increased between two- and three-fold higher than those in control animals. There was no evidence of cytochrome P-420 in microsomes from treated or control individuals, indicating no apparent photometrically detectable degradation of cytochrome P-450.

The data also suggest an increase in the specific content of cytochrome b_5. However, the average value in the treated animals was only 30% or so greater than the values in control animals, and the difference is not statistically significant.

The EROD activities, expressed per milligram of microsomal protein, were nearly 12-fold greater in the treated fish than in the control fish. The activity normalized to the apparent content of spectrally native cytochrome P-450 was fivefold greater in the treated than in the control animals. The former expression, activity/mg protein, gives an indication of the capacity of the membrane preparation to catalyze this activity. The latter expression, activity/nmole P-450, indicates that there has been a selective increase in the catalyst for this activity among the population of cytochrome P-450 forms present.

The suggested increase in specific form(s) of cytochrome P-450 was confirmed by the results of immunoblot analysis of microsomes from control and treated fish. Analysis with monoclonal antibody 1-12-3 to the hydrocarbon inducible cytochrome P-450E from the marine teleost scup showed a single cross-reacting protein in *P. stellatus*. This protein had an electrophoretic migration characteristic very close to that of P-450E, about 54,000 molecular weight. Given the specificity of this MAb 1-12-3,[33] we conclude that this protein is a counterpart to scup P-450E. As indicated in Table 9, the amount of *P. stellatus* P-450E is very strongly increased in the treated animals. Given our knowledge of the catalytic functions of scup cytochrome P-450E, we further suggest that this P-450E counterpart present in *P. stellatus* is the catalyst for EROD activity. In contrast to the results with EROD activity, the activities of estradiol 2-hydroxylase activity/mg protein were unchanged by 7,8-BF treatment. This is consistent with other information suggesting that estradiol 2-hydroxylase is not catalyzed by methylcholanthrene- or 7,8-BF-inducible P-450 forms. However, the activities expressed per nanomole P-450 showed a decline in the animals treated wtih 7,8-BF. The results also suggest that the activities of the estradiol 2-hydroxylase catalyst are unchanged in animals treated with 7,8-BF. The decrease in estradiol 2-hydroxylase activity per nanomole of total P-450 in the BNF-induced starry flounder indicates that the catalyst for this activity represents a smaller proportion of the total cytochrome P-450 in the treated animals. This would be expected if there was a large increase in a P-450, such as P-450E, that does not have E_2-hydroxylase activity, and no change in amounts of other P-450 forms.

DISCUSSION

Chlorinated Hydrocarbons in Fish

The hepatic concentrations of DDTs and their isomers were generally not significantly different between sites, which is not surprising considering the widespread contamination of the sea and atmosphere by these persistent organochlorines. However, *o,p'*-DDT had significantly elevated concentrations at Oakland and *p,p'*-DDD tended to have higher concentrations in the more urban areas of the Bay (Oakland and Berkeley).

Among the other pesticides some site differences were detected. Higher concentrations of aldrin and mirex occurred in Russian River fish and higher concentrations of chlordane and transnanochlor occurred in Oakland fish.

In the PCBs few site differences were detected in the less chlorinated biphenyls (congeners 8, 18, 28, 52, 44, 66, and 101), with elevated concentrations of congener 8 in Russian River fish and congener 101 in Oakland fish being notable exceptions. For the congeners with greater chlorine substitution there was a tendency for Oakland, and to a lesser extent Berkeley, fish to have the highest concentrations. This was true for congeners 118, 105, 153, 138, 187, 128, 180, 170, 195, 209, and ΣPCBs.

Since mean ΣPCBs in liver were 9.7 μg/g lipid in Berkeley, 13 μg/g lipid in Oakland, 5 μg/g lipid in Santa Cruz, and 3.5 μg/g lipid in Russian River fish, it appears that: (1) the San Pablo Bay and Vallejo fish, with 2–4 μg/g lipid of ΣPCB were similar to the open-coast sites, and (2) the open-coast and Oakland-Berkeley fish differed in PCB content by about a factor of two.

Hepatic Microsomal Enzymes

For the P-450 proteins and their catalytic activities, the most obvious overall pattern was that Oakland fish had the greatest indications of contaminant induction and those from the Russian River, the least. While most MFO parameters indicated little difference between sites other than Oakland, P-450E was clearly elevated in Berkeley fish (except in mature females) relative to the other sites. Relatively low values of P-450 parameters were generally evident in mature females from Berkeley, in contrast to results from past studies utilizing this site and San Pablo Bay.[2,3]

Plasma Sex Steroids

Some site differences were observed in plasma steroid concentrations. Testosterone was elevated in San Pablo Bay and Vallejo males relative to Berkeley, Oakland, and Russian River males. Estradiol was significantly elevated in Berkeley immature females, with the Russian River having the lowest concentrations.

Micronuclei

Micronuclei in circulating erythrocytes were significantly elevated in fish from all Bay sites relative to those from the Russian River site. Micronucleus frequencies were not correlated with organic contaminant concentrations. These results are consistent with the non-clastogenic properties of DDTs and PCBs.[17,46] Compounds such as mutagenic polycyclic aromatic hydrocarbons, mutagenic nitroaromatics and phenols, and carcinogenic volatile aromatic compounds that were not measured in these studies but which are present in San Francisco Bay may be more strongly correlated with micronuclei occurrence than are the chlorinated hydrocarbons measured in this study.

Relationships between Parameters

Several things seem apparent when contaminant concentrations are compared to hepatic P-450 activities. First, while EROD activity and total PCBs are correlated, AHH activity and P-450E content of microsomes, which respond similarly to induction by complex mixtures of contaminants containing PCBs, were not correlated to ΣPCBs. It should also be mentioned that other significant inducers of P-450 are present at these sites that were not measured in the liver tissue samples. Specifically, the polynuclear aromatic hydrocarbons (PAHs) are probably important inducers at many of these sites, since sediment concentrations of greater than 5 ppm (dry wt.) are commonly found in the Bay.[45] These compounds are efficiently metabolized to more polar compounds in *P. stellatus*[47] and they are not often detected in gas chromatographic analyses of tissues.

Second, P-450E, which is the major contaminant-induced isozyme in other species, is much more responsive in starry flounder to the influence of urbanization, than are either AHH or EROD activity. While P-450E abundance appears to be most responsive of the P-450 parameters to urban influences, in evaluating the usefulness of the various possible measurements in a monitoring program the implications to the organism of the presence and activities of other isozymes that may also be induced should be taken into account. For example, the relationship between AHH activity and reduced reproductive success[3] may or may not depend on P-450E abundances or activities, as other isozymes could be important in this relationship.

The possible relationship between P-450 parameters and plasma estradiol concentrations in females is interesting. The positive relationship between both P-450E and P-450 and plasma estradiol in immature females is perhaps the opposite of what would be expected based on the recent literature. Förlin and Haux[8] treated rainbow trout with 7,8-BF (a PAH-type inducer) and supplied trace amounts of radiolabeled estradiol. They found that treated trout had significantly greater radioactivity in the bile than untreated trout, suggesting prolonged induction might lead to estradiol depletion if compensatory mechanisms were overwhelmed. However, this increased excretion rate could also be interpreted as resulting just from increased rates of estradiol production. If it is accepted that this correlation in immature females *P. stellatus* is not a result of chance or other unmeasured influences then it could result from an as-yet-not-understood alteration of steroidogenesis or steroid hydroxylation by contaminants where induced fish produce or maintain more estradiol. It is not clear if there are reproductive consequences of increased estradiol in contaminant-induced immature females once they mature, especially since the reproductive endocrinology of *P. stellatus* is poorly known. In trout high concentrations of estradiol prevent final oocyte maturation.[14] This may not be the case in *P. stellatus* since individuals collected close to the time of spawning from Santa Cruz had quite high concentrations—9 to 49 ng/mL of plasma estradiol. The data from immature females do suggest the potential for contaminant induction to affect the endocrinology in feral populations of this species. Further progress on determining if such changes are significant to the reproductive success of *P. stellatus* would depend on capturing females just at the time of spawning and in carrying out some carefully designed laboratory experiments.

ACKNOWLEDGMENTS

We gratefully acknowledge the support of the Ocean Assessments Division, National Ocean Service, National Oceanographic and Atmospheric Administration for these studies. A part of this work was performed under the auspices of the U.S. Department of Energy, by the Lawrence Livermore National Laboratory under contract W-7465-Eng-48.

REFERENCES

1. Long, E. R., and M. F. Buchman, Chapter 20.
2. Spies, R. B., D. W. Rice, Jr., and J. W. Felton. "The Effects of Organic Contaminants on Reproduction of Starry Flounder, *Platichthys stellatus* (Pallas) in San Francisco Bay. I. Hepatic Contamination and Mixed-function Oxidase (MFO) Activity during the Reproductive Season," *Mar. Biol.* 98:181–190 (1988).

3. Spies, R. B., and D. W., Rice, Jr. "The Effects of Organic Contaminants on Reproduction of Starry Flounder, *Platichthys stellatus* (Pallas) in San Francisco Bay. II. Reproductive Success of Fish Captured in San Francisco Bay and Spawned in the Laboratory," *Mar. Biol.* 98:191–200 (1988).

4. Klotz, A. V., J. J. Stegeman, and C. Walsh. "Multiple Isozymes of Hepatic Cytochrome P-450 from the Marine Teleost Fish Scup (*Stenotomus chrysops*)," *Mar. Environ. Res.* 14:402–404 (1984).

5. Klotz, A. V., J. J. Stegeman, B. R. Woodin, E. A. Snowberger, P. E. Thomas, and C. Walsh. "Cytochrome P-450 Isozymes from the Marine Teleost *Stenotomus chrysops*: Their Roles in Steroid Hydroxylation and the Influence of Cytochrome b_5," *Arch. Biochem. Biophys.* 249:326–338 (1986).

6. Lu, A. Y. H. "Multiplicity of Liver Drug Metabolizing Enzymes," *Drug Metab. Rev.* 10:187–208 (1979).

7. Truscott, B., J. M. Walsh, M. P. Burton, J. F. Payne, and D. R. Idler. "Effect of Acute Exposure to Crude Petroleum on Some Reproductive Hormones in Salmon and Flounder," *Comp. Biochem. Physiol.* 75C:121–130 (1983).

8. Förlin, L., and C. Haux. "Increased Excretion in the Bile of 17β-[^{3}H]estradiol-derived Radioactivity in Rainbow Trout Treated with β-naphthoflavone," *Aquat. Toxicol.* 6:197–208 (1985).

9. Snowberger, E. A., and J. J. Stegeman. "Patterns and Regulation of Estradiol Metabolism by Hepatic Microsomes from Two Species of Marine Teleosts," *Gen. Comp. Endocrinol.* 66:256–265 (1987).

10. Freeman, H. C., G. Sangalang, and B. Flemming. "The Sublethal Effects of a Polychlorinated Biphenyl (Aroclor 1254) Diet on the Atlantic Cod (*Gadus morhua*)," *Sci. Total Environ.* 24:1–11 (1982).

11. Singh, H., and T. P. Singh. "Site of Action of Some Agricultural Pesticides in the Hypothalmo-hypophyseal-ovarian Axis of the Freshwater Catfish, *Heteropneustes fossilis* (Bloch)," in *Reproductive Physiology of Fish*, C. J. J. Richter and M. J. Th. Goos, Eds. (Poduc: Waginengen, 1982), p. 60.

12. Sangalang, G. B., and M. J. O'Halloran. "Adverse Effects of Cadmium on Brook Trout Testes and on *In Vitro* Testicular Androgen Synthesis," *Biol. Reprod.* 9:394–403 (1973).

13. Heinrichs, W. L., R. J. Gellert, J. L. Bakke, and N. L. Lawrence. DDT Administered to Neonatal Rats Induces Persistent Estrus Syndrome," *Science N.Y.* 173:642–643 (1971).

14. Theofan, G. "The *In Vitro* Synthesis of Final Maturational Steroids (*Perca flavescens*)," Ph.D. Thesis, Univ. Notre Dame, Notre Dame, Indiana, 1981.

15. Chen, T. T., and R. A. Sonstegard. "Development of a Rapid, Sensitive and Quantitative Test for the Assessment of Effects of Xenobiotics on Reproduction in Fish," *Mar. Environ. Res.* 14:429–430 (1984).

16. Schmid, W. "The Micronucleus Test for Cytogenetic Analysis," in *Chemical Mutagens: Principles and Methods for their Detection. Vol. 6.* A. Holander, Ed. (New York: Plenum Press, 1976), pp. 31–53.

17. Heddle, J. A., M. Hite, B. Kirkhart, K. Mavourin, J. T. MacGregor, G. W. Newell, and M. F. Salamone. "The Induction of Micronuclei as a Measure of Genotoxicity," *Mutation Res.* 123:61–118 (1983).

18. Hose, J. E., J. N. Cross, S. G. Smith, and D. Diehl. Elevated Circulating Erythrocyte

Micronuclei in Fishes from Contaminated Sites off Southern California," *Mar. Environ. Res.* 22:167–176 (1987).

19. Spies, R. B., D. W. Rice, Jr., P. A. Montagna, and R. R. Ireland. "Reproductive Success, Xenobiotic Contaminants and Hepatic Mixed-function Oxidase Activity in *Platichthys stellatus* Populations from San Francisco Bay," *Mar. Environ. Res.* 17:117–121 (1985).

20. Orcutt, H. G. "The Life History of the Starry Flounder *Platichthys stellatus* (Pallas)," *Calif. Fish Game Bull.* 78:1–64 (1950).

21. Spies, R. B., D. W. Rice, P. A. Montagna, R. R. Ireland, J. S. Felton, S. K. Healy, and P. R. Lewis. "Pollutant Body Burdens and Reproduction in *Platichthys stellatus* in San Francisco Bay," Final Report to NOAA, UCID-19993-84, March 1985, 95 pp.

22. Colloidi, P., M. S. Stekoll, and S. D. Rice. "Hepatic Aryl Hydrocarbon Hydroxylase Activities in Coho Salmon (*Oncorhynchus kisutch*) Exposed to Petroleum Hydrocarbons," *Comp. Biochem. Physiol.* 79C:337–341 (1984).

23. Nebert, D. W., and H. U. Gelboin. "Substrate-inducible Microsomal Aryl Hydrocarbon Hydroxylase in Mammalian Cell Culture," *J. Biol. Chem.* 243:6242–6249 (1968).

24. Spies, R. B., J. S. Felton, and L. Dillard. "Hepatic Mixed Function Oxidases in California Flatfishes are Increased in Contaminated Environments and by Oil and PCB Ingestion," *Mar. Biol.* 70:117–127 (1982).

25. Klotz, A. V., J. J. Stegeman, and C. Walsh. "An Aryl Hydrocarbon Hydroxylating Hepatic Cytochrome P-450 from the Marine Fish *Stenotomus chrysops*," *Arch. Biochem. Biophys.* 226:578–592 (1983).

26. Lowry, O. H., N. J. Rosebrough, A. L. Farr, and R. J. Randall. "Protein Measurement with the Folin Phenol Reagent," *J. Biol. Chem.* 193:265–275 (1951).

27. Förlin, L., and T. Andersson. "Storage Conditions of Rainbow Trout Liver Cytochrome P-450 and Conjugating Enzymes," *Comp. Biochem. Physiol.* 80B:569–572 (1985).

28. Klotz, A. V., J. J. Stegeman, and C. Walsh. "An Alternative 7-ethoxyresorufin O-deethylase Activity Assay: A Continuous Visible Spectrophotometric Method for the Measurement of Cytochrome P-450 Monoxygenase Activity," *Anal. Biochem.* 140:138–145 (1984).

29. Burke, M. D., S. Thompson, C. R. Elbcombe, J. Halpert, T. Haaparanta, and R. T. Mayer. Ethoxy-, Pentoxy-, and Benzoxyphenoxazones and Homologues: A Series of Substrates to Distinguish between Different Induced Cytochromes P-450," *Biochem. Pharmacol.* 34:3337–3345 (1985).

30. Kupfer, D., G. K. Miranda, and D. H. Bulger. "A Facile Assay for 2-Hydroxylation of Estradiol by Liver Microsomes," *Anal. Biochem.* 116:27–34 (1981).

31. Stegeman, J. J., and M. Chevion. "Sex Differences in Cytochrome P-450 and Mixed-function Oxidase Activity in Gonadally Mature Trout," *Biochem. Pharmacol.* 29:553–558 (1979).

32. Park, S. S., H. Miller, A. V. Klotz, P. J. Kloepper-Sams, J. J. Stegeman, and H. V. Gelboin. "Monoclonal Antibodies Against Cytochrome p-450E from the Marine Teleost *Stenotomus chrysops* (scup)," *Arch. Biochem. Biophys.* 249:339–350.

33. Kloepper-Sams, P. K., S. S. Park, H. V. Gelboin, and J. J. Stegeman. "Specificity and Cross-reactivity of Monoclonal and Polyclonal Antibodies against Cytochrome P-450E of the Marine Fish Scup," *Arch. Biochem. Biophys.* 253:268–278 (1987).

34. Towbin, H., T. Staehelin, and J. Gordon. "Electrophoretic Transfer of Proteins from Polyacrylamide Gels to Nitrocellulose Sheets: Procedure and Applications," *Proc. Natl. Acad. Sci. U.S.A.* 76:4350–4354 (1979).

35. Ozretich, R. J., and W. P. Schroeder. "Determination of Neutral Organic Priority Pollutants in Marine Sediment, Tissue and Reference Materials Utilizing Bonded-Phase Sorbent," *Anal. Chem.* 58:2041–2047 (1986).

36. Spies, R. B., H. Kruger, R. Ireland, and D. W. Rice, Jr. "Stable Isotope Ratios and Contaminant Concentrations in a Sewage-distorted Food Web," *Mar. Ecol. Prog. Ser.* 54:157–170 (1989).

37. Ballschmitter, K., and M. Zell. "Analysis of Polychlorinated Biphenyls by Glass Capillary Gas Chromatography," *Fresenius Z. Anal. Chem.* 302:20–31 (1980).

38. Mullin, M. D., C. M. Pochini, S. McCrindle, M. Romkes, S. H. Safe, and L. M. Safe. "High-resolution PCB Analysis: Synthesis and Chromatographic Properties of All 209 PCB Congeners," *Environ. Sci. Technol.* 18:468–476 (1984).

39. Capel, P. D., R. A. Rapaport, S. J. Eisenreich, and B. B. Looney. "PCBQ: Computerized Quantification of Total PCB and Congeners in Environmental Samples," *Chemosphere* 14:439–450 (1985).

40. Clark, J. U. "Structure-Activity Relationships in PCBs: Use of Principal Components Analysis to Predict Inducers of Mixed-function Oxidase Activity," *Chemosphere* 15:275–287 (1986).

41. Brown, J. F., Jr., D. L. Bedard, M. J. Brennan, J. C. Carnahan, H. Feng, and R. F. Wagner. "Polychlorinated Biphenyl Dechlorination in Aquatic Sediments," *Science* 236:709–712 (1987).

42. Yamamoto, K. "Studies on the Formation of Fish Eggs. I. Annual Cycle in the Development of Ovarian Eggs in the Flounder, *Liopsetta obscure*," *J. Fac. Sci. Hokkaido Univ. Ser. 6* 2001.1 2:363–373 (1956).

43. Rosenthal, H., and D. F. Alderdice. "Sublethal Effects of Environmental Stressors, Natural and Pollutional, on Marine Fish Eggs and Larvae," *J. Fish. Res. Bd. Can.* 33:2047–2065 (1976).

45. Rice, D. W., Jr., R. B. Spies, C. Zoffman, M. Prieto, and R. Severeid. "Organic Contaminants in Superficial Sediments of the San Francisco Bay-Delta," in preparation.

46. Green, S., J. V. Carr, K. A. Palmer, and E. J. Oswald. "Lack of Cytogenetic Effects in Bone Marrow and Spermatogonial Cells in Rats Treated with Polychlorinated Biphenyls (Arolclor 1242 and 1254)," *Bull Environ. Contam. Toxicol.* 13:14–22 (1975).

47. Varanasi, U., M. Nishimoto, W. L. Reichert, and B.-T. L. Eberhart. "Comparative Metabolism of Benzo(a)pyrene and Covalent Binding to Hepatic DNA in English Sole, Starry Flounder and Rat," *Cancer Res.* 56:3817–3824 (1986).

CHAPTER 6

Hepatic Enzymes as Biomarkers: Interpreting the Effects of Environmental, Physiological and Toxicological Variables

Braulio D. Jimenez, A. Oikari, S. M. Adams, D. E. Hinton, and J. F. McCarthy

ABSTRACT

Components of the hepatic mixed function oxidase (MFO) system are evaluated as a biomarker of contaminant exposure in fish from streams in Tennessee. The responses of several MFO components (ethoxyresorufin-O-deethylase (EROD) activity, cytochrome P-450 levels, and NADPH cytochrome c reductase activity) were comparable in fish from four pristine reference streams; EROD activity and cytochrome P-450 levels did vary over the year, with highest levels apparent in the autumn. Levels were lower in reproductively active females than in males during the same season. Fish from a stream impacted by an industrial effluent had EROD activities that were significantly higher than those from the reference stream; although the differences were greater in males during the warmer months. Reproductively active females from the contaminated stream also had higher EROD activity, but their enzyme activity was considerably lower than in males tested during the same season. Fish from the most severely impacted site did not have the highest EROD levels, possibly due to hepatotoxic damage to the liver; laboratory exposure of fish to an hepatotoxin reduced the capacity of the liver to respond to injection of benzo[a]pyrene, a known MFO inducer. MFO response in the fish was an informative biomarker of exposure, but the role of environmental, physiological, and toxicological factors on MFO response must be understood in order to properly interpret the biomarker response.

INTRODUCTION

Responses of hepatic mixed function oxidase (MFO) system in animals from areas of suspected contamination are often used as a biomarker to indicate exposure to unacceptable levels of organic chemicals such as polycyclic aromatic hydrocarbons, dibenzo-*p*-dioxins, and other higher molecular weight organics. However, in interpreting changes in various components of the MFO system as biomarkers of exposure to contaminants, care must be exercised in interpreting the data because this system is a multifunctional enzyme complex whose activities are regulated by a number of physiological and metabolic factors. The interactions of these potentially complicating factors must be understood and accounted for before we can, with any confidence, attribute changes in MFO as evidence for contaminant exposure.

This paper provides examples of several of the environmental, physiological, and toxicological factors that need to be considered in validating the use of the hepatic MFO system as a biomarker of exposure to environmental contaminants. The goal is to point out that interpretation of hepatic MFO parameters as a biomarker requires an awareness of these potentially confounding variables, and that experimental design of an environmental monitoring program that includes hepatic enzyme biomarkers must account for these factors.

We will present data from a long-term biomonitoring study in a stream in Tennessee contaminated by an industrial facility and from several uncontaminated streams in adjacent areas. The objective is to demonstrate the variability in enzyme activities and cytochrome P-450 levels as a function of season and sex. Laboratory results will be presented which confirm the role of environmental variables, such as temperature and food availability, on the responses of MFO components in fish. We will also use the field data to illustrate the potential for "false negatives"—low levels of MFO activity in fish collected from highly contaminated sites—due to hepatotoxicity resulting from exposure to complex mixtures of contaminants. In this paper, we attempt to make several points:

1. Measurements of hepatic MFO components are useful and informative biomarkers to demonstrate exposure to organic contaminants in the environment.
2. Confounding factors need to be accounted for in order to interpret the biomarker responses.
3. By monitoring responses of animals from sites of suspected contamination with those from animals at ecologically similar reference sites that are free of contaminants can account for many of the confounding factors. Furthermore, animals from the different sites should be matched by species, age, and sex, and should, if possible, exclude reproductively active organisms.
4. A *suite* of biomarkers should be measured because a single biomarker response is not sufficient to unequivocally evaluate exposure, and the responses of one biomarker can provide information that improves interpretation of other biomarkers.

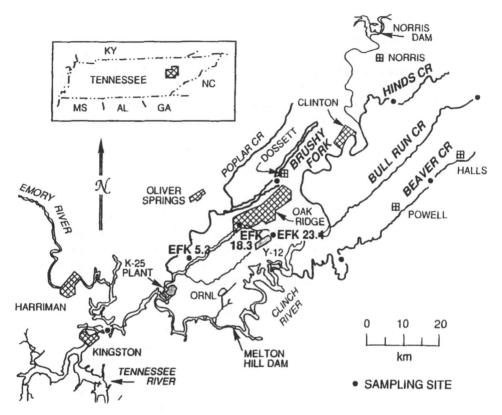

Figure 1. Map of streams around Oak Ridge, Tennessee, with stations and sampling streams of this study. See text for description.

METHODS AND MATERIALS

Field Survey of Fish from East Fork Poplar Creek

Biological markers were examined in fish as part of a Biological Monitoring and Abatement Program stipulated under the National Pollutant Discharge Elimination System permit issued to the Oak Ridge Y-12 Plant. This comprehensive monitoring program includes tasks that evaluate the toxicity of the ambient stream water, the bioaccumulation of contaminants in fish and molluscs, and the structure and function of the aquatic community, in addition to the measurement of bio-markers.[1] Effluent discharges from the Y-12 Plant enter the headwaters of East Fork Poplar Creek (EFPC; Figure 1) 23.7 km above its confluence with Poplar Creek. Effluent discharges of 388 L/s from the Y-12 Plant constitute 25% of the mean annual flow in EFPC at a gauging station at East Fork Kilometer (EFK) 5.3. The stream also receives urban runoff and some agricultural runoff between EFK 22.7 and 7.7.

Water and sediments in EFPC downstream of the Y-12 Plant contain metals, organic chemicals, and radionuclides discharged over many years of operation. Sediment samples contained ten priority pollutant organics (including seven polycyclic aromatic hydrocarbons, polychlorinated biphenyls, and total phenols) and seven metals (As, Cd, Pb, Hg, Ni, Ag, and Zr) at concentrations above background.[2] Monitoring of the discharge into EFPC identified potential toxicity problems due to copper, ammonia, residual chlorine, perchloroethylene, total nitrogen, and oil and grease. Of these, the first four could have been toxic at the maximum concentration reported, although wide variability in concentrations was observed.[1] However, monthly testing of 24-hr composite samples collected at the outfall of the Y-12 Plant (EFK 23.4) provides no evidence of chronic toxicity. This was based on bioassays that measured both survival and reproduction of *Ceriodaphnia* and survival and growth of the fathead minnow larvae.[1] Accumulation of mercury and PCBs in fish and molluscs in EFPC decreases with increasing distance from the Y-12 Plant. Furthermore, effects at the community level, measured as reduction in species diversity of benthic invertebrates or species richness of fish, is more severe at the upstream stations and improves with increasing distance downstream.[1]

This paper discusses results of measurements of hepatic enzyme biomarkers at several sites on EFPC and at four ecologically similar streams used as uncontaminated reference sites. Brushy Fork (one of the four reference streams) has no source of industrial or municipal inputs, but may receive some agricultural runoff.

Four reference streams were sampled: Brushy Fork, Beaver Creek, Hinds Creek, and Bull Run Creek (Figure 1). All of these reference streams are hard water streams draining geologically similar ridge-and-valley regions of East Tennessee. All are similar in size to EFPC. All four drain predominantly agricultural regions. Beaver Creek has a significant amount of suburban development within the watershed. None of the streams receives industrial inputs. However, headwater tributaries of Hinds Creek and Bull Run receive inputs from municipal wastewater treatment plants and some PCB contamination has been found in fish and sediments in Brushy Fork (0.6 ppm total PCBs in sunfish).

Adult redbreast sunfish (*Lepomis auritus*) were collected by electroshocking at each site. Fish were transported alive to the laboratory for dissection and analysis.

Laboratory Exposure of Fish to Benzo(a)pyrene: Effects of Temperature, Feeding Status, and Dose

Bluegill sunfish (*Lepomis machrochirus*) were obtained from Chews Fish Hatchery (Woodbury, GA). Fish were acclimated for at least 2 weeks at 4, 9, 13, and 26°C in flowing water. The fish were divided into two groups, one of which was fed ad libitum with trout chow while the other group was denied food for 2 weeks. Another group of starved fish were injected intraperitoneally with either ^{3}H-benzo[a]pyrene Aldrich (BaP) in corn oil at 0, 0.1, 10, or 20 mg BaP/kg body weight or equivalent weights of corn oil (controls). Animals were sacrificed 5 days after injection, and microsomes were isolated and assayed for enzyme activity.[4]

Laboratory Exposure of Fish to Hepatotoxins and Benzo(a)Pyrene

Fish (hybrid *Lepomis*) were dosed by gavage with allyl formate, a known hepatotoxic agent,[5,6] and 1 day later were injected intraperitoneally with benzo[a]pyrene (BaP; gold label, Aldrich) in corn oil at a dose of 8 mg/kg body weight. Vehicle controls were not treated with allyl formate, but were injected with an equivalent volume of corn oil without BaP. BaP controls received injections of BaP in corn oil but no allyl formate. Fish were sacrificed 5 days after injection, and microsomes were isolated and assayed for enzyme activities.

Measurement of MFO Components

Isolation of Hepatic Microsomes

Fish hepatic microsomes were prepared by differential centrifugation. Fish were killed by severing their spinal cords, and the livers were immediately removed and placed in ice-cold 0.1 M sodium phosphate buffer and 0.25 M sucrose at pH 7.4. The minced tissues were homogenized in 5 volumes of buffer and centrifuged at 3000 × g for 20 min with a Beckman J-21B centrifuge. The resulting supernatants were centrifuged at 105,000 × g for 60 min in a Beckman L3-50 ultracentrifuge. Microsomal pellets were resuspended in 0.1 M Tris buffer (pH 7.4), 1 mM EDTA, 20% glycerol.[4] Microsomes were frozen in liquid nitrogen and stored at $-120°C$ until assayed. No significant change in EROD activity was detected after 6 months of storage at this temperature. The activity of fish microsomes stored under these conditions has been reported to be stable for a year.[7]

Enzyme Assays

The activity of 7-ethoxyresorufin-O-deethylase (EROD) in the hepatic microsomal fraction was determined fluorometrically at 30°C by measuring the production of resorufin from 7-ethoxyresorufin (Molecular Probes, Oregon), using an Aminco SPF500 spectrofluorometer with excitation wavelength set at 530 nm and 582 nm for emission.[8] The final reaction buffer contains 150 mM Tris at pH 7.7, 200 nmol NADPH, and 1 μM 7-ethoxyresorufin in the cuvette. The activity is expressed as picomoles of resorufin per minute per milligram of microsomal protein.[4] Cytochrome P_{450} content was measured by its characteristic oxidized and reduced spectra.[9] Cytochrome P-450 samples were oxidized with carbon monoxide and reduced with sodium dithionite.[10]

NADPH-cytochrome c reductase activity was determined spectrophotometrically at 30°C and measured at 550 nm by the reduction of the electron acceptor, cytochrome c,[11] and using an extinction coefficient of 21 cm^{-1} mM^{-1}. The reaction mixture contained 0.2 M of potassium phosphate buffer (pH 7.4) 1.1 mg/mL horse heart cytochrome c, 0.125 mM NADPH and 2–10 μg of microsomal protein. Proteins were measured by the Bio-Rad (Richmond, CA) reagent method[12] on the Cobas Fara (Hoffman La Roche Instruments) using bovine serum albumin as a standard.

Statistical Analysis

For each response variable, Dunnett's test[13] was used to compare the mean from the reference stream to each of the means from the other sites along EFPC for each season and sex combination.

Histopathology of Livers from Field Fish

At necropsy, a wedge (approximately $1 \times 3 \times 2$ mm) was cut from livers of randomly selected fish and placed in Bouin's fixative within coded vials. All subsequent morphologic work and evaluation were done blindly. After analysis, individual animal codes were grouped by site with the histopathologist unaware of site designation. Histopathological findings were grouped by site and prevalence of lesions was determined for each fish. After routine dehydration in a graduated series of ethanol solutions, livers wedges were embedded in paraffin and sectioned at 5 μm. Sections were mounted on glass slides and stained with hematoxylin and eosin.

RESULTS AND DISCUSSION

MFO Responses of Fish from Uncontaminated Reference Sites

Since the responses of fish from the reference site(s) provides the basis of comparison for evaluating the degree of induction of fish from suspect sites, selection of a suitable reference site is critical for the interpretation of biomarker responses. It is appropriate, therefore, to evaluate MFO components of animals from several ecologically comparable sites to determine the variability of MFO responses in animals not exposed to known sources of pollution. Evaluation of several critical MFO parameters demonstrated that, for most responses (EROD, cytochrome P-450 levels, and NADPH cytochrome c reductase), there were no significant differences in responses of animals at the different reference sites (Table 1). The MFO parameters were also similar to those from animals raised in the laboratory in water free from any pollutants. EROD activity in laboratory-raised bluegill sunfish could vary from 10–60 pmol/min/mg protein depending on the type of food consumed and on their nutritional status. Levels of cytochrome P-450 can also fluctuate from 0.2–0.3 nmol/mg protein under various conditions.[4]

Seasonal differences in hepatic EROD activity and cytochrome P-450 levels were studied in male and female redbreast sunfish in one of the reference sites

Table 1.   Mean ± SE (n) for Biological Indicators in Redbreast Sunfish Collected at Four Different Reference Streams in the Vicinity of Oak Ridge, Tennessee.

Station	EROD[a]	P450[b]	NADPH[c]
	Females		
Brushy Fork Creek	5.4 ± 1.7 (6)	0.12 ± 0.02 (6)	59.5 ± 8.7 (6)
Beaver Creek	12.0 ± 1.7* (6)	0.15 ± 0.02 (6)	48.0 ± 2.95 (6)
Bull Run Creek	3.9 ± 0.8 (5)	0.16 ± 0.02 (5)	66.2 ± 6.5* (5)
Hinds Creek	2.0 ± 0.6 (5)	0.14 ± 0.02 (5)	52.4 ± 3.6 (5)
	Males		
Brushy Fork Creek	15.7 ± 4.4 (5)	0.22 ± 0.06 (5)	80.0 ± 7.3 (5)
Beaver Creek	41.5 ± 7.7* (5)	0.39 ± 0.07* (5)	96.4 ± 14.5 (5)
Bull Run Creek	12.0 ± 4.2 (4)	0.31 ± 0.04 (4)	93.8 ± 8.7 (4)
Hinds Creek	18.3 ± 5.2 (5)	0.24 ± 0.02 (5)	79.2 ± 4.8 (5)

[a]Ethoxyresorufin O-deethylase (pmol/min/mg protein).
[b]Cytochrome P-450 (nmol/mg protein).
[c]NADPH cytochrome c reductase (nmol/min/mg protein).
*Value is significantly different ($p < 0.05$) from other values for animals of the same sex.

(Brushy Fork; Figure 2). Higher levels of EROD activity for both sexes were observed during the fall (October) of both 1985 and 1986 (Figure 2A). However, EROD activity was low for female fish in spring (late April) 1986 and summer (July) 1986. Cytochrome P-450 also followed similar patterns for both male and female fish (Figure 2B). This pattern appears to be attributable to (1) seasonal temperature changes which increase the cytochrome level and catalytic activity of the MFO system with increasing environmental temperature (in males), and (2) the effect of reproductive status on EROD activity and cytochrome levels in females.

To demonstrate the role of seasonal changes in temperature on hepatic enzyme activity, the effects of acclimation temperature on EROD activity were measured in the laboratory. Bluegill sunfish were acclimated for several weeks at different temperatures, and assayed for hepatic EROD activity thereafter (Figure 3). Basal levels of EROD activity (assayed at 30°C) increased with increased acclimation temperature if the animals were fed ad libitum. Fish that were acclimated but then denied food for 2 weeks did not demonstrate any temperature-related increases in EROD activity[4] (Figure 3A). Comparison of the EROD activity of field-collected animals from the reference site (data in Figure 2A) with the mean water temperature at that site for the month they were collected demonstrates

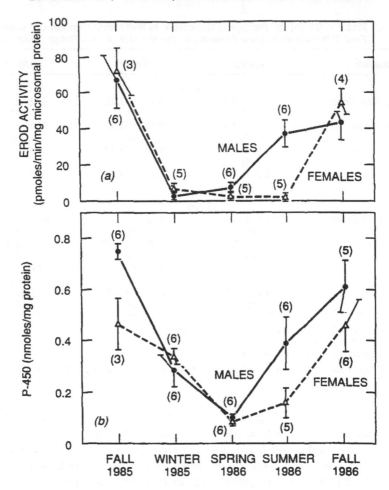

Figure 2. Changes in (a) EROD activity and (b) cytochrome P-450 levels in male (filled circles and solid line) and female (open triangle and dotted line) redbreast sunfish from Brushy Fork, a representative reference stream. The mean and standard error are indicated; the number of fish analyzed per sample is indicated in parentheses.

that the seasonal pattern of changes in field-collected animals can be attributed to changes in environmental temperature (Figure 3B).

There are two exceptions to this pattern. The male redbreasts collected in April 1986 exhibited hepatic EROD activity much lower than the well-fed laboratory animals. This may be due to the severe drought that occurred in the spring and summer of 1986. The decrease in precipitation and associated instream water flow was documented in an adjacent creek in the same drainage basin as Brushy Fork.[14] The prolonged period of low precipitation occurring before and during these periods could have affected food availability along the

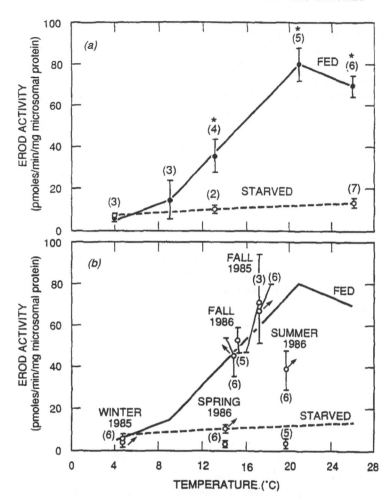

Figure 3. (a) EROD activity of laboratory bluegill sunfish acclimated to different temperatures and fed ad libitum (closed circles and solid line) or denied food for 2 weeks ("starved"; open circles and dotted line). EROD activity was assayed at 30°C. Symbols are as in Figure 2. (Modified from Jimenez, B. D., L. S. Burtis, G. H. Ezell, B. Z. Egan, N. E. Lee, J. F. McCarthy, and J. J. Beauchamp, in *Environ. Toxicol. Chem.* 7:623–634 (1988) and is reprinted with permission of the publisher.) (b) EROD activity of redbreast sunfish collected at different seasons from Brushy Fork is plotted as a function of the mean water temperature for the month the fish were sampled. The EROD data are replotted from Figure 2A. Open circle with arrow indicates a male group.

stream, since this represents a decrease of stream surface area (for the settlement of insects and other invertebrates which are sources of food along the banks of the stream) and possible crowding of organisms.

The other anomaly in Figure 3B is the very low level of EROD activity in female redbreasts collected in late April and July 1986. This reduction in EROD activity and cytochrome P-450 concentrations coincides with the reproductive season for this species (personal observations). This decrease in MFO activity of reproductively active females is consistent with correlations observed between maturity index and aryl hydrocarbon hydroxylase (AHH) activity in a marine fish (cunner).[15] In this study it was observed that levels of AHH activity in females began to decrease during the formation of eggs and reached their lowest values when gonads were enlarged and visual granulation was apparent. Furthermore, low levels of MFO activities (EROD and/or AHH) have been reported during the spawning season in several other fish species, including the freshwater vendace, *Coregonus albula L*,[16] lake trout, *Salvelinus namaycush*,[17] and rainbow trout, *Salmo gairdneri*.[18] It might also be noted that administration of 17β-estradiol caused a marked reduction of several MFO enzyme activities as well as total cytochrome P-450 content in trout.[19-21]

In summary, then, the basal level of MFO activity in fish from uncontaminated streams varies seasonally as a result of physiological changes, such as reproductive activity, and environmental variables, such as temperature and food availability.

MFO Responses of Fish from East Fork Poplar Creek (Polluted Steam)

EROD activity was measured in both male and female redbreasts in several stations of EFPC at all seasons, and compared to responses of fish from the reference stream (Brushy Fork; Figure 4). The pattern of responses in the field fish illustrate several points that are important to consider in interpreting hepatic enzymes as a biomarker. In this discussion, we assume that increases in EROD activity represent induction of new enzyme, although we did not confirm this in all cases by electrophoresis[22] or antibody probes.[23]

Effect of Season on MFO Induction

The enzyme activities observed in males collected from EFPC in summer were approximately four- to tenfold higher than those observed from the same stations in the winter. Increased temperatures in the summer may be responsible for this effect for any of several reasons, including (1) seasonal changes may mobilize contaminants in sediments or make them more available for uptake from water or from the foodchain, (2) the physiological response of these poikilothermic animals to increased temperature may increase the dose of contaminant taken up by the organism, and/or (3) the higher temperatures may increase the capacity for induction of the MFO system.

There is experimental evidence that demonstrates the effect of temperature on the degree of induction of the MFO system. Laboratory fish (bluegill sunfish) acclimated to higher water temperatures responded with higher induction of EROD activity when injected at the same dose of a known MFO inducer, benzo(a)pyrene.[21] For example, EROD activity in fish acclimated to 26°C was six-fold higher than that at 4°C when both groups were injected with 10 μg BaP/ g fish (Figure 5). Similar response of the MFO system was obtained with *Fundulus heteroclitus*.[24] Other researchers have reported a lag time of induction by the MFO system at lower temperatures.[22,25] For example, rainbow trout acclimated to 5°C reached maximum induction 3 days after injection with β-naphthoflavone, compared to 1 day for those acclimated to 17°C.[25] This suggests that fish acclimated to low temperatures can reach maximum induction given sufficient time. However, even after 18 days bluegills acclimated to 4°C and injected with BaP had reached only 8% of the activity reached by fish acclimated to 23°C.[22] These results show that the effects of temperature on the induction of the MFO system is species dependent.

In addition to the increased capacity of summer fish to respond to contaminant exposure by increasing EROD activity, the higher temperature may also increase the actual dose of contaminant to which the organism is exposed. Within the stream, higher temperatures could increase desorption of sediment-associated contaminants and thereby increase the levels of contaminants in the water. Furthermore, more contaminants may enter the foodchain in summer; for example, emergence of benthic insects may be a vector to increase availability of contaminants to the fish.

Because the higher temperatures also increase overall metabolic demand in poikilothermic animals,[26] exposure increases not only due to ingestion of larger volumes of potentially contaminated food, but also by increased exposure to chemicals dissolved in the water. Many animals respond to increased metabolic demand through compensatory increases in the volume of oxygenated (and potentially contaminated) water which is pumped over the gills. However, increases in uptake of oxygen are paralleled by increased uptake of organic contaminants. For example, changes in oxygen consumption, experimentally induced either by acute changes in temperature or by chlorine-induced gill damage, were highly correlated with changes in the uptake of polycyclic aromatic hydrocarbons and polychlorinated biphenyls by gills of rainbow trout.[27,28] In addition, the rate of uptake of BaP in water was six-fold faster in bluegills acclimated to 26°C compared to animals acclimated to 13°C.[29] Thus, the higher levels of EROD activity may reflect either increased capacity to respond to contaminant exposure, or to temperature-associated increases in the dose of chemical absorbed by the organism, or both.

Within-Season Comparisons of MFO Induction

Animals (male or female) collected from EFPC in winter clearly had much lower levels of EROD activity than did animals collected from the same sites in summer. Nevertheless, winter fish from most stations in EFPC had significantly

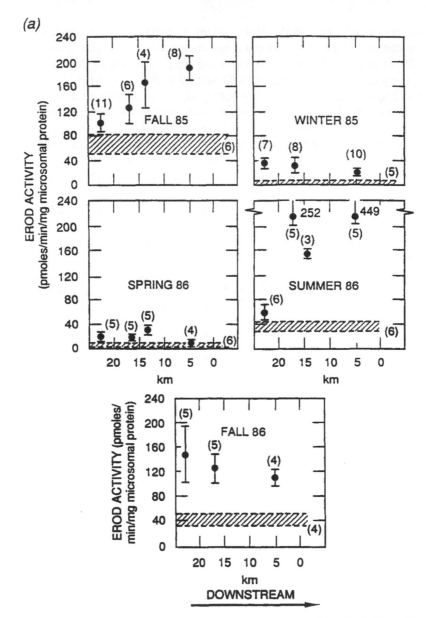

Figure 4. EROD activity of (a) male and (b) female redbreast sunfish collected from different stations along EFPC. EFK 23.7 is the station closest to the outfall of the industrial facility. Symbols are as in Figure 2. The shaded area indicates the range of the mean ± SE for redbreasts collected at the same time from the reference stream, Brushy Fork.

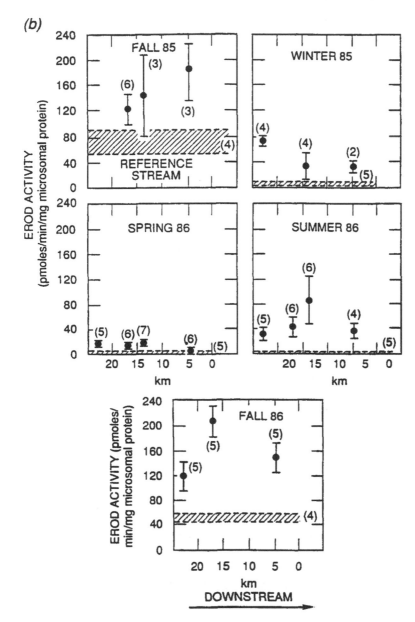

Figure 4. Continued

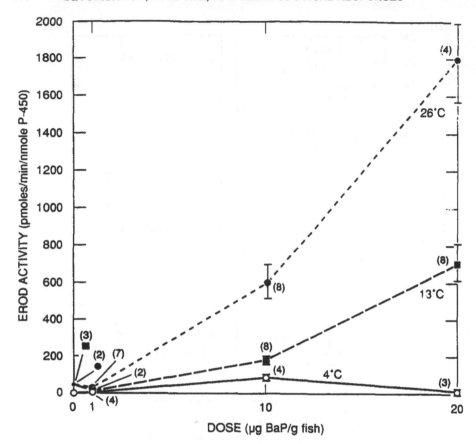

Figure 5. Effects of acclimation temperature on EROD activity in bluegill sunfish injected with different doses of BaP. Symbols are as in Figure 2. (From Jimenez, B. D., and L. S. Burtis in *Comp. Biochem. Physiol.* 93C:11–21 (1989). Reprinted with permission of the publisher.)

higher levels of EROD activity than winter fish from the reference site. Thus, although differences between reference and impacted sites may be greater during the warmer months, hepatic enzymes may be useful even in the winter, provided comparison is made with reference sites sampled during the same season.

Induction of Reproductively Active Females

EROD activity of male redbreasts collected from EFPC in summer 1986 are much higher than those of reproductively active females collected from the same sites at the same time (Figure 4A vs 4B). Nevertheless, even these reproductively active females (clutches of maturing oocytes were observed in ovaries of female redbreasts collected from the same sites throughout the summer of 1988 [M. S. Greeley, unpublished data]) demonstrated the expected induction of EROD ac-

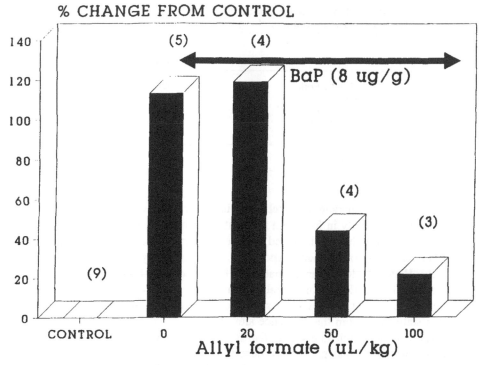

Figure 6.   Effect of allyl formate, a hepatotoxin, on BaP-induced increases in EROD activity in hybrid sunfish, expressed as a percent of the EROD activity of control fish. The numbers in parentheses represent the number of fish used in each treatment. These data were reported in McCarthy, J. F., B. D. Jimenez, L. R. Shugart, and A. Oikari, in *In Situ Evaluation of Biological Hazards of Environmental Pollutants*, S. Sandhu, Ed. (New York: Plenum Press, 1990).

tivity, compared to the very low levels of EROD activity exhibited by females from the reference stream. Other researchers have also reported induced levels of MFO activity in spawning fish from a number of fish species.[30-33]

Effect of Hepatotoxicity on Induction of EROD Activity

For both male and female redbreasts from EFPC, there are several seasons in which EROD activities in fish from the most polluted upstream station (EFK 23), while higher than those in fish from the reference stream, are much lower than those from fish further downstream. A range of endpoints including chemical analysis of sediments and water, bioaccumulation of key contaminants such as PCBs and mercury, and community-level effects such as species richness and diversity all support the intuitively logical ranking of upstream sites being more impacted by the industrial discharges than the sites further downstream.[1] Why then did the pattern of EROD activity from fish suggest that the upstream station had a lower capacity to induce MFO activity than the downstream stations?

Table 2.   Hepatic Lesions in Redbreast Sunfish from Reference Stream (Brushy
Fork) and Various Stations along East Fork Poplar Creek (EFPC).[a]

Site	% Incidence
Brushy Fork	<1
EFPC	
EFK 23.7[b]	30
EFK 18.3	10
EFK 5.3	12

[a]Fish were sampled in Spring of 1986.
[b]This site is closest to the industrial outfall.

This apparently paradoxical behavior is likely to have been the result of
hepatocellular toxicity in fish from the upstream site; i.e., the cumulative effect
of high levels of pollutants damaged the liver and impaired its capacity to respond
biochemically to contaminant exposure, as measured by increases in EROD
activity. Histopathological examination of fish livers from both reference stream
(Brushy Fork) and EFPC confirmed that there was a three-fold higher incidence
of hepatic lesions at the upstream station compared to animals from the downstream stations or from the reference site (Table 2).

Liver hispotatholologic analysis in fish from the reference stream showed structure similarities to that previously described in other freshwater teleosts.[34,35]
Hepatocytes were arranged as tubules[36,37] which, in longitudinal orientation,
appeared as a double row of hepatocytes interposed between adjacent sinusoids.
Hepatocytes contained abundant glycogen and were uniform in size. Occasional
macrophage aggregates were observed. These were small and showed hemosi-
derin pigment with no indication of melanin. Except for encysted parasites which
were associated with little to no host inflammatory response, tissues from these
fish were free of alterations.

The group of fish with the highest prevalence of lesions was collected immediately below the outfall site (EFK 23.7, Figure 1, Table 2). Most fish showed
enhanced basophilia of hepatocytes which probably was due to glycogen depletion. In addition to encysted parasites, affected livers revealed parasitic tracks
of alteration which showed liquefactive necrosis and/or varying degree of fibrotic
change. Coagulative necrosis of hepatocytes was focal to multifocal and in the
latter was associated with islands of basophilic regenerating hepatocytes. Macrophage aggregates were abundant and appeared to be larger than in fish from
the reference stream. The combined effect of liquefactive and coagulative necrosis with presence of abundant macrophages and parasites appeared to effectively reduce the volume of the liver occupied by normal-appearing, functional
parenchyma. Since the major cellular locus for cytochrome P-450 in the teleost
liver is the hepatocyte,[38,39] contaminant mixtures can produce necrosis of this
cell type, as was demonstrated in this study, consequently reducing the number
of functional cells in the liver and reducing the amount of cytochrome P-450
enzymes responsible for detoxification of organic contaminants.

To experimentally test the hypothesis of reduced hepatic MFO activity due to hepatotoxic effects, laboratory fish were treated with a known hepatotoxic agent (allyl formate) and subsequently challenged with an injection of BaP. Treatment with 50 μL/kg of allyl formate significantly reduced ($p<0.05$) the ability of the liver to induce EROD activity as a result of BaP exposure (Figure 5). The extent of liver damage was also indicated by an enzyme assay that can be considered another biomarker, levels of plasma lactic dehydrogenase activity, which increased significantly ($p<0.05$) in the animals treated with allyl formate.[40] The presence of this cytoplasmic enzyme in blood is assumed to result from liver cell damage from the hepatotoxic agent.

These observations illustrate the potential for misinterpreting a "false negative" if only EROD (or any other MFO parameter) was solely relied on to provide an indication of exposure to toxic chemicals. Low EROD may indicate the absence of an effective dose of pollutants, as it does in the reference streams, or may indicate very high hepatoxic levels of contaminants. However, comparison of several biomarker responses provides information needed to correctly interpret the MFO responses. Histopathology and serum enzyme levels were useful biomarkers that not only indicated that the organisms were exposed to toxicants, but provided information needed to correctly interpret the MFO data.

SUMMARY AND CONCLUSIONS

The utility of the hepatic MFO system as a biomarker of contaminant exposure was evaluated in sunfish from a stream impacted by effluents from an industrial facility and from uncontaminated reference streams. An iterative laboratory-field research program evaluated the effects of environmental, physiological, and toxicological factors on the responses of the hepatic enzymes and has provided a basis for interpreting the responses of this biochemical response as a biomarker of exposure to chemicals in the environment. Seasonal changes in temperature affect the basal levels of MFO activity, the magnitude of MFO induction to the same levels of exposure, and the rates of uptake of contaminants from food[41] and water.[29] Reproductively active females had lower levels of EROD activity, but demonstrated at least a limited capacity for induction as a result of exposure to environmental pollutants. Hepatotoxic effects resulting from exposure to high levels of contaminants can limit the capacity for induction of the MFO system, and could lead to misinterpretation of MFO activity as a biomarker of exposure.

While it is clear that induction of hepatic enzymes is a useful and informative biomarker of exposure to chemical agents in the environment, interpretation of the responses must take into account the role of these variable on MFO activity.

ACKNOWLEDGMENTS

This work was supported by the Exploratory Studies Program of the Oak Ridge National Laboratory and by the Oak Ridge Y-12 Plant, Department of

Environmental Management, Health, and Safety, Environment and Accountability Division. The Oak Ridge National Laboratory and the Y-12 Plant are operated by Martin Marietta Energy Systems, Inc. for the U.S. Department of Energy under contract DE-AC05-84OR21400. Publication No. 3361. Environmental Sciences Division, Oak Ridge National Laboratory.

REFERENCES

1. Loar, J. M., S. M. Adams, M. C. Black, H. L. Boston, A. J. Gatz, Jr., M. A. Huston, B. D. Jimenez, J. F. McCarthy, S. D. Reagan, J. G. Smith, G. R. Southworth, and A. J. Stewart. First Annual Report on the Y-12 Plant Biological Monitoring and Abatement Program, ORNL/TM Environmental Sciences Division Publication No. 2984, Oak Ridge National Laboratory, Oak Ridge, TN, 1989, 311 pp.
2. Tennessee Valley Authority. Instream Contaminant Study, Task 2: Sediment Characterization, Vol. I. Report to the U.S. Department of Energy, Oak Ridge Operations Office, Tennessee Valley Authority, Office of Natural Resources and Economic Development, Knoxville, TN, 1985.
3. Tennessee Valley Authority. Instream Contaminant Study, Task 2: Sediment Characterization, Vol. II—Appendices. Report to the U.S. Department of Energy, Oak Ridge Operations Office, Tennessee Valley Authority, Office of Natural Resources and Economic Development, Knoxville, TN, 1985.
4. Jimenez, B. D., L. S. Burtis, G. H. Ezell, B. Z. Egan, N. E. Lee, J. F. McCarthy, and J. J. Beauchamp. "Effects of Environmental Variables on the Mixed Function Oxidase (MFO) System in Bluegill Sunfish (*Lepomis macrochirus*)," *Environ. Toxicol. Chem.* 7:623–634 (1988).
5. Droy, B. F., M. E. Davis, and D. E. Hinton. "Mechanism of Allyl Formate-Induced Hepatotoxicity in Rainbow Trout," *Toxicol. Appl. Pharmacol.*, 98:313–324 (1989).
6. Zimmerman, H. J. "Chemical Hepatic Injury and Its Detection," in *Toxicology of the Liver,* G. Plaa, and W. R. Hewitt, Eds. (New York: Raven Press, 1982), pp. 1–45.
7. Forlin, L., and T. Andersson. "Storage Conditions of Rainbow Trout Liver Cytochrome P_{450} and Conjugating Enzymes," *Comp. Biochem. Physiol.* 3:569–572 (1985).
8. Burke, M. D., and R. T. Mayer. "Ethoxyresorufin: Direct Fluorimetric Assay of a Microsomal O-Dealkylation Which Is Preferentially Inducible by 3-Methylcholanthrene," *Drug. Metab. Dispos.* 2:583–588 (1974).
9. Omura, T. and R. Sato. "The Carbon Monoxide-Binding Pigment of Liver Microsomes," *J. Biol. Chem.* 239:2370–2378 (1964).
10. Johannesen, K. M. and J. W. DePierre. "Measurements of Cytochrome P450 in the Presence of Large Amounts of Contaminating Hemoglobin and Methemoglobin," *Anal. Biochem.* 86:725–32 (1978).
11. Phillips, A. H. and R. G. Langdon. "Hepatic Triphosphopyridine Nucleotide-cyt-c Reductase Isolation Characterization, and Kinetic Studies," *J. Biol. Chem.* 237:2652–2660 (1962).
12. Bradford, M. M. "A Rapid and Sensitive Method for the Quantitation of Protein Utilizing the Principle of Protein-Dye Binding," *Anal. Biochem.* 72:248–254 (1976).
13. Zar, J. H., in *Biostatistical Analysis,* 2nd ed. (Englewood Cliffs, NJ: Prentice Hall, Inc., 1984), pp 194.

14. Southworth, G. R., J. M. Loar, M. G. Ryon, J. G. Smith, A. J. Stewart, and J. A. Burris. Ecological Effects of Contaminants and Remedial Actions in Bear Creek. Draft ORNL/TM Report Oak Ridge National Laboratory, Oak Ridge, TN, 1990, 169 pp. plus appendices (in press).

15. Walton, D. G., L. L. Fancey, J. M. Green, J.W. Kiceniuk, and W. R. Penrose. "Seasonal Changes in Aryl Hydrocarbon Hydroxylase Activity of Marine Fish *Tautogolabrus adspersus (Walbaum)* With and Without Petroleum Exposure," *Comp. Biochem. Physiol.* 76C:247–53 (1983).

16. Lindström-Seppä, P. "Seasonal Variation of the Metabolic Enzyme Activities in the Liver of Male and Female Vendace (*Coregonus albula l*)," *Aquat. Toxicol.* 6:323–331 (1985).

17. Luxon, P. L., P. V. Hodson, and U. Borgmann. "Hepatic Aryl Hydrocarbon Hydroxylase Activity of Lake Trout (*Salvelinus namaycush*) as Indicator of Organic Pollution," *Environ. Toxicol. Chem.* 6:649–657 (1987).

18. Koivusaari, U., M. Harri, and O. Hanninen. "Seasonal Variation of Hepatic Biotransformation in Female and Male Rainbow Trout (*Salmo gairdneri*)," *Comp. Biochem. Physiol.* 70C:149–157 (1981).

19. Hansson, T., and J. A. Gustafsson. "In Vitro Metabolism of 4-Androstene-3,17-dione by Hepatic Microsomes from the Rainbow Trout (*Salmo gairdneri*): Effects of Hypophysectomy and Oestradiol-17B," *J. Endocrinol.* 90:103–112 (1981).

20. Förlin, L. and T. Hansson. "Effects of Oestradiol-17B and Hypophysectomy on Hepatic Mixed Function Oxidases in Rainbow Trout," *J. Endocrinol.* 95:245–252 (1982).

21. Vodicnik, M. J., and J. J. Lech. "The Effects of Sex Steroids and Pregnenolone-16a-carbonitrile on the Hepatic Microsomal Monooxygenase System of Rainbow Trout (*Salmo gairdneri*)," *J. Steroid Biochem.* 3:323–328 (1983).

22. Jimenez, B. D., and L. S. Burtis. "Influence of Environmental Variables on the Hepatic Mixed Function Oxidase System in the Bluegill Sunfish (*Lepomis Macrochirus*)," *Comp. Biochem. Physiol.* 93C:11–21 (1989).

23. Stegeman, J. J., F. Y. Teng, and E. A. Snowberger. "Induced Cytochrome P450 in Winter Flounder (*Pseudopleuronectes americanus*) from Coastal Massachusetts Evaluated by Catalytic Assay and Monoclonal Antibody Probes," *Can. J. Fish. Aquat. Sci.* 44:1270–1277 (1987).

24. Stegeman, J. J. "Temperature Influence on Basal Activity and Induction of Mixed Function Oxygenase Activity in *Fundulus Heteroclitus*," *J. Fish. Res. Bd. Can.* 36:1400–1405 (1979).

25. Andersson, T., and U. Koivusaari. "Influence of Environmental Temperature on the Induction of Xenobiotic Metabolism by β-Naphthoflavone in Rainbow Trout, *Salmo gairdneri*," *Toxicol. Appl. Pharmacol.* 80:43–50 (1985).

26. Breck J. E. and J. F. Kitchell. "Effects of Macrophyte Harvesting on Simulated Predator-Prey Interactions," in *Aquatic Plants, Lake Management, and Ecosystems Consequences of Lake Harvesting*, J. E. Breck, R. T. Prentki, and O. L. Loucks, Eds., (Proceedings, Conference at Madison Wisconsin, Feb. 14–15, 1979, Institute for Environmental Studies, University of Wisconsin-Madison, 1979), pp. 435.

27. Black, M. C. "Effects of Environmental and Physiological Variables on the Uptake of Hydrophobic Contaminants by the Gills of Rainbow Trout (*Salmo gairdneri*)," Ph.D. Dissertation in Ecology, University of Tennessee, 1989.

28. Black, M. C., and J. F. McCarthy. "Effects of Acute Temperature Changes on

Oxygen Consumption and Toxicant Uptake in Rainbow Trout," *Physiol. Zool.*, in press (1990).

29. Jimenez, B. D., C. R. Cirmo, and J. F. McCarthy. "Uptake, Elimination and Metabolism of Benzo[a]pyrene in Bluegill Sunfish (*Lepomis macrochirus*)," *Aquat. Toxicol.* 10:41–57 (1987).

30. Payne, J. F. and W. R. Penrose. "Induction of Aryl Hydrocarbon Benzo(a)pyrene Hydroxylase in Fish by Petroleum," *Bull. Environ. Contam. Toxicol.* 14:112–116 (1975).

31. Payne, J. F., and L. L. Fancey. "Effects of Long-Term Exposure to Petroleum on Mixed Function Oxygenases in Fish. Further Support for the Use of the Enzyme System in Biological Monitoring," *Chemosphere II*, 207–214 (1982).

32. Porter, E., L. L. Fancey, A. Rahimtula and J. F. Payne. "Induction of the Hepatic and Extra-Hepatic Mixed-Function Oxygenases in Cunner Exposed to Petroleum Hydrocarbons During Various Stages of Gonadal Maturation," *Proc. Can. Fed. Biol. Soc.* 29:140 (1986).

33. Collier, T. K., J. E. Stein, R. J. Wallace and U. Varanasi. "Xenobiotic Metabolizing Enzymes in Spawning English Sole (*Parophrys vetulus*) Exposed to Organic-Solvent Extracts of Marine Sediments from Contaminated and Reference Areas," *Comp. Biochem. Physiol.* 84C:291–298 (1986).

34. Hinton, D. E., R. L. Snipes, R. L. and M. W. Kendall. "Morphology and Enzyme Histochemistry in the Liver of Largemouth Bass, *Micropterus salmoides*," *J. Fish. Res. Bd. Can.* 29:531–534 (1982).

35. Hinton, D. E., E. R. Walker, C. A. Pinkstaff, and E. M. Zuchelkowski. "Morphological Survey of Teleost Organs Important in Carcinogenesis with Attention to Fixation," *Natl. Cancer Inst. Monogr.* 65:291–320 (1984).

36. Hamptom, J. A., R. C. Lantz, P. J. Goldblatt, D. J. Lauren, and D. E. Hinton. "Functional Units in Rainbow Trout (*Salmo gairdneri* Richardson) Liver. II. The Biliary System," *Anat. Rec.*, 221:619–634 (1988).

37. Hamptom, J. A., R. C. Lantz, and D. E. Hinton. "Functional Units in Rainbow Trout (*Salmo gairneri* Richardson) Liver. III. Morphometric Analysis of Parenchyma, Stroma, and Component Cell Types," *Am. J. Anat.*, 185:58–73 (1989).

38. Goksoyr, A., T. Andersson, T. Hansson, J. Klungsoyr, Y. Zhang, and L. Forlin. "Species Characteristics of the Hepatic Xenobiotic and Steroid Biotransformation Systems of Two Teleost Fish, Atlantic Cod (*Gadus morhua*) and Rainbow Trout (*Salmo gairdneri*)," *Toxicol. Appl. Pharmacol.* 89:347–360 (1987).

39. Miller, M. R., D. E. Hinton, J. B. Blair, and J. J. Stegeman. "Immunohistochemical Localization of Cytochrome p-450E in Liver, Gill and Heart of Scup (*Stenotomus chrysops*) and Rainbow Trout (*Salmo gairdneri*)," *Mar. Environ. Res.* 24:37–39 (1988).

40. McCarthy, J. F., B. D. Jimenez, L. R. Shugart and A. Oikari. "Biological Markers in Animal Sentinels: Laboratory Studies Improve Interpretation of Field Data," *In Situ Evaluation of Biological Hazards of Environmental Pollutants*, S. Sandhu, Ed. (New York: Plenum Press, 1990).

41. Rhead, M. M. and J. M. Perkins. "An Evaluation of the Relative Importance of Food and Water as Sources of p,p'-DDT to the Goldfish, *Carassius auratus* (L.)," *Water Res.* 18:719–725 (1984).

CHAPTER 7

Avian Mixed Function Oxidase Induction as a Monitoring Device: The Influence of Normal Physiological Functions

C. Fossi, C. Leonzio, S. Focardi and D. B. Peakall

INTRODUCTION

The induction of the mixed function oxidase (MFO) system, commonly used as a biochemical marker for xenobiotics in recent years,[1-4] is also affected by normal physiological changes. The use of this as an index of stress has contributed to our ability to detect and understand the significance of exposure to xenobiotics (both organochlorines and polynuclear aromatic hydrocarbons) in the environment.[4-7] Nevertheless, several intrinsic factors such as hormonal changes during the breeding cycle significantly modify the activity of monooxygenases.[8-11] The aim of this paper is to examine, using four specific examples of studies in wild birds, the role of the sexual cycle in the modification of MFO activity and consequently the importance of considering this aspect in planning biomonitoring.

The four examples are:

(a) the seasonal variations in the activity of hepatic aldrin epoxidase in a population of wood pigeon (*Columba palumbus*) in a relatively unpolluted area of central Italy,
(b) the seasonal variation of the same enzyme in the yellow-legged herring gull (*Larus argentatus cachinnans*) from northern Italy,
(c) changes in the activity of several MFO enzymes during the breeding cycle of

the herring gull (*Larus argentatus smithsonianus*) and a comparison with activity found during the winter in Eastern Canada, and

(d) the different levels of aldrin epoxidase activities and PCB concentrations in
the liver of the yellow-legged herring gull collected during the reproductive
and post-reproductive periods in several polluted areas of Italy.

STUDIES IN WILD BIRDS

Hepatic Aldrin Epoxidase Activity in the Wood Pigeon in Central Italy

Hepatic aldrin epoxidase activities were measured in wood pigeons collected
from open unpolluted country in central Italy using the method of Krieger and
Wilkinson.[12] This enzyme adds an epoxide group to chlorinated cyclodienes
which renders them more polar and thus increases the rate of excretion of these
compounds from the body. The data are presented in Figure 1. The activity
remains constant during the winter, but there is a statistically significant rise at
the beginning of the reproductive season.

Seasonal Variation of Hepatic Aldrin Epoxidase Activity in the Herring Gull in Northern Italy

Aldrin epoxidase activity was also measured in livers of herring gulls collected
from a lagoon in northern Italy (Figure 2). There is marked variation of the
activity of this enzyme throughout the year with a statistically significant maximum during the breeding season. The activity of aldrin epoxidase is higher than
that found in the wood pigeon, and the rise during the breeding season is more
marked.

Seasonal Variation of Several MFO Enzymes in the Herring Gull in Newfoundland

This work was carried out on the east coast of Newfoundland in an area remote
from pollution sources. The residue levels in the livers of 21-day-old chicks
were 0.53 mg/kg PCBs and 0.07 mg/kg DDE on a dry weight basis.[10] The
hepatic activities of aldrin epoxidase, epoxide hydroxylase, aminopyrine N-
demethylase (AMDM), 7-ethoxyresorufin O-deethylase (7-EROD), benzo[a]pyrene
3-hydroxylase and UDP glucuronyl transferase were determined at three different
stages of the breeding cycle. The incubation period of the herring gull is 28 days
and the fledging period about 21 days. "Early" refers to adults collected soon
after the completion of the clutch, "middle" to the end of the incubation period,
and "late" to the period when there were small young. The data for aldrin
epoxidase are given in Table 1.

Table 1. Hepatic Aldrin Epoxidase Activity in the Adult Herring Gull During the Breeding Cycle.

	Enzyme activity (nmol/mg protein/min)*		
	Early	Middle	Late
Female	0.402 ± 0.147	0.381 ± 0.134	0.529 ± 0.206
Male	0.485 ± 0.208	0.508 ± 0.205	0.476 ± 0.195

*Mean ± standard deviation, n = 10.

Relationship of Hepatic Aldrin Epoxidase Activity to PCB Concentration in the Herring Gull in Italy

Hepatic aldrin epoxidase activity and the concentration of polychlorinated biphenyls (PCBs) have been measured in herring gulls collected during the reproductive season (spring) and when the birds are sexually inactive (autumn) in central and northeastern Italy.[6] The data are summarized in Table 2. The levels of PCBs recorded in these specimens are 20 to 60 times higher than in the specimens from Newfoundland.

The high activity of aldrin epoxidase is related to the custom of these birds to feed on municipal landfills where hazardous industrial materials are also dumped. Despite the increase in PCB levels a decrease of enzyme activity outside the breeding season is noted.

SOURCE OF VARIATION OF MFO ACTIVITY

Inter-Group Variation

Even in Newfoundland, where the contaminant levels are low and relatively uniform, there is considerable within-group variation with the coefficient of variation being 0.35 to 0.45 (Table 1). The coefficients are higher in material from Italy (0.57–0.58, Table 2) and a similar gradient has been found in material from the Great Lakes of North America.[13] It would be worthwhile to examine

Table 2. Hepatic Aldrin Epoxidase Activity and PCB Concentration in Herring Gulls.

	Enzyme activity (nmol/mg protein/min)*	PCB levels (mg/kg, dry weight)*
Reproductive season, n = 12	3.18 ± 1.84	9.05 ± 7.46
Period of sexual inactivity, n = 18	1.48 ± 0.85	33.35 ± 29.14

*Mean ± standard deviation.

a sizable group from the cleanest population available to establish clearly the normal biological variation of MFOs.

Inter-Substrate Variation

The degree of induction of four MFO enzymes caused by Prudhoe Bay Crude Oil (PBCO) was examined in herring gull chicks. It was found that 0.1 mL caused a threefold increase of AMDM and a doubling of 7-EROD and aldrin epoxidase. Significant induction of benzo[a]pyrene 3-hydroxylase did not occur until the dose of PBCO was increased to 1 mL.[14] Aldrin epoxidase was maximumly induced by a dose of 0.5 mL PBCO.

Inter-Species Variation

There is considerable phylogenetic variation in MFO activity. Walker et al.[4] found, using aldrin as substrate, that activity generally decreased in the following sequence: mammals > birds and amphibia > fish. Levels in aquatic invertebrates are lower still.[15] Within the order Aves it was found[3] that fish-eating birds tended to have lower MFO activities than other species. Even among closely related species there are considerable variations in aldrin epoxidase activity; for example, Knight and Walker[16] found that while two species of the family Alcidae— razorbill, Alca torda and guillemot, Uria aalge—had similar values, the puffin, Fractercula arctica, had activities tenfold higher. The low MFO activity in some species of seabirds has been considered to be a contributing factor to the bioaccumulation of organochlorines.[17]

Inter-Seasonal Variation

It is clear from the data presented in this paper (Table 2; Figures 1 and 2) that there are considerable seasonal variations with levels of enzyme activity being higher during the reproductive season.

Inter-Sex and Age Variation

There were no consistent differences in MFO activity levels between the sexes of the herring gull when studied during the breeding season using five different substrates.[7] Only three of the possible eighteen comparisons were statistically significant. In these cases the activity in the female was lower. Knight and Walker[16] also found no clear pattern for MFO differences between the sexes for other species of seabird.

Ellenton and co-workers[13] found that hepatic AHH activity increased almost fourfold in the herring gull embryo between day 20 and day 25. MFO activity was determined in nestling gulls (2, 7, 14, and 21 days post-hatch). Significant decreases in the activity of AMDM with age were found, but no consistent changes were found with other substrates.[10]

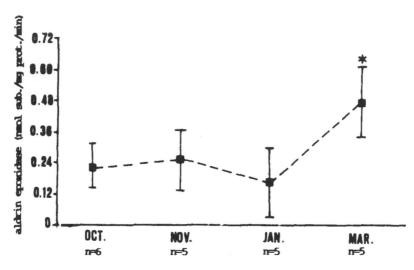

Figure 1.   Seasonal variations of hepatic aldrin epoxidase activities (mean ± standard deviation) in a population of wood pigeon (*Columba palumbus*). *: p = 0.001, Student *t*-test.

CONCLUSIONS

Despite all these variations the induction of MFO activity is a useful biomarker of exposure of organisms to xenobiotics. The system is readily induced by both organochlorines[18] and oil.[7] Seasonal validation has been shown to be a significant

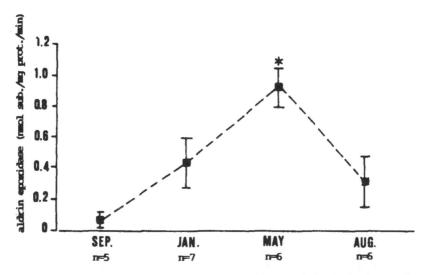

Figure 2.   Seasonal variations of hepatic aldrin epoxidase activities (mean ± standard deviation) in a population of yellow-legged herring gull (*Larus argentatus cachinnans*). *: p = 0.001, Student *t*-test.

factor and thus it is important for monitoring purposes that the reproductive status of the specimen be known. The periods immediately before and after the reproductive season should be avoided because of rapidly altering MFO activity. Quality assurance procedures are essential. This can be achieved by making an internal reference pool. We have used a pooled microsomal sample of the Sprague-Dawley rat as a reference. A sample of this reference material was included in each batch of samples analyzed. It would be valuable if all groups used the rat as a standard, running it against the most widely used substrates (AHH, 7-EROD, aldrin epoxidase). Vials from a pool of hepatic microsomal material can be frozen without loss of enzyme activity and thus standards can be included with each run. This allows quality control to be maintained within the laboratory, and would also permit comparison with the data produced by other laboratories. These procedures, routinely used by analytical chemists, should be adopted in bioeffects studies.

REFERENCES

1. Payne, J.F., L. L. Fancey, A. D. Rahimtula, and E. L. Porter. "Review and Perspective of the Use of Mixed-Function Oxidase Enzymes in Biological Monitoring," *Comp. Pharmacol. Physiol.* 86C:233–245 (1987).
2. Moore, M.N. "Cellular Responses to Polycyclic Aromatic Hydrocarbons and Phenobarbital in *Mytilus edulis,*" *Mar. Pollut. Bull.* 16:134–139 (1985).
3. Walker, C.H. "Species Variations in Some Hepatic Microsomal Enzymes that Metabolize Xenobiotics," *Prog. Drug Metabol.* 5:113–164 (1980).
4. Walker, C.H., G. C. Knight, J. K. Chipman, and J. J. Roneis. "Hepatic Microsomal Monoxygenases of Seabirds," *Mar. Environ. Res.* 14:416–419 (1984).
5. Fossi, C., C. Leonzio, and S. Focardi. "Mixed Function Oxidase Activity and Cytochrome P-450 Forms in Black-Headed Gulls Feeding in Different Areas," *Mar. Pollut. Bull.* 17:546–548 (1987).
6. Fossi, C., C. Leonzio, and S. Focardi. "Increase of Organochlorines and Mixed-Function Oxidase Activity in Water Birds in an Italian Lagoon," *Bull. Environ. Contam. Toxicol.* 37:538–543 (1986).
7. Peakall, D.B., D. A. Jeffrey, and D. Boersma. "Mixed-Function Oxidase Activity in Seabirds and Its Relationship to Oil Pollution," *Comp. Biochem. Physiol.* 88C:151–154 (1987).
8. Conney, A.H. "Pharmacological Implications of Microsomal Enzyme Induction," *Pharmacol. Rev.* 19:317–366 (1967).
9. Livingstone, D.R. "Response of the Detoxication/Toxication Enzyme System of Molluscs to Organic Pollutants and Xenobiotics," *Mar. Pollut. Bull.* 16:158–164 (1985).
10. Peakall, D.B., R. J. Norstrom, A. D. Rahimtula, and R. D. Butler. "Characterization of Mixed-Function Oxidase Systems of the Nesting Herring Gull and Its Implications for Bioeffects Monitoring," *Environ. Toxicol. Chem.* 5:379–385 (1986).
11. Fossi, C., C. Leonzio, S. Focardi, and A. Renzoni. "Seasonal Variation in Aldrin Epoxidase (MFO) Activity in Yellow-Legged Herring Gulls: The Relationship to Breeding and PCB Residues," *Bull. Environ. Toxicol. Chem.* 41:365–370 (1988).

12. Krieger, R.L., and C. F. Wilkinson. "Microsomal Mixed-Function Oxidases in Insects. I. Localization and Properties of an Enzyme System Effecting Aldrin Epoxidation in Larvae of the Southern Armyworm (*Frodenia eridania*)," *Biochem. Pharmacol.* 19:1403–1415 (1969).

13. Ellenton, J., L. Brownlee, and B. Hollebone. "Aryl Hydrocarbon Hydroxylase Levels in Herring Gull Embryos from Different Location on the Great Lakes," *Environ. Toxicol. Chem.* 4:615–622 (1985).

14. Peakall, D.B., R. J. Norstrom, and F. A. Leighton. Unpublished data.

15. Mix, M.C., S. I. Hemingway, and R. L. Schaffer. "Benzo[a]pyrene Concentrations in Somatic and Gonad Tissue of Bay Mussels," *Bull. Environ. Contam. Toxicol.* 28:46–51 (1982).

16. Knight, G.C., and C. H. Walker. "A Study of the Hepatic Microsomal Epoxide Hydrolase in Sea Birds," *Comp. Biochem. Physiol.* 73C:463–467 (1982).

17. Knight, G.C., and C.H. Walker. "A Study of Hepatic Microsomal Monoxygenases of Sea Birds and Its Relationship to Organochloride Pollutants," *Comp. Biochem. Physiol.* 73C:211–221 (1982).

18. Bend, J.R., D. S. Miller, W.B. Kinter, and D. B. Peakall. "DDE-Induced Microsomal Mixed-Function Oxidases in the Puffin (*Fractercula arctica*)," *Biochem. Pharmacol.* 26:1000–1001 (1977).

Use of Barbiturate-Induced Sleeping Time as an Indicator of Exposure to Environmental Contaminants in the Wild

Richard S. Halbrook and Roy L. Kirkpatrick

We have postulated that barbiturate-induced sleeping times can be used as an indicator of microsomal enzyme activity induced by environmental contaminants in wild populations. We have demonstrated this in a wild population inhabiting an area known to be contaminated. Muskrats collected from a region of the Elizabeth River, Va. known to be contaminated with a variety of environmental pollutants (heavy metals, polynuclear aromatic hydrocarbons, DDTs, PCBs, etc.) had a mean sleeping time (36 min) that was approximately half the mean sleeping times recorded from 2 uncontaminated areas (60 and 63 min). Laboratory studies using barbiturate-induced sleeping times to indicate changed microsomal enzyme activity resulting from exposure to specific hepatotoxins, anticholinesterases, chlorinated hydrocarbons insecticides, polychlorinated biphenyls, and polynuclear aromatic hydrocarbons have been reviewed.

In an industrial and agricultural society heavily dependent on chemicals, it is desirable to have means available for monitoring and detecting xenobiotic exposure in living organisms. Researchers are currently trying to find reliable and sensitive techniques that would indicate exposure to low levels of contaminants prior to the advent of irreversible or detrimental changes in the exposed organism. Barbiturate-induced sleeping times have been used in laboratory studies to evaluate effects of various individual organic contaminants on hepatic microsomal enzyme activity. Hepatic microsomal enzymes are induced or inhibited by several organic compounds some of which are difficult to detect by residue analyses. The purpose of this paper is to review the use of barbiturate, primarily pentobarbital (Nembutal), induced sleeping time as a means of detecting changes in microsomal enzyme metabolism due to exposure to environmental contaminants and to explore the potential for extension of its use to in situ detection of contaminant exposure in wild populations.

The history of barbiturate-induced anesthesia, the distribution, metabolism, and elimination of barbiturates, and the use of barbiturate-induced sleep times to evaluate hepatic microsome enzyme activity in laboratory animals exposed to environmental contaminants will be reviewed. One instance of use of barbiturate-induced sleeping times to evaluate exposure to hepatic microsomal enzyme inducing environmental contaminants is described. The importance of the detection of changes in hepatic microsomal enzyme activity is that these changes in activity brought on by xenobiotics, environmental contaminants included, lead to changes in metabolism not only of the xenobiotic but also of endogenous compounds such as steroids.[1-3]

BARBITURATES

History

Barbituric acid (the precursor of pentobarbital and other barbiturates) results from the reaction of urea and malonic acid and was first synthesized by Adolf von Bayer in 1867.[4] Although barbituric acid has no sedative properties, derivatives of this compound were found to have hypnotic and anesthetic effects. In 1903, Fischer and von Mering synthesized the first barbituric acid derivative with sedative properties, diethyl barbituric acid.[4] Since that time more than 100 barbituric acid derivatives have been synthesized with varying sleep producing activity.[5]

Pentobarbital (5-ethyl-5-(1-methylbutyl)barbituric acid) has been used extensively as an hypnotic and anesthetic.[6-8] Volwiler and Tabern[5] studied the sleep-producing activity of several 5,5-substituted barbituric acid derivatives in rats and found that the ethyl-(1-methylbutyl) substitution (pentobarbital) seemed to be one of the most active in sleep-producing efficiency. Relative to the other sleep-producing barbiturates, pentobarbital is considered to be short acting and to have properties intermediate between those of thiopental, an ultrashort-acting barbituate, and phenobarbital, a long-acting barbiturate. The length of activity of these barbiturates is relative to their substituted side chains and lipophilic characteristics with thiopental being the most lipophilic.[9,10] As the length of the side chains at carbon position 5 is increased from two to five carbons the onset of action shortens and the duration of action decreases. Pentobarbital with a 5-(1-methylbutyl) (five carbon) side chain has a relatively short onset and duration of activity whereas phenobarbital with a 5-phenyl (six carbon) side chain has a relatively long onset and duration of activity.

Table 1.   Distribution of Pentobarbital in Dog.

Tissues	Conc. of Pentobarbital (mg/kg)
Plasma	34.4
Plasma Water	18.7
Cerebrospinal fluid	18.2
Red cells	36.0
Liver	64.4
Brain	42.3
Muscle	27.5
Kidney	45.8
Heart	38.4
Lung	20.8
Spleen	41.4
Lumbodorsal fat	37.3

Note: The studies were made 3 hr after the intravenous administration of 0.43 g of the drug to dog weighing 10.7 kg. The pentobarbital concentration in the fatty tissue is expressed in terms of total lipid content of the tissue.

From Brodie, B. B., J. J. Burns, L. C. Mark, and P. A. Lief. "The Fate of Pentobarbital in Man and Dog and a Method for its Estimation in Biological Material," *J. Pharmacol. Exp. Ther.* 109:26–34 (1953). With permission.

Distribution, Metabolism, and Elimination

The route of administration, dose, form (ionized vs nonionized), amount of plasma binding, rate of uptake by various tissues, biotransformation, and excretion are all important influences on the fate of pentobarbital (and other drugs) in animal systems. Brodie et al.[11] reported that at plasma concentrations of 10–20 mg/L approximately 45% of the pentobarbital was bound to the nondiffusible constituents of plasma and the remaining 55% presumably represented the pharmacologically active drug in plasma. Goldstein et al.[12] reported 40% of pentobarbital bound to plasma protein in male rats following a 15 mg/kg dose i.v. Plasma concentrations in human subjects given 0.75–1 g pentobarbital sodium i.v. over 10 min decreased rapidly during the first 2–3 hr as the drug left the plasma and was distributed throughout the body. This was followed by relatively slow decreases in plasma concentration reflecting the rate of biotransformation which averaged 4%/hr and ranged from 0.5–6%/hr. The rate of detoxification of pentobarbital in dogs was reported by the same authors to be 15% of the total dose per hour.[11]

The tissue distribution at 3 hr following i.v. administration of 0.43 g pentobarbital to a 10.7 kg dog are given in Table 1. Tissue concentrations were approximately the same or somewhat higher than in the plasma and concentrations in cerebrospinal fluid were approximately the same as that found in plasma water.[11] Using in vitro methods, Kuntzman et al.[13] have shown that the metabolism of pentobarbital to the levorotatory and dextrorotatory alcohol metabolites (d and l 5-ethyl-5-(1-methyl-3-hydroxylbutyl) barbituric acid) occurs primarily

Table 2.   Recovery of Metabolites After Chromatography[a] of 0.1 ML of Urine from Dog Receiving Pentobarbital-2-C.[14]

Spot No.	R_F	CPM In Spot	Percent of Recovered Activity	Identity
1	0.86	144	0.8	Pentobarbital
2	0.63	3685	21.0	d alcohol[b]
3	0.54	7209	41.0	l alcohol
4	0.46	268	1.5	Unknown
5	0.26	182	1.0	Unknown
6	0.18	374	2.1	Unknown
7	0.12	464	2.6	Urea
8	0.07	790	4.5	Carboxylic acid[c]
9	0.03	2195	12.4	Unknown
10	0.01	2405	13.6	Glucuronide of Spot 2

[a]Developing solvent, n-butanol saturated with 1% NH_4OH.
[b]5-Ethyl-5-(3-hydroxy-1-methylbutyl)barbituric acid.
[c]5-Ethyl-5-(1-methyl-3-carboxypropyl)barbituric acid.

From Titus, E., and H. Weiss. "The Use of Biologically Prepared Radioactive Indicators in Metabolic Studies: Metabolism of Pentobarbital," *J. Biol Chem.* 214:807–820 (1955). With permission.

in the microsomal fraction of liver and to a much lesser extent in lung microsomal fractions. Homogenates of the GI tract, brain, heart, spleen, adrenals, kidney, and testes did not apparently metabolize pentobarbital (<0.1 μmol of the alcohol derivatives found).

Titus and Weiss[14] reported ten metabolites of pentobarbital in the urine of dogs following i.v. administration of pentobarbital-2-C.[14] Sixty-two percent of the recovered activity was in two diastereoisomers of 5-ethyl-5-(1-methyl-3-hydroxylbutyl) barbituric acid that had previously been identified by Maynert and Dawson,[15] and 13.6% of recovered activity was in the form of a glucuronic acid conjugate of one of these isomers (Table 2).

These and other reported findings indicate that pentobarbital is primarily metabolized by the liver by oxidation of the radicals at carbon 5 and that the metabolic products are primarily eliminated in the urine.[4,10,11,13–18] Additional historical information can be found in Keys,[19] Wright,[20] and Soma[4] and more detailed information on the distribution, metabolism, and elimination of pentobarbital can be found in Soma[4] and Harvey.[10]

LABORATORY STUDIES USING BARBITURATE-INDUCED SLEEPING TIMES

Research by Mason and Beland[21] in 1945 indicated that certain forms of toxic hepatitis resulted in a reduction in the ability of the liver to detoxify certain barbiturates. With this discovery came the concept upon which this paper is based. This concept stipulates that if the liver's ability to metabolize barbiturates is changed (either enhanced or impaired) there will be a concomitant change in the anesthetic efficiency of an administered barbiturate. Since this discovery was made, numerous laboratory studies have used barbiturate sleeping time to indicate toxic hepatic damage and chemically induced increases or decreases in hepatic enzyme activity.

One of the first of these studies involved the use of changes in hexobarbital (Evipal) sleeping time as an index of the degree of liver damage resulting from carbon tetrachloride-induced liver injury. In this study Brauer and Root[1] reported a several-fold increase in hexobarbital-induced sleeping time that was correlated with other indexes (body weight changes, plasma esterase levels) and with the degree of liver damage.

Plaa et al.[22] expanded the use of barbiturate-induced sleeping time as an index of hepatotoxicity in mice. They were able to use the prolongation of pentobarbital anesthesia that accompanied liver damage induced by several halogenated hydrocarbons (chloroform, carbon tetrachloride, 1,1,1-trichloroethane, 1,1,2-trichloroethane, syn-tetrachloroethane, trichloroethylene, tetrachloroethylene) to develop standard dose-response curves. These authors reported a 95% correlation between sleeping time and gross or microscopic hepatic lesions.

In that same year Rosenberg and Coon[23] reported that certain anticholinesterase treatments increased hexobarbital sleeping time in mice. They compared the effects of anticholinesterases that are activated (octamethylpyrophosphoramide (OMPA) and ethyl-p-nitrophenyl thionobenzene phosphonate (EPN)) or partially inactivated (malathion, chlorothion, and phostex) by liver enzymes to anticholinesterases that are not markedly activated nor rapidly inactivated (tetraethyl pyrophosphate (TEPP) and bis-dimethylamidefluorophosphate (BFP)). Malathion, phostex, and chlorothion, EPN, and OMPA all produced a prolongation of sleeping time while BFP and TEPP did not.

In 1963, two papers appeared that used changes in barbiturate sleeping times to indicate changes in liver function due either to hepatic damage or hepatic enzyme induction by environmental contaminants. Balazs and Grice[24] treated rats with various hepatotoxins (carbon tetrachloride 0.4 mL/100 g i.p., dimethylnitrosamine 2.5 and 5 mg of 2% aqueous orally, and allyl alcohol 5 mg of a 2% aqueous orally) followed the next day by a 25 mg/kg dose of sodium pentobarbital i.p. There was a significant prolongation of pentobarbital-induced sleeping time in rats treated with carbon tetrachloride and dimethylnitrosamine. Histological lesions also were observed. Allyl alcohol treated rats had minimal increases in pentobarbital-induced sleeping time and histology revealed focal periportal necrosis, which was restricted to certain hepatic lobes. The authors speculated that the remaining normal hepatic tissue compensated for the lost function of the necrotic tissue and metabolized pentobarbital at a normal rate.

In contrast to the prolongation of barbiturate-induced sleeping time resulting from hepatic injury, rats and mice treated with most chlorinated hydrocarbon insecticides usually exhibit a decrease in sleeping times. Hart and Fouts[25] treated mice with acute and rats with chronic doses of DDT and observed changes in hepatic enzyme activity measured in vivo by hexobarbital sleeping time and in vitro from side chain oxidation of hexobarbital, N-demethylation of aminopyrine, ring hydroxylation of aniline, and the reduction of the aromatic nitro group of p-nitrobenzoic acid by liver homogenates. Single i.p. doses of DDT (25 mg/kg) did not affect hexobarbital sleeping time in mice whereas single doses of chlordane (25 mg/kg), endrin (6.25 mg/kg), and trichloro 237 (25 mg/kg) did reduce sleeping times. Additional studies by Hart and Fouts[26] indicated that although a single dose of DDT and its metabolites did not stimulate drug metabolism in mice, there was stimulation of drug metabolism in rats and other species. Gabliks et al.[27] also reported that a single i.p. dose of 90 mg/kg DDT or *o,p'*DDD followed by i.p. administration of 67 mg/kg sodium pentobarbital resulted in a marked prolongation of sleeping time in mice. In vitro studies with liver homogenates from mice 18 hr after a single dose of 90 mg/kg DDT or *o,p'*DDD did indicate an inhibition of liver enzymes metabolizing pentobarbital. However, chronic exposure to 90, 135, or 180 mg/kg (i.p. daily) for 2 weeks followed by sodium pentobarbital (67 mg/kg i.p.) resulted in a marked decrease in sleeping time recorded for DDT treated mice. Mice treated with 90 mg/kg *o,p'*DDD continued to exhibit a slight prolongation of sleeping time (120% of controls), whereas mice dosed with 135 or 180 mg/kg *o,p'*DDD did not have sleeping times significantly different from controls. The authors postulated a diphasic response in mice to the administration of insecticidal hydrocarbons. They also speculated that the increased sleeping time observed in mice chronically exposed to *o,p'*DDD might be associated with an inhibition effect on the function of the adrenals. Grady et al.[28] reported no electron micrographic evidence of adrenal cortical damage in mice and that prolonged sleeping time in mice could not be reversed with cortisone. Rats placed on 500 ppm DDT diets had significant increases in in vitro drug metabolizing enzymes activity beginning at 2 weeks and hexobarbital sleeping times showed corresponding decreases at all times studied.[25]

The slight prolongation of sleeping time observed in *o,p'*DDD treated mice should not be a major concern in the use of barbiturate sleeping time as an in situ indicator of exposure to contaminants. In "real life" situations, wild populations are exposed to mixtures, not single pure chemicals, and the overall response is the index being evaluated.

During the 1970s there was an increase in laboratory studies using barbiturate (primarily pentobarbital) induced sleeping time as an indication of hepatic microsomal enzyme changes resulting from exposure to environmental contaminants. Most of these studies dealt with treatment of laboratory mice and rats but several involved avian and wild mammalian species. Using pentobarbital sleeping times, Bitman et al.[29] were able to show a difference in metabolic enzyme activity between immature and mature rats, between *o,p'*DDT and *p,p'*DDT isomers, and between rats and Japanese quail (*Coturnix coturnix*). After 1 week on a 50 ppm *o,p'* or *p,p'*DDT diet immature rats on the *p,p'*DDT diet had sleeping times that were 5% of controls while those on the o,p'DDT diet had sleeping times 30 to 40% of controls. Mature female rats placed on the same diets exhibited less of an effect with those on the p,p'DDT diet having sleeping times that were 40% of controls and those on the o,p'DDT diet having sleeping times approximately 50% of controls. These authors also reported that in contrast to the reduced pentobarbital-induced sleeping time seen in the rat, Japanese quail exposed to a diet containing 100 ppm of one of the DDT isomers or metabolites had a prolonged sleeping time compared to controls. In both male and female quail, pentobarbital-induced sleeping time was prolonged from one and one half to four times that of controls within the first few days on treated diets followed by a reduction to approximately one and one half times that of controls, with the exception of males on o,p' and p,p'DDD diets whose sleeping times returned to that of controls after 2 weeks. Grady et al.[28] reported a 26% decrease in hexobarbital sleeping time for chickens treated with 200 mg/kg p,p'DDD for 3 days. As indicated previously, the exception observed in male quail is not expected to significantly affect the anticipated use of this technique.

Litterst et al.[30] reported the effects of chronic exposure to diets containing low doses of PCBs (potent hepatic microsomal enzyme inducers) on hepatic enzyme activity in rats. They exposed rats to diets of 0.5, 5, 50, and 500 ppm Aroclor 1242, 1248, 1254, or 1260. Liver weights were increased in those rats on the 50 and 500 ppm diets of Aroclor 1248, 1254, or 1260 but not 50 ppm Aroclor 1242. An increase in cytochrome P-450 levels was observed with all PCB mixtures at dietary levels of 50 and 500 ppm and also at 5 ppm of Aroclor 1254 and 1260. A measure of the production of hydroxylated metabolites of pentobarbital indicated an increase in hydroxylation in all PCB mixtures at dietary levels of 5 ppm or greater. Based on this study, pentobarbital-induced sleeping time would be expected to be decreased in rats that were chronically exposed to as little as 5 ppm diets of PCBs with higher percentages of chlorine.

Villeneuve et al.[31] reported that rats fed 20 ppm diets of Aroclor 1254 or 1260 for 30 days had a significant decrease in sleeping time when given 40 mg/kg pentobarbital, while those fed 20 ppm Aroclor 1221 did not. However, after 240 days even the rats on the 20 ppm Aroclor 1221 diets had a significant decrease in sleeping times.

During the mid 1970s, several laboratory studies were conducted at Virginia Polytechnic Institute and State University showing an effect of organochlorines on hepatic microsomal enzyme activity using changes in pentobarbital-induced sleeping time as an index. These studies not only involved laboratory species but also several wild species. Zepp et al.[32] dosed male cottontail rabbits (*Sylvilagus floridanus*) with 45 mg/kg and female cottontail rabbits with 40 mg/kg pentobarbital after they had been on diets containing 10 ppm Aroclor 1254 for 10 and 11 weeks, respectively. Both male and female rabbits exhibited significantly shorter sleeping times compared to non-PCB treated rabbits.

Sanders and Kirkpatrick[33] reported a significant decrease in pentobarbital-induced sleeping time in male white-footed mice (*Peromyscus leucopus*) fed diets of 100, 200, or 400 ppm Aroclor 1254 for 2 weeks. They also noted that treated mice had significantly heavier livers (all treatments), altered plasma corticoid levels (at 400 ppm), smaller seminal vesicle weights (at 400 ppm), and reduced number of spermatozoa per milligram of testis (at 400 ppm) compared to controls. There was no difference in body weights, body weight changes, or paired adrenal weights between treated and control mice. These data indicate the sensitivity of pentobarbital-induced sleeping time in detecting changes associated with exposure to organic microsomal enzyme inducers. In this experiment only pentobarbital sleeping times and liver weight differences were significantly different from controls at all treatment doses. It also should be noted that mice on 200 and 400 ppm PCB diets did not have sleeping times or liver weights significantly different from each other. The authors suggested that these observations indicate a maximal level of enzyme induction after 2 weeks feeding on a 200 ppm diet of Aroclor 1254 in white-footed mice.

Montz et al.[34] measured pentobarbital-induced sleeping time in raccoons (*Procyon lotor*) after 8 days on diets containing 0 or 50 mg/kg Aroclor 1254 and 100 or 70% ad libitum diets. The pentobarbital sleeping times observed for the PCB treated raccoons were 30% lower than those of non-PCB treated raccoons but this difference was not statistically significant. Sleeping times were not significantly different between raccoons on different diets. The authors attributed the nonsignificant findings to small sample sizes (four raccoons per group) and large variability of measurements.

Many laboratory studies have shown that another group of environmental contaminants, polynuclear aromatic hydrocarbons (PAHs), are inducers of microsomal enzymes.[35] PAHs consist of two or more fused benzene rings in various arrangements. Many of these compounds are known carcinogens. In vivo studies have demonstrated the potency of PAHs as inducers of microsomal enzymes. However, the specific P450 enzymes induced by PAHs (3-methylcholanthrene (3-MC) type inducers) appear to be different from those induced by organochlorines (phenobarbital type inducers). Laboratory studies have indicated that benzo(a)pyrene (a carcinogenic PAH) metabolism occurs to a greater extent with cytochrome P450c than with other P450 isozymes, whereas hexobarbital metabolism results primarily from the action of cytochromes P450b and P450e.[36,37] This knowledge has resulted in the use of compounds other than barbiturates to indicate exposure to 3-MC type inducers. Conney et al.[38] administered 25 mg/kg benzo(a)pyrene i.p. to male rats 24 hr before an i.p. injection of zoxazolamine (100 mg/kg), a muscle relaxant that is metabolized by liver microsomal enzymes. The half-life of zoxazolamine was 10 min in treated rats compared to 9 hr in control rats. We were unable to document studies which would indicate whether mixtures of PAHs or increased concentrations of PAHs would result in induction of a greater variety of P450 isozymes or studies of barbiturate sleeping times observed in PAH-treated animals.

These laboratory studies indicate that barbiturate-induced sleeping time is a reliable and sensitive measure of change in hepatic microsomal enzyme activity resulting from exposure to a variety of hepatotoxins, anticholinesterases, chlorinated hydrocarbons insecticides, and polychlorinated biphenyls. Although there appear to be species differences in barbiturate-induced sleeping times associated with specific chemicals (for example, DDD), this should not negate the use of this technique to detect areas of increased contamination but would affect comparison between some species. Because changes are usually observed in sleeping times of treated animals and in situ situations usually will involve chronic exposure to mixtures (not single pure chemicals), we would anticipate and expect a change in sleeping time response in examined populations from contaminated areas. However, we would advise researchers designing multiple species experiments to check for species specific responses prior to comparison between species and to examine biomarkers which would indicate hepatic damage (hepatic lesions, serum glutamic oxaloacetic transaminase, serum glutamic pyruvic transaminase, orinithine carbamyl transferase, etc.).

FIELD STUDY USING BARBITURATE-INDUCED SLEEPING TIME

Although numerous laboratory studies have shown the usefulness of pentobarbital-induced sleeping time, no in situ studies have been reported. We recently used pentobarbital sleeping times to evaluate exposure to microsomal enzyme inducers in a study of the influence of contaminants on muskrat (*Ondatra zibethicus*) populations inhabiting the Elizabeth River in southeastern Virginia. The Elizabeth River is a heavily industrialized tributary emptying into the James River near its confluence with the Chesapeake Bay. Currently 48 industries discharge into the river and it serves as a center for Naval and commercial shipping activities.[39,40] The Elizabeth River estuary is considered one of the most polluted in the world and elevated levels of heavy metals and polynuclear aromatic hydrocarbons (PAHs) have been reported as well as numerous other organic compounds.[40–42]

During December 1987 through February 1988, 63 muskrats were live trapped from 2 locations in the Elizabeth River that were approximately 12 km apart (regions 1 and 2) and from 1 location in the Nansemond River (region 3), a nearby reference river (Figure 1). Region 1 of the Elizabeth River was known to be contaminated with heavy metals and PAHs[41] while the other two locations (regions 2 and 3) were not known to be contaminated. Trapped muskrats were transported to a research trailer where they were weighed and injected with sodium pentobarbital (37.5 mg/kg i.m.). The sleeping time (time from eye closing until eye opening >20 sec) was recorded.

Mean sleeping time for muskrats from the contaminated region was 39 min (N = 20), which is significantly shorter than the 60 (N = 19) and 63 (N = 24) min for those from uncontaminated regions (p = 0.03) (Table 3). Based on knowledge of contaminant concentrations in the three regions, we had hypothesized that induction of hepatic microsomes had occurred and that mean sleeping

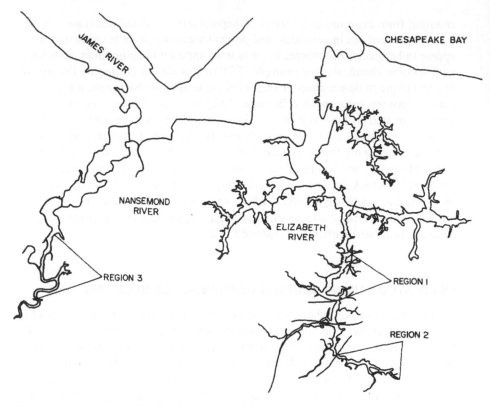

Figure 1. Muskrat trapping locations.

Table 3.   Pentobarbital Induced Sleeping Times of Muskrats Trapped from Three Study Sites in Virginia.

Location	Number	Mean ± SE
Elizabeth River		
Region 1	20	38.6 ± 6.9[a]
Region 2	19	60.1 ± 7.7[b]
Nansemond River		
Region 3	24	63.1 ± 4.0[b]

Means with different superscripts significantly different ($p = 0.03$).

times would be different. The results support our hypotheses. There were no significant differences due to sex or age. Comparisons of mean body and organ weights, blood data, necropsy data, and reproductive data failed to indicate significant differences between muskrats from the three regions. Information on these data will be published elsewhere.

Our study indicates that the rate of pentobarbital metabolism is increased in muskrats in an area known to be heavily contaminated with heavy metals and PAHs compared to muskrats inhabiting other areas. We conclude that this difference is due to increased hepatic microsomal enzyme activity resulting from induction by environmental contaminants. The actual compounds inducing the enzymes in region 1 muskrats are unknown. Storet (EPA computer data base) data indicate that PCB and DDT compounds (phenobarbital type inducers) have been reported in fish and sediment samples but at low levels. Whether PAHs which are present at high levels are involved in the observed induction is not known. It is thought that the exposure to inducing agents is not sufficient to elicit a significant response in the other parameters examined (e.g., body weights, reproduction, etc.). If this assumption is true, it is further evidence of the sensitivity and usefulness of the rate of pentobarbital metabolism, as measured by changes in sleeping time, as an in situ indicator of exposure to contaminants.

SUMMARY

We have reviewed the use of barbiturate sleeping time as a measure of changes in hepatic microsomal enzyme activity. Early studies used changes in barbiturate sleeping times to indicate the degree of liver damage due to exposure to various hepatotoxins. Damage to liver tissue resulted in a prolongation of sleeping time which was reported by some authors to be correlated with dose. These experiments were soon followed by studies that exposed laboratory and wild species to environmental contaminants. For most mammalian species, chlorinated hydrocarbons insecticides and polychlorinated biphenyls resulted in a decrease in sleeping times while anticholinesterases resulted in a prolongation in sleeping times. Although not extensively studied, avian species also have been shown to exhibit a response in barbiturate-induced sleeping time following exposure to selected organic compounds.

These studies have demonstrated the feasibility of using barbiturate-induced sleeping time as an indicator of microsomal enzyme activity resulting from exposure to a variety of environmental contaminants. We have used this concept and demonstrated for the first time its potential for detecting environmental contaminant exposure in a wild population. We believe this technique is a sensitive and reliable indicator of microsomal enzyme activity resulting from exposure to certain environmental contaminants. It is a low cost technique, does not require elaborate equipment, and does not require sacrificing the surveyed animals. We encourage additional field studies involving a variety of wild species to further test this technique and anticipate laboratory experiments which will indicate if mixtures or elevated levels of 3-MC type inducers (PAHs) can elicit a pentobarbital sleeping time change in exposed animals.

REFERENCES

1. Brauer, R. W., and M. A. Root. "The Effects of Carbon Tetrachloride Induced Liver Injury upon the Acetylcholine Hydrolyzing Activity of Blood Plasma of the Rat," *J. Pharmacol. Exp. Ther.* 88:109–118 (1946).
2. Azarnoff, D. L., H. J. Grady, and D. J. Svoboda. "The Effect of DDD on Barbiturate and Steroid-Induced Hypnosis in the Dog and Rat," *Biochem. Pharmacol.* 15:1985–1993 (1966).
3. Conney, A. H., R. M. Welch, R. Kuntzman, and J. J. Burns. "The Effects of Pesticides on Drug and Steroid Metabolism," *Clin. Pharmacol. Ther.* 8:2–10 (1967).
4. Soma, L. R. "Intravenous Anesthetic Agents," in *Textbook of Veterinary Anesthesia*, L. R. Soma, Ed. (Baltimore, MD: Williams & Wilkins, 1971), chap. 21.
5. Volwiler, E. H., and D. L. Tabern. "5, 5-Substituted Barbituric Acid," *J. Am. Chem. Soc.* 52:1676–1679 (1930).
6. Fitch, R. H., R. W. Waters, and A. L. Tatum. "The Intravenous Use of the Barbituric Acid Hypnotics in Surgery," *Am. J. Surgery* 9:110–114 (1930).
7. Lundy, J. S. "Experience with Sodium Ethyl (1-Methylbutyl) Barbiturate (Nembutal) in More than 2,300 Cases," *Surg. Clin. N. Am.* 11:909–915 (1931).
8. Sweebe, E. E., and E. C. Khuen. "Pentothol-Sodium, (Thiobarbiturate No. 8064), A New Anesthetic, *N. Am. Vet.* (January 1936), pp. 31–33.
9. Meyers, F. H., E. Jawetz, and A. Goldfien. "Sedative-Hynotics," in *Review of Medical Pharmacology* (Los Altos: Lange Medical Pub., 1978), chap. 23.
10. Harvey, S. C. "Hypnotics and Sedatives," in *The Pharmacological Basis of Therapeutics*, A. G. Gilman, L. S. Goodman, T. W. Rall, and F. Murad, Eds. (New York: MacMillan Publ., 1985), chap. 17.
11. Brodie, B. B., J. J. Burns, L. C. Mark, and P. A. Lief. "The Fate of Pentobarbital in Man and Dog and a Method for its Estimation in Biological Material," *J. Pharmacol. Exp. Ther.* 109:26–34 (1953).
12. Goldstein, A., L. Aronaw, and S. M. Kalman. *Principles of Drug Action: The Basis of Pharmacology* (New York: Harper & Row, 1968), pp. 884.
13. Kuntzman, R., M. Ikeda, M. Jacobson, and A. H. Conney. "A Sensitive Method for the Determination and Isolation of Pentobarbital-C^{14} Metabolites and its Application to *In Vitro* Study of Drug Metabolism," *J. Pharmacol. Exp. Ther.* 157:220–226 (1967).
14. Titus, E., and H. Weiss. "The Use of Biologically Prepared Radioactive Indicators in Metabolic Studies: Metabolism of Pentobarbital," *J. Biol. Chem.* 214:807–820 (1955).
15. Maynert, E. W., and J. M. Dawson. "Ethyl(3-hydroxy-1-methylbutyl) Barbituric Acids as Metabolites of Pentobarbital," *J. Biol. Chem.* 195:389–395 (1952).
16. Maynert, E. W. "The Alcohol Metabolites of Pentobarbital and Amobarbital in Man," *J. Pharmacol. Exp. Ther.* 150:118–121 (1965).

17. Maynert, E. W., and H. B. van Dyke. "The Metabolic Fate of Pentobarbital," *J. Pharmacol. Exp. Ther.* 98:174–179 (1950).
18. Maynert, E. W., and H. B. van Dyke. "The Isolation of a Metabolite of Pentobarbital," *Science* 110:661–662 (1949).
19. Keys, T. E. *The History of Surgical Anesthesia* (New York: Dover Pub., 1963), pp. 200.
20. Wright, J. G. "Anaesthesia in Animals: A Review," *Vet. Rec.* 76:710–713 (1964).
21. Mason, C. M. C., and E. Beland. "Influence of the Liver and the Kidney on the Duration of Anesthesia Produced by Barbiturates," *Anesthesia* 6:483–491 (1945).
22. Plaa, G. L., E. A. Evans, and C. H. Hine. "Relative Hepatotoxicity of Seven Halogenated Hydrocarbons," *J. Pharmacol. Exp. Ther.* 123:224–229 (1958).
23. Rosenberg, P., and J. M. Coon. "Increase of Hexobarbital Sleeping Time by Certain Anticholinesterases," *Proc. Soc. Exp. Biol. Med.* 98:650–652 (1958).
24. Balazs, T., and H. C. Grice. "The Relationship Between Liver Necrosis and Pentobarbital Sleeping Time in Rats," *Toxicol. Appl. Pharmacol.* 5:387–391 (1963).
25. Hart, L. G., and J. R. Fouts. "Effects of Acute and Chronic DDT Administration on Hepatic Microsomal Drug Metabolism in the Rat," *Soc. Exp. Bio. Med.* 114:388–392 (1963).
26. Hart, L. G., and J. R. Fouts. "Further Studies on the Stimulation of Hepatic Microsomal Drug Metabolizing Enzymes by DDT and its Analogs," *Naunyn-Schmiedeberg Arch. Exp. Pathol.* 249:486–500 (1965).
27. Gabliks, J., and E. Maltby-Askari. "The Effect of Chlorinated Hydrocarbons on Drug Metabolism in Mice," *Pesticides Symposia*, W. B. Deichmann, Ed. (Miami: Halos and Associates, 1970), pp. 27–31.
28. Grady, H. J., D. L. Azarnoff, and D. J. Svoboda. "Species Differences in DDD-Induced Changes in Barbiturate Hypnosis," *Fed. Proc. (Abstr.)* 25:417 (1966).
29. Bitman, J., H. C. Cecil, S. J. Harris, and G. F. Fries. "Comparison of DDT Effects on Pentobarbital Metabolism in Rats and Quail," *J. Arg. Food Chem.* 19:333–338 (1971).
30. Litterst, C. L., T. M. Farber, A. M. Baker, and E. J. van Loon. "Effect of Polychlorinated Biphenyls on Hepatic Microsomal Enzymes in the Rat," *Toxicol. Appl. Pharmacol.* 23:112–122 (1972).
31. Villeneuve, D. C., D. I. Grant, and W. E. J. Phillips. "Modification of Pentobarbital Sleeping Times in Rats Following Chronic PCB Ingestion," *Bull. Environ. Contam. Toxicol.* 7:264–269 (1972).
32. Zepp, R. L., O. T. Sanders and R. L. Kirkpatrick. "Reduction of Pentobarbital-Induced Sleeping Times in PCB-Treated Cottontail Rabbits," *Bull. Environ. Contam. Toxicol.* 12:518–521 (1974).
33. Sanders, O. T., and R. L. Kirkpatrick. "Effects of a Polychlorinated Biphenyl (PCB) on Sleeping Times, Plasma Corticosteroids, and Testicular Activity of White-Footed Mice," *Environ. Physiol. Biochem.* 5:308–313 (1975).
34. Montz, W. E., W. C. Card, and R. L. Kirkpatrick. "Effects of Polychlorinated Biphenyls and Nutritional Restriction on Barbiturate-Induced Sleeping Times and Selected Blood Characteristics in Raccoons (*Procyon lotor*)," *Bull. Environ. Contam. Toxicol.* 28:578–583 (1982).
35. Conney, A. H. "Induction of Microsomal Enzymes by Foreign Chemicals and Carcinogenesis by Polycyclic Aromatic Hydrocarbons: G. H. A. Clowes Memorial Lecture," *Cancer Res.* 42:4875–4917 (1982).

36. Elangbam, C. S., C. W. Qualls, and M. Bauduy. "Induction of Hepatic Cytochrome P-450 Activity in Wild Cotton Rats (*Sigmodon hispidus*) by Phenobarbital and 3-Methylcholanthrene," *Bull. Environ. Contam. Toxicol.* 42:716–720 (1989).

37. Conney, A. H. "Induction of Microsomal Cytochrome P-450 Enzymes: The First Bernard B. Brodie Lecture at Pennsylvania State University," *Life Sci.* 39:2493–2518 (1986).

38. Conney, A. H., C. Davison, R. Gastel, and J. J. Burns. "Adaptive Increases in Drug-Metabolizing Enzymes Induced by Phenobarbital and Other Drugs," *J. Pharmacol. Exp. Ther.* 130:1–8 (1960).

39. Lu, M. Z. "Organic Compound Levels in a Sediment Core from the Elizabeth River of Virginia," M.S. Thesis, College of William and Mary, Williamsburg, VA (1982).

40. "The Elizabeth River: An Environmental Perspective," Virginia State Water Control Board, Basis Data Bull. No 61 (1983).

41. Huggett, R. J., M. E. Bender, and M. A. Unger. "Polynuclear Aromatic Hydrocarbons in the Elizabeth River, Virginia," in *Fate and Effects of Sediment-bound Chemicals in Aquatic Systems*, K. L. Dickson, A. W. Makie, and W. A. Brumgs, Eds. (Pergamon Press, in press).

42. Bieri, R. H., C. Hein, R. J. Huggett, P. Shou, C. Smith, and C. Su. "Toxic Organic Compounds in Surface Sediments from the Elizabeth and Patapsco Rivers and Estuaries," Virginia Institute of Marine Sciences Publication.

CHAPTER 9

Stress Proteins: Potential as Multitiered Biomarkers

Brenda Sanders

INTRODUCTION

The biomarker concept involves the use of biochemical, cellular, and physiological parameters as diagnostic screening tools in environmental monitoring. Although biomarkers have been used in a variety of contexts this article deals specifically with the application of biomarkers for two different purposes: to diagnose sublethal stress in an organism, designated as a tier I biomarker, and to detect exposure to specific contaminants, a tier II biomarker.

Important criteria for evaluating the utility of tier I biomarkers for diagnosing sublethal stress would include: (1) its ability to be used in a broad range of organisms when exposed to a wide variety of stress conditions in their environment, (2) that it correlates with decreased physiological function and survival of the organism, and (3) that, in practicality, it can be easily measured in a cost efficient manner. Tier II biomarkers, used to identify exposure to specific contaminants, should be detected in organisms exposed to a particular class of contaminants in their environment and be easily measured. By integrating both kinds of biomarkers into a multitiered approach to environmental monitoring, one could develop a series of assays in which organisms are initially screened with biomarkers to detect general stress and, if the results were positive, could be assayed with an array of tier II biomarkers, each of which identifies exposure to a particular class of contaminants or physical conditions. Such a strategy would provide a comprehensive overview of both the extent of biological damage and the "culprits" responsible.

A major advantage of the biomarker approach is that biochemical and cellular events tend to be more sensitive, less variable, more highly conserved and often easier to measure than stress indices commonly examined at the organismic level such as inhibition of growth, changes in rate of development, and reduced reproductive potential. The major disadvantage is that it can be more difficult to relate these biochemical responses to the health of the organism and to adverse effects on the population, the type of information which is often the bottom line in environmental monitoring. We can best overcome this disadvantage if we select as biomarkers cellular and biochemical events which are intimately involved in protecting and defending the cell from environmental insults. It has recently become apparent that one of the earliest cellular responses to environmental stress involves changes in differential gene expression which are part of the cell's attempt to protect itself.[1] In this paper we focus specifically on the potential use of these transcriptional changes as the basis for developing a series of biomarkers for environmental monitoring.

Since biomarkers for general stress should be related to adverse effects on the organism, the mechanisms by which cells actively respond to environmental stress are ideal candidates. However, it is important to keep in mind that one of the most difficult aspects of identifying such primary responses to environmental stress has been the difficulty in distinguishing them from secondary, symptomatic events caused by adverse conditions in the cell as a result of the stressed condition. For example, it is difficult to determine if a shift in energy metabolism is a consequence of the stress event, i.e., the result of the inactivation of key respiratory pigments, or a proactive strategy initiated by the cell to protect itself.

Only very recently have cell biologists begun to understand the molecular mechanisms underlying the physiology of stressed cells.[2] Serious attention to this important cellular phenomena has occurred largely because of the discovery that all cells dramatically alter their gene expression in response to environmental stress. This alteration in transcriptional activity appears to be an attempt to protect the cell from damage and to repair existing damage.[3] This response, initially referred to as the heat shock response because it was discovered upon exposure to a heat shock, is now more commonly referred to as the cellular stress response since it can be elicited by a variety of physical and chemical stressors. Changes in gene expression associated with the stress response are extremely rapid and result in the induced synthesis and accumulation of what is now referred to as stress proteins. Although a few stress proteins appear unique to stressed cells, most are found in much lower concentrations in nonstressed cells where they play a role in normal cellular function.[4] Although the induction of some of these stress proteins is independent of the nature of the stressor, others are quite stressor specific.

Molecular biologists have taken advantage of this heat shock, or stress response, as a model system for studying the regulation of gene expression because the response can be easily turned on and off by manipulating environmental factors. Much is known about the multiple levels of regulation of stress protein synthesis.[5] Many of the stress proteins and the genes which code for them have been sequenced in a wide range of organisms and found to be remarkably conserved. Less is known about the biochemical role stress proteins play in stress physiology. However, new biochemical functions are being identified rapidly and there is ample evidence that, in concert, these proteins are involved in protecting the cell from subsequent stress and repairing cellular components.[4,6] Although bacteria, yeast, *Drosophila*, and mammalian cell lines have been used most extensively as model systems, the highly conserved nature of this response allows for broad extrapolation to other organisms. Further, many of the cDNA probes and antibodies developed for these model systems have been found to broadly cross react across phyla.[7]

Stress proteins satisfy many of the conditions of ideal candidates for developing a multitiered biomarker strategy for environmental monitoring.[8] They are part of the primary cellular protective response from environmental stress, are induced by a wide variety of environmental stressors, and are highly conserved in all organisms from bacteria to man.[3] Since a thorough understanding of the cellular function of any potential general stress indicator is essential to relating them to the physiological state of the organism this paper reviews what is currently known about the biochemical and physiological function of the various stress proteins, discusses their role in protection from and adaptation to environmental stress, and evaluates the feasibility of their use as biomarkers for general stress. It also discusses how some stress proteins might be used to determine the chemical and physical nature of the stressor.

THE CELLULAR STRESS RESPONSE

The stress protein response is often broken into two major groups: the "classic" heat shock proteins (hsps) whose synthesis is dramatically increased by heat and a variety of other stressors, and the glucose regulated proteins (grps) whose synthesis is increased in cells deprived of either glucose or oxygen.[4] These two stress protein groups are closely related, having similar biochemical and immunological characteristics and considerable homology exists between families. Interestingly, in many instances the synthesis of these two groups seems to be inversely regulated, i.e., cells deprived of glucose have increased synthesis of grps and a concomitant decrease in the synthesis of hsps.[9] Each stress protein is comprised of a multigene family in which some proteins, termed cognates, are constitutively expressed and others are highly inducible in response to environmental stressors. The stress protein cognates play a role in the cell's basic physiology and are found in unstressed cells.

In this paper we designate a third group of stress proteins, those induced by a specific set of chemical or physical conditions, the stressor-specific stress proteins. Although these proteins may not be related to one another, they share in common the characteristic that their synthesis is dependent upon the chemical or physical nature of the stressor and is not substantially increased by heat.

The Stress Proteins Induced by Heat: Heat Shock Proteins

This group of stress proteins, the heat shock proteins (hsps), has a high potential as biomarkers for general stress. As it turns out, what had originally been referred to as the "heat shock response", because it was discovered in response to heat, is induced by a variety of other stressors as well. This heat shock response is a fundamental aspect of cellular physiology in which exposure to a stressor results in a dramatic redirection of metabolism such that this suite of stress proteins is rapidly synthesized and the synthesis of other cellular proteins is repressed.[5] The heat shock response has been found in all organisms examined to date and the genes which encode these stress proteins, and the proteins themselves, are remarkably conserved, from bacteria to man.[3]

The hsps are induced by a wide variety of stressors including heavy metals,[10,11] xenobiotics,[12] oxidative stress,[13] anoxia,[14] salinity stress,[15] teratogens,[16,17] and hepatocarcinogens.[18]

The Classes of HSPs and Their Cellular Function

Although the number of hsps induced by heat shock and their exact size are both tissue and species specific, five "universal" stress proteins are found in all eucaryotes. Four of these are referred to by their apparent molecular weight on SDS-polyacrylamide gels: hsp90, hsp70, hsp58, and the low molecular weight hsp20-30. In eucaryotes each hsp is comprised of a multigene family, the members of which are regulated by different promoters and code for closely related protein isoforms.[5,19,20] The fifth hsp is an 8 kDa protein, called ubiquitin. Most of these proteins are synthesized at high levels in stressed cells. However, with the exception of the 72 kDa protein, a highly inducible member of the hsp70 family, all of these proteins are also present in much lower concentrations in unstressed cells.[4] The initial observations that many hsps are found in "normal" cells and that hsp20-30 are developmentally induced in larval systems lead to the suggestion early on that hsps play a role in normal cellular activities. In this section we will discuss what is currently known about the function of each of these protein families in the nonstressed cell, the roles they may play in the physiology of cells experiencing stress, and their potential use as biomarkers.

hsp90. The hsp90 protein is also referred to as hsp83 or hsp89 depending upon the species under study. In mammalian cell culture hsp90 has been found to associate with a number of cellular proteins, including steroid receptors,[21,22] several kinsases,[23] and a number of retrovirus encoded oncogene proteins many of which are tyrosine specific protein kinases.[4,24] The unifying functional theme in each of these cases appears to be that the binding of hsp90 to these cellular components regulates their activity by preventing them from carrying out their normal functions. In mammals hsp90 is abundant in cells under normal conditions and its synthesis increases about three- to fivefold upon exposure to stress.[4] In light of its normal function, this protein may participate in redirecting cellular metabolism in stressed cells through such mechanisms as the alteration of signal transduction. Given the normal abundance of hsp90 and its limited induced synthesis upon exposure to stress, it may not, by itself, have a great deal of potential as a biomarker.

hsp70. There are two major members of this most highly conserved stress protein, the hsp70 family, each of which is present in multiple isoforms.[5] The larger protein of the two, 73 kDa in mammals, is often referred to as the hsp cognate because it is found in unstressed cells and also exhibits a marked increase in synthesis upon exposure to various stressors. The smaller protein, 72 kDa, is only synthesized upon exposure to stress and is not found in the cell under normal conditions. Although these two hsps are closely related and have similar biochemical properties, they are distinct gene products.

As for the functional aspects of these important proteins, a general picture is now emerging in which hsp70 acts to either stabilize or solubilize a target protein.[4] Under normal conditions such binding may serve a "chaperone" function for newly synthesized secretory and organellular proteins by helping them to translocate across a membrane.[25,26] Another member of the hsp70 family, called BiP or grp78 (discussed below), is also found under normal conditions and has recently been shown to be transported into the endoplasmic reticulum where it may perform a similar chaperone function for proteins transported into this compartment.[27]

The highly inducible hsp72 in conjunction with the other hsp70 proteins may perform a similar role in cells experiencing stress. A major feature of stressed cells is the loss of integrity of the nucleolus and the associated inhibition of rRNA synthesis and ribosomal assembly.[4] Under stress hsp72 rapidly migrates to the nucleolus where it is speculated to resolubilize denatured pre-ribosomal complexes and help restore nucleolar function during recovery from stress.[28] During recovery it migrates to the cytoplasm and associates with ribosomes and polyribosomes where it is speculated that it may bind to denatured proteins and in an ATP dependent manner, facilitate their resolubilization.[28,29] Since the hsp70 family accounts for much of the translational activity in stressed cells and it is one of the most highly conserved proteins known in biology, it is an excellent candidate for a biomarker for general stress.

hsp60. The hsp60 protein, found in the mitochondria, is believed to be another "chaperoning" protein which facilitates the translocation and assembly of oligomeric proteins into that compartment.[30] It is homologous to the bacterial hsp GroEL and the Rubisco-binding protein of chloroplasts,[31] forms large aggregates in the matrix of the mitochondria, and is essential for the assembly of oligomeric complexes imported into the mitochondria. Its synthesis is increased in stressed cells where it is believed to perform similar functions. Because it is highly conserved and its synthesis is increased in stressed cells, it is also a good candidate as a biomarker.

hsp20-30.   The low molecular weight stress proteins (hsp20-30) are the least conserved of the stress proteins. They show homology to the alpha crystalline lens protein, and also share with that protein the tendency to form higher ordered structures.[32] These stress proteins are highly species specific. One protein of this class is found in yeast (26 kDa) and mammals (28 kDa), *Drosophila* has four (28, 26, 23, 22 kDa), and some plants have been found to have up to 20 proteins in this class.[5] Considerable variation exists even within the same class of organisms. For example, within the molluscs *Mytilus edulis* has two low molecular weight stress proteins (29, 32 kDa; Figure 1), *Collisella scabra* has at least five (16, 17.5, 18.5, 19, 24 kDa; Figure 2) and *Collisella pelta* has two (17.5, 18.5 kDa).[33] Although low molecular weight hsps are regulated during development and differentiation, and by the hormones estrogen, progesterone, and ecdysone, we know very little about the function of these proteins in cells under normal conditions.[4,5,34,35] However, they do localize within the nucleus upon heat shock and return to the perinuclear region of the cytoplasm upon recovery.[36] Since they are highly species specific and regulated by a number of factors besides exposure to stressors, their use as a biomarker of general stress may be misleading and of limited value.

Ubiquitin.   Ubiquitin is a small molecular weight (7 KDa) protein found in all eucaryotic cells. Under normal conditions it is involved in the nonlysosomal degradation of intracellular proteins.[20] When ubiquitin is conjugated to proteins by a ubiquitin-protein ligase system these proteins are selectively degraded. Ubiquitin synthesis increases with exposure to heat and is an essential component of the stress response.[37] In stressed cells the function of this protein would complement the resolubilization and stabilization function of hsp70 by targeting denatured proteins for degradation and removal. In yeast the polyubiquitin gene which is transcribed in stressed cells has been shown to be essential for resistance to heat shock, starvation, and other forms of physiological stress.[37] Because of the inability to detect proteins of such low molecular weight using standard electrophoretic gels, little data are available that would aid in evaluating this stress protein as a biomarker. However, given that the generation of denatured proteins is a common problem in stressed cells, increased synthesis of this stress protein is probably ubiquitous in eucaryotic systems making it a good candidate as a biomarker for general stress. Further, since such damage is quite localized it is unlikely that its synthesis will be regulated by extracellular signals.

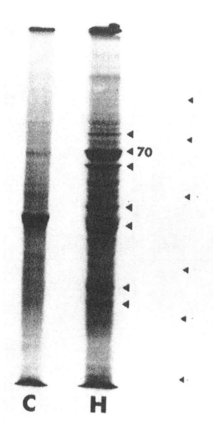

Figure 1. The stress protein response in gill tissue of *Mytilus edulis.*[8] The control (C) was maintained at 17°C, while the heat shock (H) sample was exposed in vivo to a heat shock for 1 hr at 31°C. Tissues were then incubated in ^{35}S methionine, homogenized and equal protein was loaded for one-dimensional electrophoresis (SDS-PAGE). The gels were subsequently fluorographed. Arrows at the far right indicate molecular weight markers of 130, 75, 50, 39, 27, and 17 kDa, respectively. Arrows next to the heat shocked sample indicate stress proteins of 80, 74, 72, 60, 47, 43, 32, and 29 kDa, respectively.

The Glucose-Regulated Proteins

The major glucose regulated proteins (grps) are of 100, 80, and 75 kDa and are structurally and functionally related to the heat shock group. The 75 and 80 kDa proteins are homologous to the hsp70 family,[38] and the 100 kDa protein is homologous to hsp90.[40] This group of stress proteins is present in unstressed cells and shows increased synthesis in cell cultures deprived of glucose or oxygen, exposed to elevated lead,[40] and agents that perturb calcium homeostasis.[4] Since

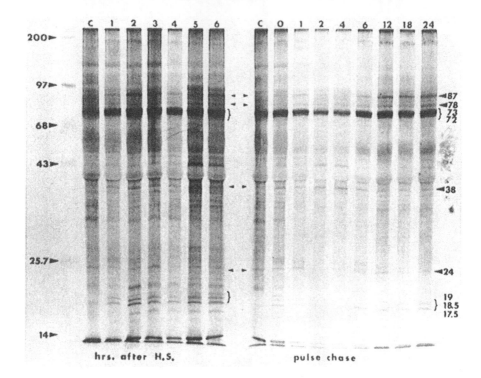

Figure 2. Fluorograph of *Collisella scabra* mantle tissue from induction and half-life experiments. Samples were processed as described in Figure 1. Arrows at the right of the fluorograph identify hsps of 87, 78, 73, 38, 24, 19, 18.5, and 17.5 kDa, respectively. Molecular weight markers are shown on the left. *Left:* Incorporation of ^{35}S methionine into proteins in *C. scabra* mantle tissue at different points in time after a 1-hr, 31°C heat shock. Limpets were heat shocked in vivo and incubated for 1 hr at the times indicated to examine how long the response was induced after heat shock. The control (C) was not heat shocked. Increased synthesis of stress proteins can be seen for at least 6 hr after heat shock. *Right:* Incorporation of ^{35}S methionine into proteins in *C. scabra* mantle tissue during a pulse chase experiment to examine the half-life of the stress proteins. Limpets were heat shocked for 31°C for 1 hr, immediately pulsed with ^{35}S methionine for 1 hr, and incubated in cold methionine over time. The numbers above the sample represent the hours that the tissue was incubated in cold methionine after heat shock. The control (C) was not heat shocked. Stress proteins remain prominent 24 hr after heat shock.

very little is known about the induction of synthesis of these stress proteins in whole organisms, their potential as biomarkers is unclear. Given the specific conditions of increased synthesis of the grps, they appear to be poor candidates as biomarkers for general stress. However, they may have potential as biomarkers to identify anoxic conditions. Also, if the synthesis of grps is increased in starved organisms they may prove useful as biomarkers for evaluating nutritional state.

100 kDa. This grp is found in the endoplasmic reticulum and golgi.[9] That it is homologous to hsp90 suggests it too may be involved in regulating the activity of various cellular components within these cellular compartments under normal conditions. Its amino acid sequence indicates that it is a transmembrane bound protein with much of the protein found on the cytosolic side of the membrane.[41]

80 kDa. This protein, as mentioned earlier, is also referred to as BiP or grp78. It is abundant in normal cells within the endoplasmic reticulum. Its homology with hsp70 suggests that it is involved in the ATP-dependent stabilization, folding, and assembly of proteins transferred into this compartment.[42]

75 kDa. Grp75 is a recently discovered stress protein which is located in the mitochondria. It is another ATP binding protein homologous to hsp70 which probably serves the same protein stabilization and assembly function as the other members of this family.[43]

The Protective Role of Stress Proteins

Collectively these two groups of stress proteins appear to be involved in the protection, enhanced survival, and restoration of normal cellular activities in stressed cells.[44] The induction of hsps by a mild conditioning stress enhances the tolerance of the cell to subsequent, more severe stress situations, a phenomenon often referred to as thermotolerance or, when other stressors are involved, "acquired tolerance".[45-50] The induction, expression, and decay of acquired tolerance correlates with the induction, accumulation, and degradation of stress proteins.[46,51-55] The fact that other studies have demonstrated thermotolerance in the absence of stress protein synthesis,[56-58] has resulted in considerable confusion regarding the role of hsps in thermotolerance. However, Welch and Mizzen[6] have shown that most of these stress proteins are abundant in unstressed cells and rapidly relocate to other subcellular compartments upon heat shock. This observation suggests that even in the absence of protein synthesis these constitutive hsps can be called upon to perform new protective and repair roles under such situations.[4] Although the molecular bases for acquired tolerance is not known, both RNA processing[59] and translation activity[60] appear to be protected by stress proteins. Also lesions caused by heat-shock treatment are prevented or repaired more quickly in cells that have been made thermotolerant by hsp synthesis.[6]

Both the hsp70 class and the low molecular weight stress proteins have been implicated in the enhanced survival of cells exposed to stress.[45,61,62] Hsp70 plays a role in the recovery of nucleoli after heat damage[28,63] and may be involved in the resolubilization of proteins damaged by heat.[29] The accumulation of hsp70 is also correlated with a decrease in the inhibition of protein synthesis which occurs upon severe heat-shock.[60] Mammalian cell lines selected for survival at high temperatures constitutively synthesize hsp70 at high levels while temperature sensitive mutants are unable to elicit the stress protein response.[5] The hsp20-30 family correlates with acquired tolerance in *Drosophila*[47] and sorghum,[64] and accumulates in desert succulents acclimatized to high temperatures.[65] Also, a temperature sensitive mutant of *Dictoselium* is unable to synthesize low molecular weight stress proteins.[61]

In a recent study we examined the heat shock response in two closely related limpet species, *Collisella scabra* and *Collisella pelta*, which have very different tolerances to elevated temperatures that correspond to different microhabitats in the intertidal.[33] We found both quantitative and qualitative differences in the heat shock response. The temperature range for induction of the response reflected the temperature mortality curve for each species, i.e., temperature tolerant *C. scabra* elicited the response and synthesized hsps at higher temperatures than temperature sensitive *C. pelta*. Further *C. scabra* synthesized the highly inducible hsp70 at a higher rate and had many more low molecular weight hsps than *C. pelta*. Boon-Niermeijer et al.[66] have demonstrated that synthesis of stress proteins correlates with acquired tolerance (as assayed by survival) in the mollusc *Lymnaea stagnalis* lending further support to the premise that stress proteins may be involved in enabling these marine invertebrates to survive in the environmental extremes of the intertidal zone.

Stressor Specific Stress Proteins

This third class of stress proteins has in common the fact that its synthesis is induced only under specific chemical or physical conditions. These proteins appear to participate in specific biochemical pathways involved in the metabolism of chemicals, metabolites, or harmful by-products that are the result of a particular chemical or physical condition rather than being part of the cell's protective system in response to general cellular damage. As a consequence, this class does not encompass a homologous group of proteins and is not related functionally or structurally to either of the other two groups of stress proteins. Although numerous inducible enzyme systems (i.e., cytochromes) could technically fall within such a broad definition, this paper will focus on proteins whose kinetics of induction and recovery are rapid (minutes to hours) and are thus part of the cell's immediate stress response. Novel stressor-specific stress proteins are being discovered at a rapid rate; however, most of these are currently only identified by their apparent molecular weight on an SDS gel. We know little about their structure or function. Therefore, this section will be limited to those stressor specific stress proteins which are best characterized, heme oxygenase and metallothionein.

Heme oxygenase

In the last few years a 32 kDa stress protein that was inducible by metals, sodium aresenite, and thiol-reactive agents had been reported in the literature.[11,40] This protein has recently been isolated and identified as heme oxygenase, an enzyme essential for heme catabolism that cleaves heme to form biliverdin, which is subsequently reduced to bilirubin.[67,68] Although it is considered by some to be a minor hsp because of limited evidence that it can be induced to some extent by heat in rats and humans,[68–70] others report that heat treatment has no effect on heme oxygenase activity or mRNA levels.[71,72] It is most highly inducible by a variety of stressors which cause oxidative damage, such as UVA radiation, sodium arsenite, and hydrogen peroxide.[68] It has been suggested that since breakdown products of heme can readily react with peroxyl radicals they may play an important role in protecting cells from oxidative damage as free radical scavengers in concert with glutathione.[68] Interestingly, Cd and other metals such as Cu, Zn, Pb, sodium arsenite, and gold have been shown to be particularly effective inducers of heme oxygenase and it is the most prominent stress protein induced by these metals upon in vitro exposure.[11,40,68]

Metallothionein

Metallothionein is a low molecular weight (≥ 10 kDa) cysteine rich metal binding protein whose synthesis is induced upon trace metal exposure.[73] The kinetics of induction tend to be slower than that of the other "classic" stress proteins. For example in trout, metallothionein synthesis is increased within 24 hr of exposure to transition metals while stress protein synthesis increases within 30 min.[74] Although its synthesis is not coordinately induced with the other stress proteins I have chosen to include a discussion of metallothionein in this section because of its role in protection from metal toxicity[73–75] and because its potential in environmental monitoring has been studied extensively.[76–78]

This metal-binding ligand appears to be part of a cellular compartmentalization/sequestration system which evolved to regulate the uptake and tissue distribution of essential trace metals such as the transition elements Zn and Cu.[79] Since the nonessential metal Cd is also a transition element with a similar chemistry to Zn and Cu, it also interacts with this system.[80] Under normal conditions this compartmentalization system provides the cell with ready access to essential metals. However, transition metals are highly toxic and capable of damaging cell function by nonspecific binding to metalloenzymes and other cellular components. When cells are exposed to elevated metals this system also protects the cell by either sequestering the metals in membrane-bound vesicles or on soluble metal binding ligands such as metallothionein and glutathione.[81]

Initial studies suggested that the induction of metallothionein might be a successful biomarker for metal exposure and perhaps even for toxic effects in cases when metals were the major stressor.[78] However, for several reasons metallothionein should be used cautiously as a monitoring tool and is, perhaps, most useful when used in association with other biomarkers. First, it is now clear that, at least in mammals, insects, and crustaceans, metallothionein is induced under many other conditions besides metal exposure, including extracellular signals such as steroid and peptide hormones, interleukin II, and interferon.[82] These and other observations suggest that in many organisms metallothionein may be used by the cell in a proactive manner to regulate metal compartmentalization and accomodate metabolic needs. As a consequence, in higher organisms the increased synthesis or accumulation of metallothionein does not necessarily reflect a reaction to increased metal exposure. Secondly, since metallothionein is only one component of the compartmentalization/sequestration system which regulates metal metabolism, there is no mechanistic basis to assume that measuring metallothionein accumulation without examining the other essential components that regulate metal metabolism (and the metals themselves) will provide reliable information on either metal exposure or the physiological condition of the organism. However, an examination of metal uptake and subcellular distribution has proven quite useful in evaluating organismal stress in the laboratory and in situ.[83–86]

CONSIDERATIONS FOR EVALUATING STRESS PROTEINS AS BIOMARKERS

Organisms can be exposed simultaneously to many stressors, both natural and anthropogenic, and have only a finite capacity to adapt to them. Yet in environmental monitoring we often look only at one potential stressor out of context with the multiple stressors with which the organism must cope in its environment. Tier I stress responses, those that occur regardless of the nature of the stressor, can be used to evaluate this integrated stress load and determine its overall impact on the physiological state of the organism. However, due to the lack of specificity general responses make it difficult to determine what factors are responsible for the observed stress to the organism. The second tier of stress responses are those which are stressor specific in that they only respond to a specific stressor, or group of stressors, which share chemical or physical characteristics. Although these later responses are less integrative, they facilitate the establishment of cause effect relationships between biological impact and exposure regimes in the environment.

A monitoring strategy using biomarkers would be most effective if it incorporated tier I, general stress responses for initial screening, with tier II, specific stress responses to help identify the stressors. As discussed in previous sections stress proteins have the potential for providing both types of information. Accumulation of the highly inducible hsp72 and hsp60 may provide a useful tier I biomarker which would reflect the integrated stress load on the organism regardless of the type or number of stressors that might be involved. When the accumulation of one of these stress proteins is calibrated to organismal indices of stress, such as inhibition of growth or reproduction, it could be highly predictive of adverse impacts on the population. Once assays based on antibodies raised against these stress proteins are developed to measure stress protein concentration, organisms could be screened rapidly to determine if, and to what degree, they are stressed. If stress proteins have not accumulated, no further testing would be necessary. Accumulation of one or more of the grps may also be used as a supplemental assay to screen for nutritional imbalances which might exacerbate the adverse effects of environmental stress on organismal physiology.

The hsps, grps, and stressor-specific stress proteins all have the potential to play a role in the tier II assays which would examine the nature of the stressors involved. By identifying unique patterns of accumulation of the various stress proteins induced by key classes of contaminants we could develop a series of antibody based assays that would allow us to screen for exposure to each type of stressor. As with the general stress proteins, laboratory calibrations between accumulation of particular proteins and organismal stress would provide insight into the relative contributions of each stressor to the entire stress load.

Since the criteria for evaluating biomarkers will be different for each purpose some stress proteins will be more suitable for one application than another. In this section we will examine what is currently known about the use of stress proteins as biomarkers and discuss the kinds of experiments that are needed in order for us to accurately evaluate their potential in environmental monitoring.

Tier I: Stress Proteins as Biomarkers for General Stress

Environmental Relevance

In order to evaluate the potential of using stress proteins as the basis for developing a screening strategy in biomonitoring several characteristics of the response need to be determined. Since an ideal tier I biomarker would evaluate the impact of environmental stress on native organisms and populations in situ, elevated stress proteins must accumulate in organisms under realistic environmental conditions in response to a variety of stressors. Thus, the response should be sustained over time in a variety of organisms upon exposure to sublethal concentrations of stressors.

Unfortunately, to date we know little about the environmental relevance of the stress response and much research needs to be focused in this area. With the exception of metallothionein, much of the research on the stress response has involved exposure of cells in culture to perturbations which are often extreme and unlikely to occur in the environment.[6,62,87] Metabolic labeling studies in our laboratory have demonstrated induction of the stress response at concentrations as low as 7.5 ppt (ng/L) for tributyl tin (TBT) and 10^{-12} M free cupric ion activity following a 4-hr in vitro exposure of M. edulis hemolymph (Figure 3). These concentrations are well within the ranges measured in the environment. Tributyltin, the active component of some antifouling paints, has been reported to range from <0.04–0.35 ppb near the entrance to a harbor in England.[88] It has been shown that molluscs are particularly sensitive to TBT; concentrations as low as 1 ppt results in 'imposex', the induction of male sexual characteristics

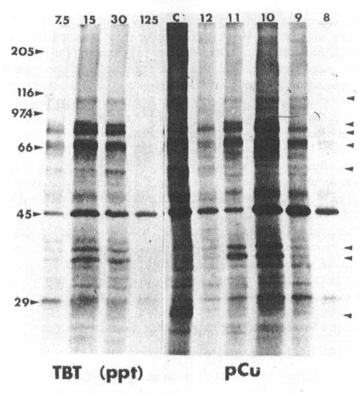

Figure 3. Induction of the stress response in hemolymph of *M. edulis* exposed to tributyltin and copper. Samples were exposed in vitro to the concentrations indicated for 4 hr and then incubated for 4 hr in ^{35}S methionine. Samples were processed as described in Figure 1. Arrows at the left are for the molecular weight markers. Those on the right indicate stress proteins of approximately 110, 85, 78, 74, 72, 60, 38, 36, and 27 kDa, respectively. TBT = tributyltin, ppt = parts per trillion, pCu = −log of free cupric ion concentration.

in females[89] and the 15-day larval LC_{50} for tributyltin oxide is 0.1 ppb.[90] Free cupric ion activities have been measured as high as $10^{-10.5}$ M in several estuaries in North Carolina, a concentration at which growth is inhibited in crab larvae.[83]

Further, recent studies have indicated that the stress response is induced in vivo at concentrations found in polluted environments. Irby et al.[12] have demonsrated several-fold increases in accumulation of hsp60 in rotifers after 96-hr exposures to concentrations as low as 15 ppb (μg/L) of TBT and 10 ppb (μg/L) Cu. In experiments recently completed in collaboration with scientists at the EPA/ERL laboratory in Narragansett, Rhode Island, we found significant increases in hsp60 concentrations in *M. edulis* mantle tissue after a 7-day in vivo exposure to Cu low as 3.2 ppb (μg/L).[33] Thus, it does appear promising that the stress response is elicited under environmentally relevant concentrations across a broad range of organisms.

If it is to be useful for native organisms exposed in situ the stress response must also persist over time. Although reports in the literature are contradictory most researchers now estimate that hsp70 has a half-life of at least 4 days in most organisms.[5] Pulse chase experiments in this laboratory have shown stress proteins to be stable for at least 72 hr in marine molluscs (Figure 2).[33]

Data on the kinetics of induction and recovery of the stress response suggest differences which are specific to the type of environmental insult and severity of the stress. In mammalian cells a 42°C heat shock induces the response within a few minutes and recovery, as determined by a reversion back to translational patterns similar to controls, occurs 4 hr after removal of the heat shock.[70] Although the kinetics of induction are similar in response to sodium arsenite, recovery is not apparent until 8 days after removal of the stressor, presumably because the stressor is still present in the cell. Surprisingly, marine invertebrates which are notorious for their slow metabolism, display similar induction and recovery kinetics. In the mollusc *C. scabra* hsps are rapidly synthesized within minutes of a 31°C heat shock and the organism continues to synthesize stress proteins for at least 6 hr after removal from elevated temperatures (Figure 2).

In a broad range of organisms continuous exposure to a contaminant which results in moderate stress appears to elicit a transient response, while continuous exposure to a contaminant which results in a more severe stress results in a sustained response.[5] One concern in terms of the usefulness of the stress response in biomonitoring is that if chronic environmental exposure resulted in a less severe stress, and thereby a transient expression of the stress response, the organism may be experiencing stress yet stress proteins may not be significantly elevated over time. If this was the case it would limit their usefulness as bio-markers for in situ exposure of native organisms.

It is important, however, to distinguish between the stress response, which is defined as the increased rate of synthesis of stress proteins and inhibition of the rate of synthesis of normal proteins as detected through translational patterns, and significant increases in accumulation of stress proteins which are persistent in time. Mizzen and Welch[60] have demonstrated that hsp70 expression is regulated in such a way that regardless of the number and severity of heat shock treatments its synthesis is increased until a threshold concentration of the protein accumulates in the cell. After this point no amount of stress will result in an increase in hsp70 concentrations. Given what we now know about the function of this protein it is not surprising that it is the total pool of this protein which is being regulated. As a result of this regulation it is also quite likely that under chronic stress conditions elevated stress proteins persist in time by balancing the rate of synthesis against the rate of degradation. If this is the case elevated stress proteins would persist even though the translational patterns characteristic of the stress response would not be observed.

Although numerous laboratory and field experiments are needed before we will be able to evaluate this aspect of the response, Kee and Noble[65] have demonstrated that low molecular weight stress proteins remain abundant in desert succulents even after completion of high temperature acclimation. In *M. edulis* we have observed sustained elevated accumulation of hsp60 for at least 1 week upon exposure to Cd and to Cu.[33] Clearly long term experiments need to be carried out to determine if elevated accumulation of stress proteins is a chronic condition upon exposure to sublethal stress.

Relationships to Physiological State of the Organism

A major advantage of the stress response is that because it is involved in protecting the cell from environmental damage, it provides a direct measure of the cellular physiological state. Thus it has the potential to be more sensitive than existing organismal indices for stress, yet can be correlated to adverse physiological conditions in the organism. Further, as a quantitative response, it could provide the added benefit of evaluating the extent to which an organism is stressed. To evaluate this aspect of the stress response more research is needed that examines the relationships between stressful conditions at a tissue level and the physiological condition of the organism.

A few studies have attempted to begin to examine these relationships. In collaboration with EPA scientists we have attempted to correlate Cu exposure in *M. edulis* to tissue specific increases in hsp60 concentration and to scope for growth (SFG), a common organismal stress index based on bioenergetic changes which measures clearance rate, respiration rate, and assimilation efficiency.[33] In these experiments, groups of organisms were exposed to Cu concentrations of 0, 3.2, 10, 32, and 100 ppb for 7 days. One half of each treatment (n = 8) was then used for scope for growth measurements and the other half for hsp60 accumulation. The SFG index showed no significant differences among organisms exposed to 0, 1, 3.2, and 10 ppb Cu. However, for those exposed to 32 and 100 ppb the SFG was significantly lower than that for controls, a condition indicative of growth inhibition. Dramatic reductions in clearing rates across the gills seemed to be a major factor in this reduction. When we measured the accumulation of hsp60 in mantle tissue we found no difference in its accumulation between controls and 1 ppb Cu. However, hsp60 concentrations were significantly higher at all other Cu concentrations. There was a linear relationship between the log of the Cu concentration and the log of hsp60 concentration in tissues exposed to 3.2 to 100 ppb Cu. Further, we could detect stress at the tissue level, defined by a significant increase in hsp60, at a Cu concentration one order of magnitude lower than could be detected with scope for growth.

Assays to Measure the Stress Response

On a practical level an ideal biomarker should be easy to use, rapid and inexpensive and would measure a response which is so highly conserved that a single probe could be used with a broad range of organisms, vertebrates, invertebrates, and plants alike, so that native organisms could be examined in a site specific manner.

The techniques most frequently used in the study of stress proteins have involved metabolic labeling and specific antibodies or cDNA probes. In metabolic labeling studies tissues are incubated with an amino acid tagged with a radio-isotope (i.e., ^{35}S, ^{14}C, ^{3}H). The tissue is then homogenized and the proteins are separated by one- or two-dimensional electrophoresis, and autoradiographed to examine incorporation of the radioisotope into specific proteins (Figures 1, 2, 3). This technique provides information on the entire translational profile in response to a stressor and can be particularly useful for identifying new inducible proteins. However, under continuous exposure to moderate stress conditions these dramatic changes in translational patterns are transient (approximately 18 hr in *Mytilus* exposed to a mild heat shock) and translational activity reverts to patterns similar to those found in controls. Therefore, it appears likely that it will be the stress protein concentration which will be most reflective of stress in organisms exposed to moderate perturbations over long periods of time. Since there is currently no simple quantitative assay for measuring any of the stress proteins these techniques will need to be developed before extensive research on the quantitative aspects of the stress response can be carried out.

Assays based on antibodies raised against the stress proteins have the best potential for environmental monitoring because they will allow us to directly quantify stress protein accumulation. Of all the stress proteins the hsp70 family is an ideal candidate for an assay of general stress because it is highly inducible and is found in high concentrations in tissues of organisms exposed to environmental perturbations. Further, since it is so highly conserved there is the potential of developing antibodies which may crossreact in a broad range of organisms. A number of antibodies have been raised against the two hsp70 gene products isolated in mammals, *Drosophila*, yeast, and bacteria. However, the extent to which they might crossreact with other organisms is not clear. We have been encouraged by the fact that a monoclonal antibody against mammalian hsp70 crossreacts with both major members of the hsp70 family in *Mytilus* (Figure 4). Most promising, we found that by Western blotting with this antibody we could detect the highly inducible hsp72 in mantle tissue exposed to heat shock, Cu, and TBT but not in the control.

Although perhaps not as higlhy inducible as hsp70, another good candidate for a tier I biomarker is the mitochondrial hsp60. Our own experiments have examined the accumulation of this stress protein because of the availability of an antibody probe with excellent crossreactivity for aquatic organisms. This hsp60 antibody was made against hsp60 from a moth[91] and has been shown to crossreact with hsp60 from a particularly wide variety of species ranging from the groEL gene product in *E. coli* to the hsp60 in rotifers,[12] *Mytilus* (Figure 5) and fathead minnow.[33]

Tier II: Stress Proteins as Stressor Specific Indices

In addition to providing information on an organism's integrated stress load, the relative synthesis of some stress proteins appears to be specific for different classes of stressors. These differences are particularly apparent in metabolic

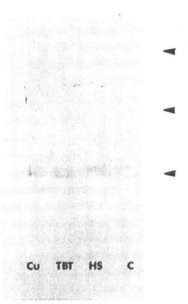

Figure 4.   Western blot of *M. edulis* mantle tissue against a monoclonal antibody raised against mammalian hsp70 (from W. Welch). Tissues were exposed for 8 hr in vitro to copper, tributyltin, heat shock for 1 hr at 31°C and a control sample from left to right, respectively. Samples of equal protein were run on a 12% SDS gel before blotting. Arrows at the right designate prestained markers of approximately 194, 111, and 60 kDa, respectively. Cu = copper exposed sample, TBT = tributyltin exposed sample, HS = heat shock at 31°C, C = control.

labeling studies under acute conditions. The synthesis of hsp70 relative to other proteins, for example, is much greater in response to heat shock than in response to chemical inducers (Figures 2, 3). Also the inhibition of synthesis of other cellular proteins is less dramatic with chemical inducers than with heat shock. Although it is unclear if these differences will be of any practical use in environmental monitoring, it will at least allow us to determine if heat is a major stressor in specific acute situations or if organisms were accidently heat shocked during sample collection.

More promising as biomarkers for identifying exposure is the observation that certain contaminants elicit the synthesis of unique proteins.[92] The two most studied stressor specific stress proteins, heme oxygenase and metallothionein, have been discussed previously. In this section we will consider the potential of using unique translational patterns which correlate to different classes of stressors as biomarkers. At least initially, this approach would need to involve broad scale screening to examine changes in gene expression in response to different classes of environmental stressors. Cell lines from a variety of organisms would be exposed to individual contaminants and examined for increase synthesis of transcriptional or translational products. Once unique gene products were identified

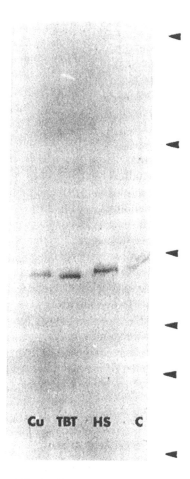

Figure 5. Western blot of *M. edulis* mantle tissue against a polyclonal antibody raised against hsp60 in moth.[91] Tissues and electrophoresis are described in Figure 4. Arrows represent prestained markers of approximately 194, 111, 60, 35, 26, 20, and 16 kDa, respectively. Cu = copper exposed sample, TBT = tributyltin exposed sample, HS = heat shock at 31°C, C = control.

in response to a class of stressor, an antibody could be raised against the unique protein for detecting in situ exposure.

This approach has been used extensively in facultative microorganisms where a set of genes and gene products that respond to a particular stress is referred to as a stimulon.[93] Although there is often overlap between stress proteins induced by such stressors as heat shock, anaerobiosis, oxidative stress, and starvation, unique proteins can also be attributed to each stressor.[14]

A few studies in eucaryotes have taken this approach to date and have generated promising results. *Neurospora* cells elicit different translational patterns when exposed to heat shock, arsenite, and oxidative stress.[13] A pattern of 11 proteins is translated in response to heat shock, whereas in response to arsenite only the hsp70, hsp80, and a unique arsenite specific protein of approximately 40 kDa are translated. Hydrogen peroxide exposure leads to the synthesis of hsp70 and an oxidative stress-responsive protein (osp80).

Differential translation patterns are also found in both human and murine melanoma cells where hypothermia induces the synthesis of four stress proteins of 100, 90, and a doublet at 70 kDa.[11] In contrast, the heavy metals Cu and Cd, and thiolreactive agents induce the synthesis of these four proteins plus the 32 kDa stress protein, heme oxygenase. Lead induces an entirely different stress response in rat fibroblasts and epithelial cells.[40] Lead glutamate results in the increased synthesis of three stress proteins, grp100, grp80, and heme oxygenase, and does not induce hsp70 or metallothionein. In barley seedlings, salinity stress induces the synthesis of unique sets of low molecular weight stress proteins from the hsp20-30 family in a tissue-specific manner.[15] Other stress proteins, including hsp70, are not induced. Exposure of salivary gland cells to low-frequency electromagnetic fields alters translational patterns in a unique manner.[94] In addition to the five stress proteins which are also induced by heat shock, different subsets of stress proteins are induced in a frequency-dependent manner.

From the data currently available it appears that a limited number of specific genes may be activiated in a stressor specific manner resulting in unique patterns of translational products. Massive screening will of course be needed before we can accurately evaluate the potential of these unique patterns as biomarkers for exposure.

CONCLUSIONS

Several points need to be emphasized regarding the use of stress proteins as biomarkers. These proteins represent changes in gene expression in response to environmental variables which are part of the cellular strategy to protect itself from potential damage. Further, since cellular responses that would confer protection would be highly selected for in evolution, it is not surprising that stress responses are so highly conserved in procaryotes and eucaryotes alike. From a mechanistic viewpoint the strategy for selecting potential biomarkers which are based on the molecular mechanisms underlying protection from stress is a sound one with high potential for identifying environmentally relevant biomarker assays.

Much is known about regulation of gene expression of these stress responses and we are beginning to understand their role in the physiology of normal and stressed cells. However, very little is known about the environmental relevance of these responses in whole organisms exposed to stressors in their enviroment. Although preliminary data suggest that the accumulation of stress proteins can provide useful information for environmental monitoring and toxicological screening, much more research will be required before their usefulness can be accurately evaluated. The response needs to be studied in a broad range of environmental contaminants at environmentally realistic concentrations. Research on the stress response is particularly needed in fish, aquatic invertebrates, and plants. Also the relationships between tissue level stress responses and impairment of function at the organismic level will be particularly important for this evaluation. The persistence of elevated stress protein concentrations in tissues of organisms exposed to contaminants in their environment needs to be fully explored under a wide range of environmental conditions. Finally, field studies must be undertaken to examine relationships between stress protein accumulation in organisms, other organismal and population stress indices, and chemical data on contaminant exposure.

ACKNOWLEDGMENTS

I wish to thank L. Martin and V. Pascoe for technical assistance and help in manuscript preparation. I am also grateful to S. Miller and W. Welch for furnishing heat shock protein antibodies. This work was partially supported by a cooperative agreement (CR-814323-01-0) with the U.S. Environmental Protection Agency and by a Public Health Service grant (GM42214-01) from the National Institute of General Medical Sciences.

REFERENCES

1. Atkinson, B.G., and D.B. Walden, Eds. *Changes in Eukaryotic Gene Expression in Response to Environmental Stress*, (Orlando: Harcourt Brace Jovanovich, 1985), p. 379.
2. Welch, W.J., and J.P. Suhan. "Cellular and Biochemical Events in Mammalian Cells during and after Recovery from Physiological Stress," *J. Cell. Biol.* 103:2035–2053 (1986).
3. Schlesinger, M.J., M. Ashburner, and A. Tissieres. *Heat Shock. From Bacteria to Man* (Cold Spring Harbor, NY: Cold Spring Harbor Laboratory, 1982), p. 440.
4. Welch W.J. "The Mammalian Stress Response: Cell Physiology and Biochemistry of Stress Proteins," in *The Role of the Stress Response in Biology and Disease*, R. Moromoto, and A. Tissieres, Eds. (Cold Spring Harbor, NY: Cold Spring Harbor Laboratory, 1990), in press.
5. Lindquist, S. "The Heat Shock Response," *Ann. Rev. Biochem. 55:1151–1191 (1986)*.
6. Welch, W.J., and L.A. Mizzen. "Characterization of the Thermotolerant Cell II. Effects on the Intracellular Distribution of Heat Shock Protein 70, Intermediate Filaments and Small Ribonucleoprotein Complexes," *J. Cell Biol.* 106:1117–1130 (1988).
7. Greenberg, S.G., P.F. Drake, and R.J. Lasek. "Differential Synthesis of Heat Shock Proteins by Connective Cells and by Neurons of *Aplysia californica*," *J. Cell Biol.* 97:152a (1983).
8. Sanders, B.M. "The Role of the Stress Proteins Response in Physiological Adaptation of Marine Molluscs," *Mar. Envir. Res.* 24:207–210 (1988).

9. Welch, W.J., J.G. Garrels, G.P. Thomas, J.J. Lin, and J.R. Feramisco. "Biochemical Characterization of the Mammalian Stress Proteins and Identification of Two Stress Proteins as Glucose and Ca^{+2} Ionophore Regulated Proteins," *J. Biol. Chem.* 258:7102–7111 (1983).

10. Hammond, G.L., Y.K. Lai, and C.L. Market. "Diverse Forms of Stress Lead to New Patterns of Gene Expression through a Common and Essential Metabolic Pathway," *Proc. Natl. Acad. Sci. U.S.A.* 79:3485–3488 (1982).

11. Caltabiano, M.M., T.P. Koestler, G. Poste, and R.G. Greig. "Induction of 32 & 34-kDa Stress Proteins by Sodium Arsenite, Heavy Metals, and Thiol-reactive agents," *J. Biol. Chem.* 261:13381–13386 (1986).

12. Cochrane, B.J., R. Irby, and T.W. Snell. "Stress protein synthesis in response to toxin exposure in two species of rotifers," paper presented at the 13th Aquatic Toxicity and Risk Assessment Symposium, Atlanta, Ga, April 16–18, 1989.

13. Kapoor, M., and J. Lewis. "Alteration of the Protein Synthesis Pattern in *Neurospora crassa* cells by Hyperthermal and Oxidative Stress," *Can. J. Microbiol.* 33:162–168 (1987).

14. Spector, M.P., Z. Aliabadi, T. Gonzalez, and J.W. Foster. "Global Control in *Salmonella typhimurium*: Two-Dimensional Electrophoretic Analysis of Starvation-, Anaerobiosis-, and Heat Shock-Inducible Proteins," *J. Bacteriol.* 168:420–424 (1986).

15. Ramagopal, S. "Salinity Stress Induced Tissue-Specific Proteins in Barley Seedlings," *Plant Physiol.* 84:324–331 (1987).

16. Bournias-Vardiabasis, N., R.L. Teplitz, G.F. Chernoff, and R.L. Seecof. "Detection of Teratogens in the *Drosophila* Embryonic Cell Culture Test: Assay of 100 Chemicals," *Teratology* 28:109–122 (1983).

17. Bournias-Vardiabasis, N., and C.H. Buzin. "Developmental Effects of Chemicals and the Heat Shock Response in *Drosophila* Cells," *Teratogen. Carcinogen. Mutagen.* 6:523–536 (1986).

18. Carr, B.I., T.H. Huang, C.H. Buzin, and K. Itakura. "Induction of Heat Shock Gene Expression without Heat Shock by Hepatocarcinogens and during Hepatic Regeneration in Rat Liver," *Cancer Res.* 46:5106–5111 (1986).

19. Schlesinger, M.J. "Heat Shock Proteins: The Search for Functions," *J. Cell Biol.* 103:321–325 (1986).

20. Schlesinger, M.J. "Function of Heat Shock Proteins," *Atlas Sci. Biochem.* 161–164 (1988).

21. Catelli, M.G., N. Binart, I. Jung-Testas, J.M. Renoir, E.E. Baulieu, J.R. Feramisco, and W.J. Welch. "The Common 90KD Protein Component of Nontransformed '8S' Steroid Receptors is a Heat Shock Protein," *EMBO J.* 4:3131–3137 (1985).

22. Ziemiecki, A., M.G. Catelli, I. Joab, and B. Moncharont. "Association of the Heat Shock Protein HSP 90 with Steroid Hormone Receptors and Tyrosine Kinase Oncogene Products," *Biochem. Biophys. Res. Commun.* 138:1298–1307 (1986).

23. Rose, D.W., W.J. Welch, G. Kramer, and B. Hardesty. "Possible Involvement of the 90kDa Heat Shock Protein in the Regulation of Protein Synthesis," *J. Biol. Chem.* 264:6239–6244 (1989).

24. Brugge, J.S. "Interaction of the Rous Sarcoma Virus Protein pp60 src with the Cellular Proteins pp50 and pp90," in *Curr. Topics Microbiol. Immunol.* 123:1–22 (1986).

25. Chirico, W.J., M.G. Waters, and G. Blobel. "70K Heat Shock Related Proteins Stimulate Protein Translocation into Microsomes," *Nature* 333:805–810 (1988).

26. Deshaies, R.J., B.D. Koch, M. Weiner-Washiburne, E. Craig, and R. Schekman. "A Subfamily of Stress Proteins Facilitates Translocation of Secretory and Mitochondrial Precursor Polypeptides," *Nature* 332:800–805 (1988).

27. Craig, E.A., J. Kramer, and J. Kosic-Smithers. "SSC1, a Member of the 70kDa Heat Shock Protein Multigene Family of *Saccharomyces cerevisiae* is Essential for Growth," *Proc. Natl. Acad. Sci. U.S.A.* 84:4156–4160 (1987).

28. Welch, W.J., and J.R. Feramisco. "Nuclear and Nucleolar Localization of the 72,000 Dalton Heat Shock Protein in Heat Shocked Mammalian Cells," *J. Biol. Chem.* 259:4501–4510 (1984).

29. Pelham, H.R.B. "Coming in from the Cold," *Nature* 332:776–777 (1988).

30. Cheng, M.Y., F-U. Hartl, J. Martin, R.A. Pollack, F. Kalousek, W. Neupert, E.M. Hallberg, R.L. Hallberg, and A.L. Norwich. "Mitochondrial Heat Shock Protein HSP 60 is Essential for Assembly of Proteins Imported into Yeast Mitochondria," *Nature* 337:620–624 (1989).

31. Reading, D.S., R.L. Hallberg, and A.M. Myers. "Characterization of the Yeast HSP60 Gene Coding for a Mitochondrial Assembly Factor," *Nature* 337:655–659 (1989).

32. Arrigo, A.P., and W.J. Welch. "Characterization and Purification of the Small 28,000-Dalton Mammalian Heat Shock Protein," *J. Biol. Chem.* 262:15359–15369 (1987).

33. Sanders, B.M. unpublished results (1989).

34. Ireland, R., and E. Burger. "Synthesis of Low Molecular Weight Heat Shock Proteins Stimulated by Ecdysterone in a Cultured *Drosophila* Cell Line," *Proc. Natl. Acad. Sci. U.S.A.* 79:855–859 (1982).

35. Edwards, D.P., D.J.. Adams, N. Savage, and W.L. McGuire. "Estrogen Induced Synthesis of Specific Proteins in Human Breast Cancer Cells," *Biochem. Biophys. Res. Commun.* 93:804–812 (1980).

36. Arrigo, A.P., J.P. Suhan, and W.J. Welch. "Dynamic Changes in the Structure and Intracellular Locale of the Mammalian Low-Molecular-Weight Heat Shock Protein," *Mol. Cell. Biol.* 8:5059–5071 (1988).

37. Finley, D., E. Ozkaynak and A. Varshavsky. "The Yeast Polyubiquitin Gene is Essential for Resistance to High Temperatures, Starvation, and Other Stresses," *Cell* 48:1035–1046 (1987).

38. Munro, S., and H.R.B. Pelham. "An HSP-70 Like Protein in the ER: Identify with the 78kD Glucose-Regulated Protein and Immunoglobulin Heavy Chain Binding Protein," *Cell* 46:291–300 (1986).

39. Sargan, D.R., M.J. Tsai, and B.W. O'Malley. "HSP 108, a Novel Heat Shock Inducible Protein of Chicken," *Biochem.* 25:625–6259 (1986).

40. Shelton, K.R., J.M. Todd, and P.M. Egle. "The Induction of Stress-Related Proteins by Lead," *J. Biol. Chem.* 261:1935–1940 (1986).

41. Mazzarella, R.A., and M. Green. "ERp 99, an Abundant Conserved Glycoprotein of the Endoplasmic Reticulum is Homologous to the 90kDa Heat Shock Protein (hsp90) and the Glucose Regulated Protein (grp 94)," *J. Biol. Chem.* 262:8875–8883 (1987).

42. Kassenbrock, C.K., P.D. Garcia, P. Waller, and R.B. Kelley. "Heavy-Chain Binding

Protein Recognizes Aberrant Polypeptides Translocated *in vitro*," *Nature* 333:90–93 (1988).

43. Mizzen, L.A., C. Chang, J.G. Garrels, and W.J. Welch. "Identification, Characterization and Purification of Two Mammalian Stress Proteins Present in Mitochondria: One Related to HSP 70, the other to GroEL," submitted to *J. Biol. Chem.* (1989).

44. Subjeck, J.R., and T.-T. Shyy. "Stress Protein Systems of Mammalian Cells," *Cell. Physiol.* 19:C1–C17 (1986).

45. Dean, R.L., and B.G. Atkinson. "The Acquisition of Thermal Tolerance in Larvae of *Calpodes ethlius* (Lepidoptera) and the *in situ* and *in vitro* Synthesis of Heat-Shock Proteins," *Can. J. Biochem. Cell. Biol.* 61:472–479.

46. Landry, J., D. Bernier, P. Chretien, L.M. Nicole, R.M. Tanguay, and N. Marceau. "Synthesis and Degradation of Heat Shock Proteins during Development and Decay of Thermotolerance," *Cancer Res.* 42:2457–2461 (1982).

47. Berger, H.M., and M.P. Woodward. "Small Heat Shock Proteins in *Drosophila* May Confer Thermal Tolerance," *Exp. Cell Res.* 147:437–442 (1983).

48. Stephanou, G., S.N. Alahiotis, C. Christodoulou, and V.J. Marmaras. "Adaptation of *Drosophila* to Temperature: Heat-Shock Proteins and Survival in *Drosophila melanogaster*," *Dev. Genet.* 3:299–308 (1983).

49. Roberts, P.B. "Growth in Cadmium-Containing Medium Induces Resistance to Heat in *E. Coli*," *Int. J. Radiat. Biol.* 45:27–31 (1984).

50. Mirkes, P.E. "Hyperthermia-Induced Heat Shock Response and Thermotolerance in Postimplantation Rat Embryos," *Dev. Biol.* 119:115–122 (1987).

51. Subjeck, J.R., J. Sciandra, and R.J. Johnson. "Heat Shock Proteins and Thermotolerance: Comparison of Induction Kinetics," *Br. J. Radiol.* 55:579–584 (1982).

52. Nickells, R.W., and L.W. Browder. "Region-Specific Heat-Shock Protein Synthesis Correlates with a Biphasic Acquisition of Thermotolerance in *Xenopus laevis* Embryos," *Dev. Biol.* 112:391–395 (1985).

53. Tomasovic, S.P., and T.M. Koval. "Relationship between Cell Survival and Heat-Stress Protein Synthesis in a *Drosophila* Cell Line," *Int. J. Radiat. Biol.* 48:635–650 (1985).

54. Mosser, D.D., J. van Oostrom, and N.C. Bols. "Induction and Decay of Thermotolerance in Rainbow Trout Fibroblasts," *J. Cell. Physiol.* 132:155–160 (1987).

55. Mosser, D.D., and N.C. Bols. "Relationship between Heat-Shock Protein Synthesis and Thermotolerance in Rainbow Trout Fibroblasts," *J. Comp. Physiol. B.* 158:457–467 (1988).

56. Landry, J., and P. Cretien. "Relationship between Hyperthermia Induced Heat Shock Proteins and Thermotolerance in Morris Hepatoma Cells," *Can. J. Biochem.* 61:428–437 (1983).

57. Carper, S.W., J.J. Duffy, and E.W. Gerner. "Heat Shock Protein in Thermotolerance and Other Cellular Processes," *Cancer Res.* 47:5249–5255 (1987).

58. Easton, D.P., P.S. Rutledge, and J.R. Spotila. "Heat Shock Protein Induction and Induced Thermal Tolerance are Independent in Adult Salamanders," *J. Exp. Zool.* 241:263–267 (1987).

59. Yost, H.J., and S. Lindquist. "RNA Splicing is interrupted by Heat Shock and is Rescued by Heat Shock Protein Synthesis," *Cell* 45:185–193 (1986).

60. Mizzen, L.A., and W.J. Welch. "Characterization of the Thermotolerant Cell. I.

Effects on Protein Synthesis Activity and the Regulation of Heat-Shock Protein 70 Expression," *J. Cell Biol.* 106:1105–1116 (1988).

61. Loomis, W.F., and S. Wheeler. "Heat Shock Response of *Dictyostelium*," *Dev. Biol.* 79:399–408 (1980).

62. Heuss-La Rosa, K., R.R. Mayer, and J.H. Cherry. "Synthesis of Only Two Heat Shock Proteins is Required for Thermoadaptation in Cultured Cowpea Cells," *Plant Physiol.* 85:4–7 (1987).

63. Pelham, H.R.B. "Speculations on the Functions of the Major Heat Shock and Glucose-Regulated Proteins," *Cell* 46:959–961 (1986).

64. Ougham, H.J., and J.L. Stoddart. "Synthesis of Heat-Shock Protein and Acquisition of Thermotolerance in High-Temperature Tolerant and High-Temperature Susceptible Lines of *Sorghum*," *Plant Sci.* 44:163–167 (1986).

65. Kee, S.C., and P.S. Noble. "Concomitant Changes in High Temperature Tolerance and Heat-Shock Proteins in Desert Succulents," *Plant Physiol.* 80:596–598 (1986).

66. Boon-Niermeijer, E.K.,, M. Tuyl, and H. van der Scheur. "Evidence for Two States of Thermotolerance," *Int. J. Hyperthermia* 2:93–105 (1986).

67. Caltabiano, M.M., G. Poste, and R.G. Greig. "Induction of the 32-kD Human Stress Protein by Auranofin and Related Triethylphosphine Gold Analogs," *Biochem. Pharmacol.* 37:4089–4093 (1988).

68. Keyse, S.M., and R.M. Tyrrell. "Heme Oxygenase is the Major 32-kDa Stress Protein Induced in Human Skin Fibroblasts by UVA Radiation, Hydrogen Peroxide, and Sodium Arsenite," *Proc. Natl. Acad. Sci. U.S.A.* 86:99–103 (1989).

69. Shibahara, S., R. Muller, and H. Taguchi. "Transcriptional Control of Rat Heme Oxygenase by Heat Shock," *J. Biol. Chem.* 262:12889–12892 (1987).

70. Shuman, J., and A. Przybyla. "Expression of the 31-kD Stress Protein in Rat Myobasts and Hepatoytes," *DNA* 7:475–482 (1988).

72. Caltabiano, M.M., T.P. Koestler, G. Poste, and R.G. Greig. "Induction of MamOxygenase cDNA and Induction of *its* mRNA by Hemin," *Eur. J. Biochem.* 171:457–of Rheumatoid Arthritis," *Biochem. Biophys. Res. Commun.* 138:1074–1080 (1986).

72. Caltabiano, M.M., T.P. Koestler, G. Poste, and R.G. Greig. "Induction of Mammalian Stress Proteins by a Triethylphosphine Gold Compound Used in the Therapy of Rheumatoid Arthritis," *Biochem. Biophys. Res. Commun.* 138:1074–1080 (1986).

73. Hamer, D.H. "Metallothionein," *Ann. Rev. Biochem.* 55:913–951 (1986).

74. Heikkila, J.J., G.A. Schultz, K. Iatrou, and L. Gedamu. "Expression of a Set of Fish Genes Following Heat or Metal Ion Exposure," *J. Biol. Chem.* 257:12000–12005 (1982).

75. Kagi, J.H.R., and M. Nordberg, Eds. *Metallothionein: Proceedings of the First International Meeting on Metallothionein and other Low Molecular Weight Metal-Binding Proteins.* Zurich, July 17–22, 1978 (Basel: Birkhauser, 1979), p. 378.

76. Jenkins, K.D., D.A. Brown, P.S. Oshida, and E.M. Perkins. "Cytosolic Metal Distribution as an Indicator of Toxicity in Sea Urchins from the Southern California Bight," *Mar. Poll. Bull.* 13:413–421 (1982).

77. Roch, M., J.A. McCarter, A.T. Matheson, M.J.R. Clark, and R.W. Olafson. "Hepatic Metallothionein in Rainbow Trout (*Salmo gairdneri*) as an Indicator of Metal

Pollution in the Campbell River System," *Can. J. Fish. Aquat. Sci.* 39:1596–1601 (1982).

79. Viarengo, A., S. Palmero, G. Zanicchi, R. Capelli, R. Vaissiere, and M. Orunesu. to Heavy Metals by Rainbow Trout (*Salmo gairdneri*). II. Held in a Series of Contaminated Lakes," *Comp. Biochem. Physiol.* 77C:77–82 (1984).

79. Viarengo, A., S. Palmero, G. Zanicchi, R. Capelli, R. Vaissiere, and M. Orunesu. "Role of Metallothioneins in Cu and Cd accumulation and Elimination in the Gill and Digestive Gland Cells of *Mytilus Galloprovincialis* Lam," *Mar. Environ. Res.* 23–26 (1985).

80. Viarengo, A., M.N. Moore, G. Mancinelli, A. Mazzucotelli, R.K. Pipe, and S.V. Farrar. "Metallothioneins and Lysosomes in Metal Toxicity and Accumulation in Marine Mussels: The Effect of Cadmium in the Presence and Absence of Phenanthrene," *Mar. Biol.* 94:251–257 (1987).

81. George, S.G., and B.J.S. Pirie. "The Occurrence of Cadmium in Sub-cellular Particles in the Kidney of the Marine Mussel, *Mytilus edulis*, Exposed to Cadmium: The Use of Electron Microprobe Analysis," *Biochim. Biophys. Acta* 580:234–244 (1979).

82. Karin, M. "Metallothioneins: Proteins in Search of Function," *Cell* 41:9–10 (1985).

83. Sanders, B.M., K.D. Jenkins, W.G. Sunda, and J.D. Costlow. "Free Cupric Ion Activity in Seawater: Effects on Metallothionein and Growth in Crab Larvae," *Science* 222:53–55 (1983).

84. Sanders, B.M., and K.D. Jenkins. "Relationships between Free Cupric Ion Concentrations in Sea Water and Copper Metabolism and Growth in Crab Larvae," *Biol. Bull.* 167:704–712 (1984).

85. Jenkins, K.D., and B.M. Sanders. "Relationships between Free Cadmium Ion Activity in Seawater, Cadmium Accumulation and Subcellular Distribution, and Growth in Polychaetes," *Environ. Health Pers.* 65:205–210 (1986).

86. Jenkins, K.D., D.M. Brown, G.P. Hershelman, and W.C. Meyer. "Contaminants in White Croakers *Genyonemus lineatus* (Ayres, 1855) from the Southern California Bight. I. Trace Metal Detoxification/Toxication," in *Physiological Mechanisms of Marine Pollution Toxicity*, Vernberg, W.B., A. Calabrese, F.P. Thurberg, and F.J. Vernberg, Eds. (New York: Academic Press, 1982) p. 177–196.

87. Krause, K.W., E.M. Hallberg, and R.L. Hallberg. "Characterization of a *Tetrahymena thermophila* Mutant Strain Unable to Develop Normal Thermotolerance," *Molec. Cell. Biol.* 6:3854–3861 (1986).

88. Cleary, J.J., and A.R.D. Stebbing. "Organotin and Total Tin in Coastal Waters of Southwest England," *Mar. Poll. Bull.* 16:350–355 (1985).

89. Bryan, G.W., P.E. Gibbs, L.G. Hummerstone, and G.R. Burt. "The Decline of the Gastropod *Nucella labillus* Around South-west England: Evidence for the Effect of Tributyltin from Antifouling Paints," *J. Mar. Biol. Ass. U.K.* 66:611–640 (1986).

90. Beaumont, A.R., and M.D. Budd. "High Mortality of the Larvae of the Common Mussel at Low Concentrations of Tributyltin," *Mar. Poll. Bull.* 15:402–405 (1984).

91. Miller, S.G. "Association of a Sperm-Specific Protein with the Mitochondrial F1F0-Atpase in *Heliothis*," *Insect Biochem.* 17:417–432 (1987).

92. Pipkin, J.L., J.F. Anson, W.G. Hinson, E.R. Burns, and D.A. Casciano. "Mi-

croscale Electrophoresis of Stress Proteins Induced by Chemicals during the *In Vivo* Cell Cycle," *Electrophoresis* 7:463–471 (1986).

93. Neidhardt, F.C., V. Vaughn, T.A. Phillips, and P.L. Block. "Gene-Protein Index of *Escherichia coli* K-12," *Microbiol. Rev.* 47:231–284 (1983).

94. Goodman, R., and A.S. Henderson. "Exposure of Salivary Gland Cells to Low-Frequency Electromagnetic Fields Alters Polypeptide Synthesis," *Proc. Natl. Acad. Sci. U.S.A.* 85:3928–3932 (1988).

CHAPTER 10

Macrophage Responses of Estuarine Fish as Bioindicators of Toxic Contamination*

B. A. Weeks, R. J. Huggett, J. E. Warinner and E. S. Mathews

Fish exposed to toxic pollutants both in the wild and in the laboratory show significant changes in the immune activity of kidney macrophages. Decreases in the chemotactic and phagocytic responses and increases in neutral red uptake and melanin accumulation were observed in several species of fish captured in the Elizabeth River, Virginia, which is highly contaminated with polynuclear aromatic hydrocarbons (PAHs). The chemiluminescent response of macrophages was inhibited by laboratory exposure of fish to toxic sediments and by in vitro exposure of macrophages to tributyltin (TBT). Results of these assay techniques are presented with the aim of providing reliable bioindicators of the effects of environmental contamination on fish health.

INTRODUCTION

Portions of the Chesapeake Bay, the second largest estuary in the world, are known to be among the most contaminated waters found to date. Studies conducted at the Virginia Institute of Marine Science have shown that one of the bay tributaries, the Elizabeth River, is highly contaminated by polynuclear aromatic hydrocarbons (PAHs).[1] Chemical analyses utilizing gel permeation chromatography, high pressure liquid chromatography, glass capillary-gas chromatography and glass capillary-gas chromatography-mass spectrometry show a gradient

---

*VIMS Contribution Number 1518.

of concentrations of PAHs in sediments increasing in an upstream direction. Over 300 aromatic compounds were identified in sediment samples, with concentrations of organic compounds as high as 440 μg/g dry weight at some sampling sites. About half of the resolved concentration sum was accounted for by 14 PAHs which are characteristic combustion products.[1,2]

In addition, tributyltin (TBT), an antifouling ingredient in marine paints, is known to leach from paint films and accumulate to hazardous levels in the waters of harbors and marinas. In the Chesapeake Bay, the presence of increasing numbers of recreational, commercial, and military vessels has resulted in measurable levels of TBT as high as 0.1 μg/L in the vicinity of marinas and in open waters.[3] Extremely low levels of TBT (<100 ng/L) have been shown to be toxic to some marine and estuarine species.[4]

The occurrence of these and other toxicants in fish and shellfish habitats can have acute effects, i.e., the concentration of toxicants may be high enough to cause injury or death in a short period of time. In addition, there is mounting evidence that chronic exposure to sublethal concentrations of toxicants predisposes finfish to diseases caused by bacterial, fungal, and viral pathogens.[5,6] A high proportion of the fish collected from the Elizabeth River exhibits cataracts, fin rot, and integumental lesions.[7] Fish exposed to Elizabeth River sediments in the laboratory develop signs ranging from fin loss, severe skin lesions, cataracts, gill hyperplasia, and liver and pancreatic necrosis to death.[8]

Immune Functions of Fish

Fish, like other vertebrates, possess immune systems that function to maintain their health in the presence of disease-causing agents. It is apparent from the literature that fish have evolved immune mechanisms of a high degree of complexity which, for the most part, are not fundamentally different from those present in higher vertebrates.[9] A highly developed cell-mediated immune system has been demonstrated in teleosts, with aspects of homology to the mammalian major histocompatibility complex (MHC).[10] Nonspecific immune mechanisms as well as a humoral antibody system occur in all classes of fish.[11]

Macrophages are an important part of the cellular immune system of fish and function as the first line of defense by ingesting foreign material including disease-causing agents. The specific uptake commonly referred to as endocytosis includes the processes of chemotaxis, the response to a chemical stimulus; phagocytosis, the ingestion of particulate material; and pinocytosis, the uptake of liquid droplets. Other aspects of immune activity include accumulation of melanin granules and chemiluminescence (CL). Aggregates of macrophages containing melanin have commonly been found in the spleen, liver, and kidney of higher fish. Melanin and related pigments are considered to play a defensive role in many organisms in their capacity for hydrogen peroxide production as a bactericidal system. Therefore increases in the numbers of these melano-macrophage aggregates might serve as indicators of a change in the health status of fish.[12] The chemiluminescent response is a by-product of phagocytosis. During phagocytosis, fish macrophages show a burst of oxygen consumption associated with the production of microbicidal reactive oxygen species that emit detectable light upon relaxation.

Biological Monitoring

The immune system, in its capacity to destroy foreign material and protect the host against disease, can serve as a useful sentinel of the health status of environmentally stressed organisms, as well as provide regulatory agencies with additional means of assessing the extent of pollution. Biological measures of pollution effects such as assays of immune function are potentially superior to chemical analyses as indicators of exposure hazards because they provide direct evidence of the link between environmental challenge and health status. In addition, biomarkers offer a more sensitive and reliable assessment of exposure risks than ambient monitoring, and can provide regulatory agencies with indisputable evidence of impaired health status in exposed animals.

In its infancy, the concept of immunological biomonitoring was limited because of the scarcity of well-established methodologies for obtaining specific data on the physiological functioning of feral animals. Workers were restricted to making broad generalizations concerning the health status of organisms. However, in recent years, because of the remarkable growth of the field of comparative immunology, many new methods of determining the functional capacity of specific components of the immune system have been developed. With an increased understanding of normal immune function in a wide variety of organisms, it has become possible to assume a more mechanistic approach to biomonitoring, i.e., to relate a particular toxicant to impairment of a specific aspect of the immune response. Hence, we are approaching the point at which we can detect very subtle but potentially devastating alterations in immunological functioning that might never have been discovered using more general markers of stress.

Bioindicators have been developed to evaluate many aspects of immune function and status. Among them are responses such as lymphocyte mitogenesis,[13,14] antibody-producing cell formation,[15] antibody production[16] and nonspecific macrophage activity.[17-21] These and other elements of the immune system have been shown to be affected (depressed or stimulated) by exposure to toxicants. Therefore measurement of immune function may be used to provide information about recent exposure to environmental stress.

As efforts to monitor and control the contamination of valuable finfish habitats continue, there is an increasing need for information on the organismic effects of aquatic toxicants. The assays of immune function reported here were adapted and tested for their potential use as biomarkers of exposure of fish to environmental contamination. These results have been previously published.[17-21]

Table 1.   Effect of Environment on Macrophage Activity.

Assay	Spot		Hogchoker	
	Control River	Elizabeth River	Control River	Elizabeth River
% Chemotaxis at 90 min	55 ± 5[a]	33 ± 2*	85 ± 5	56 ± 5*
% Phagocytosis at 120 min	74 ± 5	19 ± 3*	88 ± 4	32 ± 5*
NR uptake at 120 min (μg NR/10^6 cells)	1.58 ± 0.12	1.48 ± 0.27	0.56 ± 0.04*	1.25 ± 0.20

[a]Each value represents the mean + SEM of 4 experiments. Results were evaluated by the Student t test. Significant differences are indicated by *.

METHODS AND RESULTS

Specimens

Spot (*Leiostomus xanthurus*), hogchoker (*Trinectes maculatus*), flounder (*Paralichthys dentatus*), oyster toadfish (*Opsanus tau*), and croaker (*Micropogonias undulatus*) were captured from the York River, a relatively nonpolluted river, and from the contaminated Elizabeth River. Individual kidneys from toadfish or kidneys pooled from six to ten specimens of spot, hogchoker, flounder, and croaker were dissected and macerated. Kidney cell types were separated by discontinuous density gradient centrifugation in isoosmotic Percoll®* to yield single-cell suspensions of macrophages for use in in vitro studies.[17] Macrophages were identified by light and electron microscopy and by their ability to phagocytize *Escherichia coli*.

Chemotaxis Assay

Chemotactic activity of spot and hogchoker macrophages was measured by a modification of the Boyden double-chamber filter technique.[17] Macrophages were allowed to migrate from the upper chamber through a millipore filter to the lower chamber, which contained heat-killed *E. coli* as the chemotactic stimulus. Chemotactic activity was measured microscopically by counting macrophages on the upper and lower surfaces of the filter. Chemotactic migration reached a maximum within 90 min and was markedly reduced in Elizabeth River fish (Table 1).

Phagocytosis Assay

Phagocytic activity of spot and hogchoker macrophages was determined using a light microscopic assay of bacterial ingestion.[18] Macrophage suspensions were

*Registered trademark of Pharmacia Fine Chemicals, Uppsala, Sweden.

incubated with heat-killed *E. coli* at 20°C for 120 min. Phagocytosis of bacteria was measured by the microscopic enumeration of the proportion of phagocytes that contained intracellular bacteria and expressed as percent phagocytosis.

Phagocytic activity increased with time, reaching maximal values within 120 min. The phagocytic efficiency of macrophages isolated from the kidneys of spot and hogchoker was significantly lower in Elizabeth River fish as compared to York River controls (Table 1).

Assay of Neutral Red Uptake

The uptake of neutral red (NR) dye by kidney macrophages from spot and hogchoker was determined by means of a spectrophotometric assay.[19] Macrophages were incubated with neutral red dye for 120 min at 17°C. At intervals samples were removed and lysed, and the concentration of neutral red in the lysate was determined. Characteristic kinetics of uptake were observed for each species. It was found that the uptake activity of macrophages from spot exposed to Elizabeth River pollutants did not differ from control values. However, there was a twofold increase in uptake in hogchoker obtained from the Elizabeth River (Table 1).

Melanin Accumulation

Preliminary transmission electron microscopic studies in this laboratory have shown that flounder macrophages accumulate melanin granules by a process that appears to be similar to phagocytosis. Macrophages from flounder taken from the polluted Elizabeth River contain significantly larger numbers of melanin granules than do macrophages in fish from nonpolluted waters (Figure 1).

Chemiluminescence Assay

Chemiluminescence emitted during phagocytosis was measured by incubating macrophages in mini-vials with zymosan (yeast cell wall preparation) and luminol.[20] Vials were counted in a liquid scintillation counter in the out-of-coincidence mode. Macrophages from spot captured in the Elizabeth River were markedly deficient in generating a zymosan-induced CL response, as were macrophages from York River spot exposed to Elizabeth River sediments in the laboratory (Table 2).

In vitro exposure of hogchoker, toadfish, and croaker macrophages to TBT resulted in significant decreases in the CL response at TBT levels approximating those found in fish muscle tissues in market surveys[21] (Table 3). Cell viability was not affected by exposure to TBT at these levels (data not shown).

CONCLUSION

Several immunological methods have been described that are rapid and reproducible and have the necessary qualities to monitor the health of fish. These

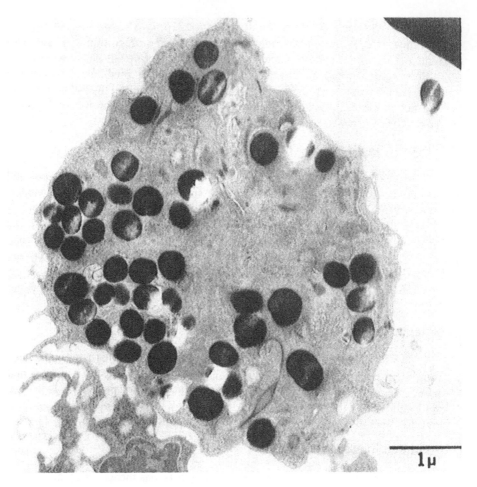

Figure 1.   Transmission electron micrographs of representative flounder macrophages from (a) the Elizabeth River and (b) a nonpolluted river.

techniques have been used extensively to test the immune responses of fish captured from contaminated sections of the Elizabeth River, and have consistently shown that these fish have altered (depressed or elevated) macrophage function. Further it has been shown that there are species differences in response to toxic sediments in the Elizabeth River. The phagocytic activity of macrophages from Elizabeth River spot and hogchoker returned to normal after the fish were held in clean water for 3 weeks, indicating that the decreased activity was related to exposure to Elizabeth River pollutants but may be reversible. Since sediments from the Elizabeth River contain high levels of polynuclear aromatic hydrocarbons (PAHs), it is possible that these chemicals are responsible for the observed immunomodulation. Further research will be required to determine the causative

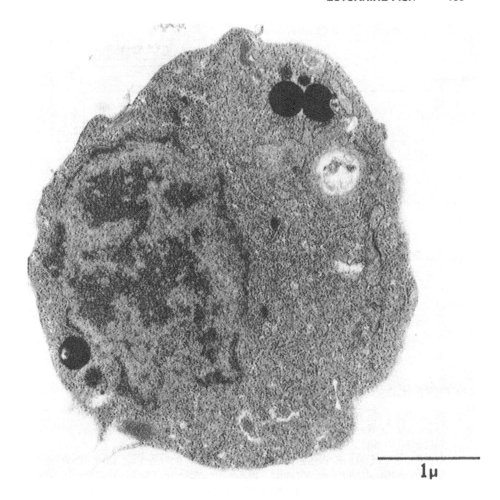

Figure 1. Continued

Table 2.   Chemiluminescent Response of Spot Exposed to Polluted Sediments.

	Peak Amplitude (cpm)	
Sediment Exposure	Control River	Elizabeth River
Spot exposed to sediments in the wild	350,000[a]	15,000
Spot exposed to sediments in the laboratory	540,000	220,000

[a]Values are given for a representative experiment.

Table 3.   Effect of TBT on CL Response.

Species	n[b]	Peak Amplitude of CL Response[a]		
		0 µg/L[c]	40 µg/L	400 µg/L
Toadfish	10	9.8 ± 3.2[d]	10.2 ± 2.4	1.1 ± 0.5*
Hogchoker	8	13.0 ± 3.0	8.3 ± 2.6	1.4 ± 0.7*
Croaker	7	16.7 ± 4.3	1.8 ± 1.1	0.4 ± 0.1*

[a]Values shown are cpm × 10^{-3} per thousand phagocytes.
[b]n = number of experiments.
[c]TBT concentration.
[d]Values shown are means ± SEM.
Results were evaluated by ANOVA and Dunnett's test. Significant differences are indicated by *.

agent(s) of impaired immune function and the mechanism of the toxic effect. In addition, the effects of age, nutritional status, and other water-borne contaminants remain to be determined. It is proposed that these assays can serve as useful biomarkers of the immune status of fish, and that careful choice of method and interpretation of results can give important information for evaluating the health of fish exposed to environmental contamination.

REFERENCES

1. Bieri, R. H., C. Hein, R. J. Huggett, P. Shou, H. Slone, C. Smith, and C. W. Su. "Polycyclic Aromatic Hydrocarbons in Surface Sediments from the Elizabeth River Subestuary," *Int. J. Environ. Anal. Chem.* 26:97–113 (1986).
2. Huggett, R. J., M. E. Bender, and M. A. Unger. "Polynuclear Aromatic Hydrocarbons in Elizabeth River, Virginia," In *Fate and Effects of Sediment-Bound Chemicals in Aquatic Systems*, K. L. Dickson, A. W. Maki, and W. A. Brungs, Eds. (New York: Pergamon Press, 1987), chap. 21.
3. Huggett, R. J., M. A. Unger, and D. J. Westbrook, "Organotin Concentrations in the Southern Chesapeake Bay," in *Oceans 86 Proceedings, Organotin Symposium* (New York: IEEE Publishing Services, 1986), p. 1262.
4. Bryan, G. W., P. E. Gibbs, L. C. Hummerstone, and G. R. Burt. "The Decline of the Gastropod *Nucella lapidus* around South-West England: Evidence for the Effect of Tributyltin from Antifouling Paints," *J. Mar. Biol. Ass. UK* 66:611–40 (1986).
5. Sindermann, J. "An Examination of Some Relationships between Pollution and Disease," *Rapp. P.-V. Reun. Cons. Int. Explor. Mer.* 182:37–43 (1983).
6. Snieszko, S. F. "The Effects of Environmental Stress on Outbreaks of Infectious Diseases of Fish," *J. Fish Biol.* 6:197–208 (1974).
7. Hargis, W. J., Jr., and D. E. Zwerner. "Effects of Certain Contaminants on Eyes of Several Estuarine Fishes," *Mar. Environ. Res.* 24:265–70 (1988).
8. Hargis, W. J., Jr., M. H. Roberts, and D. E. Zwerner. "Effects of Contaminated Sediments and Sediment-Exposed Effluent Water on Estuarine Fish: Acute toxicity," *Mar. Environ. Res.* 14:337–54 (1984).

9. Cushing, J. E. "Immunology of Fish," in *Fish Physiology*, W. S. Hoar, and D. J. Randall, Eds. (New York: Academic Press, 1970), chap. 12.

10. Jurd, R. D. "Specialisation in the Teleost and Anuran Immune Response: A Comparative Critique," in *Fish Immunology*, M. J. Manning and M. F. Tatner, Eds. (London: Academic Press, 1985), p. 9.

11. Corbel, M. J. "The Immune Response in Fish: A Review," *J. Fish Biol.* 7:539–563 (1975).

12. Wolke, R. E., C. J. George, and V. S. Blazer. "Pigmented Macrophage Accumulations (MMC; PMB): Possible Monitor of Fish Health," in *Parasitology and Pathology of Marine Organisms in the World Ocean*, W. J. Hargis, Jr., Ed., NOAA Technical Report NMFS25 (1984), p. 93.

13. Laudenslager, M. L., S. M. Ryan, R. C. Drugan, R. L. Hyson, and S. F. Maier. "Coping and Immunosuppression: Inescapable but not Escapable Shock Suppresses Lymphocyte Proliferation," *Science* 221:568–569 (1983).

14. Spitsbergen, J. M., K. A. Schat, J. M. Kleeman, and R. E. Peterson. "Interactions of 2,3,7,8-tetrachlorodibenzo-p-dioxin (TCDD) with Immune Response of Rainbow Trout," *Vet. Immunol. Immunopathol.* 12:263–280 (1986).

15. Anderson, D. P., B. Merchant, O. W. Dixon, C. F. Schott, and E. F. Lizzio. "Flush Exposure and Injection Immunization of Rainbow Trout to Selected DNP Conjugates," *Dev. Comp. Immunol.* 7:261–268 (1983).

16. O'Neill, J. G. "Effects of Intraperitoneal Lead and Cadmium on the Humoral Immune Response of *Salmo trutta*," *Bull. Environ. Contam. Toxicol.* 27:42–48 (1981).

17. Weeks, B. A., J. E. Warinner, P. L. Mason, and D. S. McGinnis. "Influence of Toxic Chemicals on the Chemotactic Response of Fish Macrophages," *J. Fish Biol.* 28:653–8 (1986).

18. Weeks, B. A., and J. E. Warinner. "Effects of Toxic Chemicals on Macrophage Phagocytosis in Two Estuarine Fishes," *Mar. Environ. Res.* 14:327–35 (1984).

19. Weeks, B. A., A. S. Keisler, J. E. Warinner, and E. S. Mathews. "Preliminary Evaluation of Macrophage Pinocytosis as a Technique to Monitor Fish Health," *Mar. Environ. Res.* 22:205–13 (1987).

20. Warinner, J. E., E. S. Mathews, and B. A. Weeks. "Preliminary Investigations of the Chemiluminescent Response in Normal and Polluant- Exposed Fish," *Mar. Environ. Res.* 24:281–4 (1988).

21. Wishkovsky, A., E. S. Mathews, and B. A. Weeks. "Effects of Tributyltin on the Chemiluminescent Response of Phagocytes from Three Species of Estuarine Fish," *Arch. Environ. Contam. Toxicol.* 1989 (in press).

SECTION FOUR

Genotoxic Responses

CHAPTER 11

Biological Monitoring: Testing for Genotoxicity

Lee R. Shugart

BACKGROUND

Effective environmental management requires knowledge of the fate of contaminants in natural systems. Often these contaminants are anthropogenic in origin and are present in low concentrations. An important consideration is whether these substances will have an adverse biological impact on the biota present. Unfortunately, under these circumstances a long latent period frequently exists between initial exposure and the subsequent expression of a pathological state. An approach that can help document the exposure of a living organism to contaminants and possibly indicate the potential for deleterious impact is biological monitoring. This technique makes use of biological endpoints in living organisms that are indicators (biomarkers) of environmental insults. The organism functions as an integrator of exposure, accounting for abiotic and physiological factors that modulate the dose of toxicant taken up from the environment. The subsequent magnitude of the response can then be used to estimate the severity of exposure, hopefully in time to take preventive or remedial measures. Monitoring organismal responses as a means of evaluating exposure and effect of contaminants is a recognized concept for use in environmental health research.[1]

The presence of a contaminant in the environment of a living organism does not by itself constitute a hazard. However, once exposure has occurred and the substance has become bioavailable, a sequence of biological responses may progress. Whether or not the well-being of the organism is eventually affected will depend upon many factors, some of which are intrinsic such as the age, sex, health, and nutritional status of the organism, and some of which are extrinsic such as the dose, duration, and route of exposure to the contaminant and the presence of other chemicals. All these factors represent barriers to the study of exposure and to the assessment of subsequent risk from that exposure. However, biological monitoring can help circumvent these problems to a large extent by focusing on relevant molecular events that occur subsequent to exposure and metabolism.

Reliable and sensitive analytical methods are currently being used to clarify the relationship between exposure to xenobiotic compounds and effect. This is particularly evident in the field of environmental genotoxicity.

BIOLOGICAL MONITORING FOR GENOTOXICITY

Introduction

DNA damage has been proposed as a useful parameter for assessing the genotoxic properties of environmental pollutants.[1,2] Many of these pollutants are chemical carcinogens and mutagens with the capacity to cause various types of DNA damage, and if the damage goes uncorrected, it is not unreasonable to expect subsequent adverse effects to the organism. The exposure of an organism to toxic chemicals may result in the induction of a cascade of events starting with an initial insult to the DNA and culminating in the appearance of an overt pathological disease.

By using this rationale, genotoxicity testing of environmental species has been attempted, and a few investigations are listed in Table 1. It should be noted that this list is intended only to be instructional and does not represent a comprehensive review of the scientific literature.

The approaches taken in Table 1 (detecting primary DNA damage and chromosomal damage) are based on our understanding of the mode of action of genotoxic compounds. Primary DNA damage, in the form of adducts, occurs because cellular detoxification mechanisms are not always complete, and highly reactive electrophilic metabolites are formed that can attack nucleophilic macromolecules. The interaction of these moieties with DNA may result in adduct formation where the compound becomes covalently bound.[3] The ^{32}P-postlabeling analysis technique[4] has allowed for very sensitive detection of certain classes of chemicals in the DNA of living organisms. Dunn et al.[5] have employed this technique to detect putative, carcinogenic DNA adducts in wild fish from polluted waterways and to provide information on their past exposure history. Martineau et al.,[6] using a fluorescence/HPLC analysis technique,[7] demonstrated the feasibility of detecting a specific carcinogen (benzo[a]pyrene) adducted to the DNA of beluga whales.

Ensuing changes to the DNA that occur subsequent to adduct formation can result in chromosomal aberrations. McBee and Bickham[8] used flow cytometry to detect increased dispersion of cellular DNA content in spleen cells of wild

Table 1. Genotoxicity Testing of Environmental Species.

Test	Environment and Species	Suspected Contaminant	Ref.
DNA content spleen cells flowcytometry	Terrestrial: white-footed mouse	Complex mixture organics and metals at waste site	8
^{32}P-postlabeling analysis of DNA	Aquatic: catfish	Sediment-bound PAHs	5
Micronuclei frequencies	Aquatic: white croker and kelp bass	DDT, PCBs and PAHs	10
Direct DNA adduct determination	Aquatic: beluga whale	Complex mixture organics and PAHs	6

rodents. It was postulated that this type of chromosomal damage was due to clastogenic agents present in the environment of the animals. Micronuclei formation is another form of chromosomal aberration.[9] Hose et al.[10] found elevated frequencies of circulating erythrocyte micronuclei in two marine fish species from contaminated areas relative to fishes from less contaminated sites.

Other endpoints of DNA damage on the continuum from exposure to effect are being studied and range from oncogene-activation[11] to the appearance of malignant cells. It is obvious that the limitation of present biological knowledge and analytical technology dictate the application of biomarkers in the area of environmental genotoxicity.

DNA Integrity as a Biomarker

From the preceding discussion it is evident that the integrity of DNA can be affected by genotoxic agents. The modulation of DNA integrity occurs subsequent to exposure and is amenable to detection. Some of these changes in DNA integrity are depicted in Figure 1.

DNA is present in the nucleus of a cell as a functionally stable, double-stranded entity without discontinuity (strand breaks). As such it is considered to have high integrity.[12] DNA damage can occur as the result of wear and tear of normal metabolic events (pathway 1); physical agents such as ultraviolet light and ionizing radiation (pathway 2); and chemical agents (pathway 3). These various processes can give rise to strand breaks in the DNA and the resulting, transient population of DNA would have low integrity. Ionizing radiation and some chemicals work through free-radical mechanisms and cause strand breaks directly, while ultraviolet light and numerous mutagens and carcinogens produce modifications (i.e., dimerization of pyrimidines and adducts). Abasic sites (sugars without purine or pyrimidine bases) are frequent lesions that occur as a result of random thermal collisions (normal wear and tear) or chemically unstable adducts. Even the normal cellular process of replication (pathway 5) produces

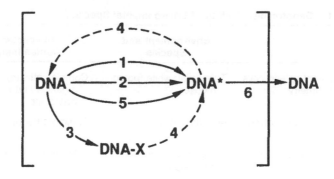

INSULT	REPAIR	SYNTHESIS
1. Normal wear and tear	4. Incision	5. Replication
2. UV and ionizing radiation (γ and X irradiation)	excision resynthesis	6. Postreplication modification
3. Chemical	ligation	

DNA: Normal double-stranded DNA with no strand breaks.
DNA-X: Chemically modified DNA.
DNA*:   DNA with strand breaks.

Figure 1. DNA integrity: a schematic representation of the status of DNA in relation to insults that disrupt DNA integrity and cellular processes that maintain DNA integrity.

DNA with strand breaks.[12] Nevertheless, all DNA with low integrity is subject to repair by the cell (pathway 4). The integrity of DNA is important for survival, and the rigid maintenance of this integrity is reflected in the low mutational rate observed in living organisms, which has been estimated to be in the order of one mutation per average gene per 200,000 years.[12,13]

Methods for Analysis of DNA Integrity

It is the inability (permanent or transient) of an organism to cope with DNA damage and to maintain DNA integrity that provides the investigator the opportunity to test for genotoxicity. This chapter will describe analytical techniques that detect and quantify DNA damage that we have been using in the Environmental Sciences Division at the Oak Ridge National Laboratory to evaluate the relation between exposure, metabolism, and subsequent biological effects of genotoxic chemicals in the environment. In relation to Figure 1, these techniques include methods that (1) quantify the formation of stable adducts between genotoxic chemicals and DNA (pathway 3); (2) detect excess strand breakage within the DNA polymer (DNA*); and (3) measure the aberrant nucleotide composition

Table 2.   Effect of Temperature, Dose, and Prior Exposure to Chemicals on DNA
Adduct Formation in Liver of Bluegill Sunfish Exposed to Benzo[a]pyrene.

Temperature (°C)	Dose[a] (μg/g)	DNA Adducts[b] (ng/g)
10[c]	20	360 ± 90
	20[d]	740 ± 175
13	5	340 ± 25
	10	773 ± 190
20	5	928 ± 119

[a]Six fish at each dose level were exposed by i.p. injection and maintained at temperature specified. The experimental protocol has been previously detailed.[19]
[b]Adduct formation expressed as ng of tetrol I-1 per gram of DNA ± s.e.m.[7]
[c]Adduct formation determined 144-hr postexposure for fish exposed at 10°C and 72 hr postexposure for all other temperatures.
[d]Fish preexposed to 20 μg of 3-methylcholanthrene 6 days prior to benzo[a]pyrene exposure.[46]

of DNA (pathway 6). The development of these methods is detailed below. An important consideration has been the validation of the techniques under controlled laboratory conditions.

DNA Adducts

Because genotoxic agents may exert their activity through irreversible reactions with DNA to form adducts, the detection of such reaction products should provide a reliable basis upon which to determine exposure. This approach is being actively pursued in the monitoring of human exposure to environmental pollutants.[14,15]

Polycyclic aromatic hydrocarbons (PAH) are a class of ubiquitous environmental chemicals known to have tumorigenic and carcinogenic properties. This genotoxicity correlates with reactive cellular metabolites of these compounds that covalently bind to DNA to form adducts.[16]

Shugart et al.,[7,17] have developed a fluorescence/HPLC method for the detection of benzo[a]pyrene, a representative PAH, subsequent to its metabolism and adduction to DNA. Essentially the technique consists of the acid-induced removal of the adduct from DNA as the strongly fluorescent, free tetrols, which are then separated and quantified. The technique allows fmol quantities of benzo[a]pyrene adducts to be detected.

Initial laboratory experiments indicated that we can detect, identify, and quantify, subsequent to several routes of exposure, the binding of benzo[a]pyrene with cellular DNA of mice[7,18] and fish.[19]

The level of adduct formation in fish exposed to benzo[a]pyrene in the laboratory can be modulated experimentally by various factors. Examples are given in Table 2. The data indicate that the temperature at which the fish are maintained during exposure will affect the level of adduct formation. For example, fish maintained at 20°C and receiving 5 μg of benzo[a]pyrene per gram of wet weight have greater than 2.5 times as many adducts as those receiving the same dose but maintained at 13°C. There is also a dose-dependent response. Fish injected at 13°C with 10 μg benzo[a]pyrene per gram of wet weight show twice the level of adducts as those exposed to one-half that dose. Finally, prior exposure to 3-methylcholanthrene, a PAH that induces detoxifying enzymes in fish,[20] will increase the level of adduct formation. It is obvious that information gained from exposures conducted under rigid laboratory conditions will help in the correct interpretation of data from field sampling.

As pointed out in Table 1 the fluorescence/HPLC technique has been used to determine whether beluga whales had been exposed to benzo[a]pyrene by determining adduct levels in their DNA.[6] In addition, the technique has been applied to the detection of benzo[a]pyrene adducts in hemoglobin of terrestrial animals exposed environmentally.[21] Biomonitoring of hemoglobin-bound metabolites represents a novel approach to the estimation of exposure to potentially harmful chemicals. Although these adducts have no putative mechanistic role in carcinogenesis, they do relate quantitatively to exposure and to activation since they may approximate the systemic dose (in vivo dose) of a chemical. They can be a measure of the chemical's genotoxic potency as well. Furthermore, since the adducts which form with hemoglobin are stable and have life-spans equal to that of the circulating erythrocyte, quantitation of these adducts can be used to integrate the dose obtained from chronic low level types of exposure which most likely occur in chemically polluted environments.[18,21]

DNA Strand Breaks

Alkaline unwinding is a sensitive analytical technique that has previously been used in cells in culture to detect and quantify DNA strand breaks induced by physical and chemical carcinogens.[22-25] To assess the level of strand breaks of DNA in environmental species, existing methods were modified to allow for the isolation of intact, highly polymerized DNA and the subsequent estimation of strand breaks.[26,27]

DNA isolation is accomplished by homogenizing the intact liver of the sunfish in 1 N NH$_4$OH/0.2% Triton® X-100. The DNA is further purified by differential extraction with chloroform/isoamyl alcohol/phenol (24/1/25-v/v) and passage through a molecular sieve column (Sephadex G50). DNA strand breaks are measured in the isolated DNA by an alkaline unwinding assay.[24,26,27] The technique is based on the time-dependent partial alkaline unwinding of DNA followed by determination of the duplex:total DNA ratio (F value). Since DNA unwinding takes place at single-strand breaks within the molecule, the amount of double-stranded DNA remaining after a given period of alkaline unwinding will be inversely proportional to the number of strand breaks present at the initiation of the alkaline exposure, provided renaturation is prevented. The amounts of these two types of DNA are quantified by measuring the fluorescence that results with bisbenzimidazole (Hoechst dye #33258), a specific DNA-binding dye that fluoresces with double-stranded DNA at about twice the intensity as it does with single-stranded DNA.[24,26,27]

Rydberg[28] has established the theoretical background for estimating strand breaks in DNA by alkaline unwinding, which is summarized by the equation

$$\ln F = -(K/M)(t^b)$$

where K is a constant, t is time, M is the number average molecular weight between two breaks, and b is a constant less than 1 that is influenced by the conditions for alkaline unwinding.

The relative number of strand breaks (N value) in DNA of sunfish from sampled sites can be compared to those from reference sites as follows:[24,27]

$$N = (\ln F_s/\ln F_r) - 1$$

where F_s and F_r are the mean F values of DNA from the sampled sites and referenced site respectively. N values greater than zero indicate that DNA from the sampled sites has more strand breaks than DNA from the reference site; an N value of 5, for example, indicates five times more strand breakage.

The alkaline unwinding technique has been used to examine the integrity of DNA from two aquatic organisms—the fathead minnow (*Pimephales promelas*) and the bluegill sunfish (*Lepomis macrochirus*)—chronically exposed to benzo[a]pyrene in their water at a concentration of 1 μg/L. The assay detected in both fish an increase in strand breaks in the benzo[a]pyrene-exposed populations.[26,27]

Current laboratory research with this technique has focused on the effect that simultaneous exposure to nongenotoxic-type stress conditions and benzo[a]pyrene will have on DNA integrity. An example of this approach is depicted in Figure 2. The exposure conditions were as detailed previously[27] except that the benzo[a]pyrene was introduced without carrier by passing water through a generator column containing glass beads coated with an excess of the chemical.[29] Nonstressed condition means the fish were acclimated slowly from 10°C to the exposure temperature of 20°C over a period of several weeks before commencing exposure. Stressed condition means the fish underwent the change from 10°C to 20°C over a period of 2 hr. Analyses for effect of these conditions on DNA integrity indicate that "temperature-stressed" fish exposed to benzo[a]pyrene had more strand breaks in their DNA than fish that had been either exposed to the genotoxic agent but not temperature stressed or had been temperature stressed only. For the purpose of computing the N value (see definition above), the control population of fish were those that were not temperature stressed and not exposed to benzo[a]pyrene. The data indicate that a sudden and significant increase in environmental temperature prevented fish from maintaining the integrity of their

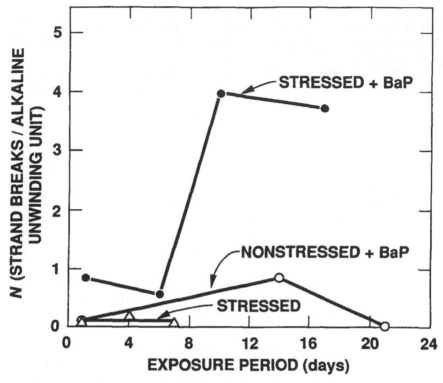

Figure 2. The effect of "temperature stress" on the kinetics of appearance of strand breaks (N) in the DNA from livers of sunfish (*Lepomis macrochirus*) exposed to a chronic dose (1 μg/L) of benzo[a]pyrene via their water.[27]

DNA when exposed concomitantly to a genotoxic agent. Preliminary experiments with hepatotoxic agents demonstrate the same phenomenon (i.e., potentiation of genotoxicity).[30]

The feasibility of utilizing this technique on environmental species as a general biomarker for pollution-related genotoxicity is currently being evaluated.[21,30] In this regard, a preliminary report of its application to turtles has been presented.[31] In this study, DNA strand breakage was determined in two populations of turtles (*Pseudemys scripta* and *Chelydra serpentian*) from a chemically contaminated settling basin and from an uncontaminated reference site. DNA damage was significantly greater in both turtle species from the contaminated site.

The detection of strand breaks by various analytical methodologies is being investigated in humans, rodents, and aquatic species as a measure of DNA damage that results from exposure to genotoxic agents.[14,32–36]

DNA Minor Nucleoside Composition

Loss of DNA methylation can lead to heritable abnormalities in gene expression.[33–39] Chemical carcinogens have been postulated to alter normal patterns of DNA methylation as a result of their effect on the maintenance and initiation (de novo) of DNA methyltransferase activities in cells.[40,41] The direct inactivation of the methyltransferases is one possible mode of action of carcinogens. In addition, carcinogenic agents cause DNA-damaging events such as strand breaks, adduct formation, induction of apurinic sites, and cross-linking that can prevent methylation (i.e., interfere with the enzymatic transfer of methyl groups to hemi-methylated DNA). With benzo[a]pyrene the inhibition of methylation has been shown in vitro[42] to be proportional to the degree of adduct formation present in the DNA.

In eukaryotic DNA, 5-methyl deoxycytidine (m^5-dCyd) is generally the only methylated deoxyribonucleoside present. The m^5-dCyd content of DNA has been measured by several established analytical techniques and is present in vertebrates between 0.7 and 2.8 mol%.[43]

From the above discussion it is obvious that the maintenance of DNA methylation is a measure of DNA integrity, and it was therefore of interest to determine whether hypomethylation occurred in the DNA of sunfish exposed to a genotoxic agent. Isolated DNA from sunfish exposed to benzo[a]pyrene[27] for a period of 40 days was subjected to analyses for m^5-dCyd by cation exchange liquid chromatography.[44] It was observed[45] that hypomethylation of the DNA occurred shortly after exposure commenced, it progressed throughout the exposure period, and it persisted after termination of exposure. During this time frame the m^5-dCyd content of DNA decreased by approximately 38% from that at day zero. These data tend to confirm that hypomethylation may occur as a biological response to genotoxic agents, and such a response may prove suitable as a biological marker. Additional laboratory and field studies are being performed to assess this assumption.

SUMMARY AND CONCLUSIONS

Biological monitoring is an approach of considerable interest to scientists investigating the effects of hazardous substances in our environment. Responses in living organisms, usually detected at the molecular level, are evaluated for their potential to identify exposure to dangerous substances and to define or to predict subsequent deleterious effects. DNA damage has been proposed as a useful biomarker for assessing the genotoxic properties of environmental contaminants. Many of these substances are chemical carcinogens and mutagens with the capacity to cause various types of DNA damage. Genotoxic chemicals may form stable adducts with DNA or cause strand breaks. Additional damage may occur through the inhibition of normal DNA repair. Each type of damage may contribute toward the eventual transformation of the cell. The feasibility of monitoring DNA damage in an organism as a biomarker of environmental genotoxicity by detecting and quantifying changes in DNA integrity was discussed.

If biological monitoring is to become a viable scientific tool for such areas as environmental research,[1,14] then continued effort and emphasis must be placed on

1. research concerned with those basic mechanisms by which substances elicit their responses in a living organism,
2. development of new analytical techniques, and
3. interpretation of data from field studies.

All three efforts are important, interrelated components of a successful biological monitoring program.

ACKNOWLEDGMENTS

The Oak Ridge National Laboratory is operated by Martin Marietta Energy Systems Inc., under contract DE-AC05-84OR21400 with the U.S. Department of Energy. Support for this research was funded in part by the ORNL Director's R & D Program. Environmental Sciences Division Publication No. 3508.

REFERENCES

1. Committee on Biological Markers of the National Research Council. "Biological Markers in Environmental Health Research," in *The Role of Biomarkers in Reproductive and Developmental Toxicology, Environ. Health Pers.* 74:3–9 (1987).
2. Kohn, H. W. "The Significance of DNA-Damaging Assays in Toxicity and Carcinogenicity Assessment," *Ann. N.Y. Acad. Sci.* 407:106–118 (1983).
3. Wogan, G. N., and N. J. Gorelick. "Chemical and Biochemical Dosimetry to Exposure to Genotoxic Chemicals," *Environ. Health Pers.* 62:5–18 (1985).
4. Randerath, K., M. Reddy, and R. C. Gupta. "^{32}P-Postlabeling Analysis for DNA Damage," *Proc. Natl. Acad. Sci. U.S.A.* 78:6126–6129 (1981).
5. Dunn, B., J. Black, and A. Maccubbin. "^{32}P-Postlabeling Analysis of Aromatic DNA Adducts in Fish from Polluted Areas," *Cancer Res.* 47:6543–6548 (1987).
6. Martineau, D., A. Legace, P. Beland, R. Higgins, D. Armstron, and L. R. Shugart. "Pathology of Stranded Beluga Whales (*Delphinapterus leucas*) from the St. Lawrence Estuary, Quebec, Canada," *J. Comp. Pathol.* 98:287–311 (1988).
7. Shugart, L. R., J. M. Holland, and R. Rahn, "Dosimetry of PAH Carcinogenesis: Covalent Binding of Benzo[a]pyrene to Mouse Epidermal DNA," *Carcinogenesis* 4:195–199 (1983).
8. McBee, K., and J. W. Bickham. "Petrochemical-Related DNA Damage in Wild Rodents Detected by Flow Cytometry," *Bull. Environ. Contam. Toxicol.* 40:343–349 (1988).
9. Schmid, W. "The Micronucleus Test for Cytogenetic Analysis," in *Chemical Mutagens: Principles and Methods for Their Detection, Vol. 6*, A. Hollaender, Ed. (New York: Plenum Press, 1976), pp. 31–53.
10. Hose, J. E., J. N. Cross, S. C. Smith, and D. Diehl, "Elevated Circulating Erythrocyte Micronuclei in Fishes from Contaminated Sites off Southern California," *Mar. Environ. Res.* 22:167–176 (1987).
11. Bos, J. L. "The *ras* Gene Family and Human Carcinogenesis," *Mutat. Res.* 195:255–271 (1988).
12. Alberts, B., D. Bray, J. Lewis, M. Raff, K. Roberts, and J. D. Watson. *Molecular Biology of the Cell*, 2nd. ed. (New York: Garland Publishing Inc., 1989), pp. 220–227.

13. Sancar, A., and G. B. Sancar. "DNA Repair Enzymes," *Ann. Rev. Biochem.* 57:29–67 (1988).

14. Bartsch, H., K. Hemminki, and I. K. O'Neill, Eds. *Methods for Detecting DNA Damaging Agents in Humans: Application in Cancer Epidemiology and Prevention* (New York: IARC Scientific Publication No. 89, Oxford University Press, 1988).

15. Soileau, S. D., and T. G. Gen, Eds. *Carcinogen-DNA Adducts* (Las Vegas, NV: Environmental Protection Agency Publication-600/4-87-005, January 1987).

16. Harvey, R. C. "Polycyclic Hydrocarbons and Cancer," *Am. Sci.* 70:386–393 (1982).

17. Rahn, R., S. Chang, J. M. Holland, and L. R. Shugart "A Fluorometric-HPLC Assay for Quantitating the Binding of Benzo[a]pyrene Metabolites to DNA," *Biochem. Biophys. Res. Commun.* 109:262–269 (1982).

18. Shugart, L. R., and J. Kao. "Examination of Adduct Formation *in vivo* in the Mouse Between Benzo[a]pyrene and DNA of Skin and Hemoglobin of Red Blood Cells," *Environ. Health Pers.* 62:223–226 (1985).

19. Shugart, L. R., J. F. McCarthy, B. D. Jimenez, and J. Daniel. "Analysis of Adduct Formation in the Bluegill Sunfish (*Lepomis macrochirus*) between Benzo[a]pyrene and DNA of the Liver and Hemoglobin of the Erythrocyte," *Aquat. Toxicol.* 9:319–325 (1987).

20. Kleinow, K. M., M. J. Melancon, and J. J. Lech. "Biotransformation and Induction: Implications for Toxicity, Bioaccumulation and Monitoring of Environmental Xenobiotics in Fish," *Environ. Health Pers.* 71:105–119 (1987).

21. Shugart, L. R., S. M. Adams, B. D. Jimenez, S. S. Talmage, and J. F. McCarthy. "Biological Markers to Study Exposure in Animals and Bioavailability of Environmental Contaminants," in *ACS Symposium Series No. 382, Biological Monitoring for Pesticide Exposure: Measurement, Estimation, and Risk Reduction*, R. G. M. Wang, C. A. Franklin, R. C. Honeycutt, and J. C. Reinert, Eds. (Washington, DC: American Chemical Society, 1989), pp. 86–97.

22. Ahnstrom, G., and K. Erixon. "Measurement of Strand Breaks by Alkaline Denaturation and Hydroxylapatite Chromatography," in *DNA Repair, Vol. 1, Part A*, E. C. Friedberg, and P. C. Hanawalt, Eds. (New York: Marcel Dekker, Inc., 1980), pp. 403–419.

23. Kanter, P. M., and H. S. Schwartz. "A Hydoxylapatite Batch Assay for Quantitation of Cellular DNA Damage," *Anal. Biochem.* 97:77–84 (1979).

24. Kanter, P. M., and H. S. Schwartz. "A Fluorescence Enhancement Assay for Cellular DNA Damage," *Mol. Pharmacol.* 22:145–151 (1982).

25. Daniel, F. B., D. L. Haas, and S. M. Pyle. "Quantitation of Chemically Induced DNA Strand Breaks in Human Cells via an Alkaline Unwinding Assay," *Anal. Biochem.* 144:390–402 (1985).

26. Shugart, L. R. "An Alkaline Unwinding Assay for the Detection of DNA Damage in Aquatic Organisms," *Mar. Environ. Res.* 24:321–325 (1988).

27. Shugart, L. R. "Quantitation of Chemically Induced Damage to DNA of Aquatic Organisms by Alkaline Unwinding Assay," *Aquat. Toxicol.* 13:43–52 (1988).

28. Rydberg, B. "The Rate of Strand Separation in Alkali of DNA of Irradiated Mammalian Cells," *Radiat. Res.* 61:274–285 (1975).

29. Billington, J. W., G-L Huang, F. Szeto, F. W. Y. Shiu, and D. McKay. "Preparation of Aqueous Solutions of Sparingly Soluble Organic Substances: I. Single Component Systems," *Environ. Toxicol. Chem.* 7:117–124 (1988).

30. McCarthy, J. F., B. D. Jimenez, L. R. Shugart, and A. Oikari. "Biological Markers

in Animal Sentinels: Laboratory Studies Improve Interpretation of Field Data," in *First Symposium on in situ Evaluation of Biological Hazards of Environmental Pollutants*, S. Sandhu, Ed. (New York: Plenum Press, Environmental Research Series, 1989), in press.

31. Meyers, L. J., L. R. Shugart, and B. T. Walton. "Freshwater Turtles as Indicators of Contaminated Aquatic Environments," paper presented at the 9th Annual Meeting of the Society of Environmental Toxicology and Chemistry, Arlington, VA, November 15, 1988.

32. Walles, S. A. S., and K. Erixon. "Single-Strand Breaks in DNA of Various Organs of Mice Induced by Methyl Methanesulfonate and Dimethylsulfoxide Determined by the Alkaline Unwinding Technique," *Carcinogenesis* 5:319–323 (1984).

33. Batel, R., N. Bihari, B. Kurelec, and R. K. Zahn. "DNA Damage by Benzo[a]pyrene in the Liver of Mosquite Fish *Gambusia affinis*," *Sci. Total Environ.* 41:275–283 (1985).

34. Singh, N. P., M. T. McCoy, R. R. Tice, and E. L. Schneider. "A Simple Technique for Quantitation of Low Levels of DNA Damage in Individual Cells," *Exp. Cell Res.* 175:184–191 (1988).

35. Nelson, M. A., and R. J. Bull. "Induction of Strand Breaks in DNA by Trichloroethylene and Metabolites in Rat and Mouse Livers *in vivo*," *Toxicol. Appl. Pharmacol.* 94:45–54 (1988).

36. Carlo, P., R. Finollo, A. Ledda, and G. Brambilla. "Absence of Liver DNA Fragmentation in Rats Treated with High Oral Doses of 32 Benzodiazepine Drugs," *Fundam. Appl. Toxicol.* 12:34–41 (1989).

37. Razin, A., and A. D. Riggs. "DNA Methylation and Gene Function," *Science*. 210:604–609 (1980).

38. Holliday, R. "The Inheritance of Epigenetic Defects," *Science* 238:163–170 (1987).

39. Cedar, H. "DNA Methylation and Gene Activity," *Cell* 53:3–4 (1988).

40. Boehim, T. L., and D. Drahovsky. "Alteration of Enzymatic Methylation of DNA Cytosines by Chemical Carcinogens: A Mechanism Involved in the Initiation of Carcinogenesis," *J. Natl. Cancer Inst.* 71:429–433 (1983).

41. Wilson, V. L. and P. A. Jones. "Inhibition of DNA Methylation by Chemical Carcinogens *in vitro*," *Cell* 32:229-246 (1983).

42. Pfeifer, G. P., D. Grungerger, and D. Drahovsky. "Impaired Enzymatic Methylation of BPDE-Modified DNA," *Carcinogenesis* 5:931–935 (1984).

43. Ehrlich, M., and R. Y.-H. Yang. "5-Methylcytosine in Eukaryotic DNA," *Science* 212:1350–1357 (1981).

44. Uziel, M., C. K. Koh, and W. E. Cohn. "Rapid Ion-Exchange Chromatographic Microanalysis of Ultraviolet-Absorbing Materials and Its Application to Nucleosides," *Anal. Biochem.* 25:77–98 (1965).

45. Shugart, L. R. "5-Methyl Deoxycytidine Content of DNA from Bluegill Sunfish (*Lepomis macrochirus*) Exposed to Benzo[a]pyrene," *Environ. Toxicol. Chem.*, 9:205–208 (1990).

46. McCarthy, J. F., D. N. Jacobson, L. R. Shugart, and B. D. Jimenez. "Pre-exposure to 3-Methylcholanthrene Increases Benzo[a]Pyrene Adducts on DNA of Bluegill Sunfish," *Mar. Environ. Res.* in press.

CHAPTER 12

DNA Adducts in Marine Mussel *Mytilus galloprovincialis* Living in Polluted and Unpolluted Environments

B. Kurelec, A. Garg, S. Krča, and R. C. Gupta

ABSTRACT

We have used a generally applicable ^{32}P-postlabeling assay to examine for the presence of DNA adducts in mussels experimentally exposed to known carcinogens and in mussels collected from sites impacted by wastewaters. Mussels exposed to seawater artificially polluted with 2-aminofluorene showed exclusively one adduct which was identified to be dG-C8-2-aminofluorene. Under the same experimental conditions, Diesel-2 oil did not induce any detectable adducts. When mussel digestive gland DNA was collected and analyzed from one unpolluted site, two moderately impacted sites, and one site heavily impacted by cannery wastewaters, mussel DNA from the unpolluted and only one moderately polluted site showed the presence of 6 to 10 adducts. This indicates they were not related to the pollution. This was further supported by the absence of dose-related adducts. Clear evidence for the presence of pollution-related DNA adducts was, however, found in juvenile mussels collected from an oil refinery site. One major and three minor adducts were detected in these mussels with no adducts detected in juvenile mussels from an unpolluted site.

Our results indicate that while mussels are capable of metabolizing 2-aminofluorene to a DNA binding species, the environmental pollutants may be metabolized only selectively. Further, the mussel digestive gland DNA also contained DNA modifications ("natural adducts") which are unrelated to pollutants and their presence appeared to be sex and season dependent.

In genotoxic risk assessment studies in the aquatic environment, biomonitoring, i.e., measurement of xenobiotics and their metabolites in body compartments of an aquatic organism, has been generally considered more relevant than measuring concentration of xenobiotics in the water. This kind of chemical dosimetry corrects for interindividual variation in absorption, metabolism and excretion, and also integrates exposure from all sources, and therefore can be used as a monitoring tool for use in exposure assessment from multiple chemicals. Measurement of free carcinogens in body compartments, however, does not measure actual DNA adducts or related phenomena, like protein adducts. Since interindividual variations in the metabolism of carcinogens is 10- to 200-fold, a more accurate indicator of biologically relevant exposure to carcinogens would be measurement of DNA adducts in the target tissue of the exposed organism.[1] Such biomarkers would be expected to possess the potential to predict the adverse, pathobiological effects of that exposure. It is well established that the majority of chemical carcinogens and mutagens covalently bind to DNA only after they are metabolised to the electrophiles.[2] This binding or formation of DNA adducts has been considered a key step in the initiation of carcinogenesis.[3] At the same time, there is a good correlation between the concentration of DNA adducts and induction of gene mutation,[4] as well as induction of chromosomal aberrations.[5] Since real life exposure is continuous, or intermittant on a regular basis, measurement of the *status presens* of carcinogen DNA adducts does provide an estimate of the genotoxic risk at the time of sampling. This is because the levels measured are the net result of the dynamic state of adduct formation and adduct losses by DNA repair and/or dilution of cell cycle.[6] Thus, analysis of DNA adducts in field organisms may be used both as a measure of the extent of exposure to genotoxic chemicals and as a predictor of pathobiological consequences of that exposure.

In the assessment of genotoxic risks in an aquatic environment, measurement of carcinogen-related DNA adducts may have implication only if the analytical method fulfills two basic requirements—extreme sensitivity and applicability to all (even unknown) carcinogens, irrespective of their chemical nature. These criteria have been largely satisfied by recently developed ^{32}P-postlabeling assay.[7-11] Due to its general applicability and high sensitivity of adduct detection (1 adduct/10^{10} nucleotides), enhanced versions of the ^{32}P assay[9,10] were employed in the present survey of the aromatic and/or hydrophobic DNA adducts in mussel *Mytilus galloprovincialis*. Mussels have been shown to possess the prerequisite needed for the induction of DNA adducts—the potential to bioactivate precarcinogens to metabolites that bind to DNA. Because of their lack of or low cytochrome P-450-dependent monooxygenases,[12-16] and because of their very active FAD-containing monooxygenase,[15] this potential is rather selective; it activates precarcinogenic aromatic amines, but not the precarcinogenic polycyclic aromatic hydrocarbon benzo[a]pyrene, to *Salmonella typhimurium* TA 98 mutagens,[17–19] or to metabolites that form DNA adducts.[20] The main purpose of this investigation was to ascertain whether chronic exposure to the polluted or unpolluted ("natural") environments on the East coast of Northern Adriatic induced exposure-related DNA adducts in mussels.

MATERIALS AND METHODS

Materials

Proteinase K was from Boehringer-Mannheim, and RNase A, RNase T_1, nuclease P_1, and α-amylase (Type I-A) were purchased from Sigma Chemical Co. T4 polynucleotide kinase was from New England Biolabs. [τ-^{32}P]ATP was prepared in the laboratory from ^{32}Pi (ICN) by an enzymatic procedure,[21] as described.[8,9] The sources of all other biochemicals and chromatographic materials have been described.[8,9]

Field Sites

In this survey, populations of mussels inhabiting a spatial pollution gradient were examined. This study was conducted in the aquatorium of the town of Rovinj, on the West Coast of Istra Peninsula, Northern Adriatic. In addition, a population of mussels inhabiting a site near the INA Refinery at Urinj, Rijeka Bay, Northern Adriatic was also examined. At the Rovinj site, there is a pollution gradient ranging from an unimpacted reference site at the Channel of Lim through gradually more impacted sites, namely Tobacco factory and Hospital sites, and finally the Institute-Pier site which is heavily impacted by the wastes of a local cannery. These were characterized by the following biological paramters. First, the pollution was defined by the frequency of chromosomal aberrations (FCA) in mussels native to these sites. FCA in mussels from different sites were as follows: reference site 2.9%, Tobacco factory site 5.3%, Hospital site 6.1%, and Institute-Pier site 9.6%.[22] Second, the state of induction of benzo[a]pyrene monooxygenase (BaPMO) activity in the livers of *Blennius pavo*, a fish species with restricted territorial range, revealed values of 1.6 (reference site), 8.9 (Tobacco factory site), 5.1 (Hospital site), and 36.7 (Institute-Pier site) pmoles of BaPOH/min/mg protein.[23] Third, i.p. administration of hexane-seawater extracts (HESW) from 1 L of seawater collected at Institute-Pier site to experimental juvenile carp, induced their BaPMO activity from 2.1 to 46.2 pmoles BaPOH/min/mg. In contrast, equivalent HESW from a reference site did not induce the BaPMO activity in the experimental carp.[24] Finally, of the HESWs tested from all sites, only HESW from the Institute-Pier site increased the number of *Salmonella typhimurium* TA 100 his$^+$ revertants in Ames-microsomal mutagenicity test by 60%.[24]

Similarly, the state of pollution at Urinj Harbor, located 200 m downstream from the INA Refinery wastewater outlet, and the unimpacted reference site at 12 km distance from Urinj Harbor at Bakarac, were characterized by the extent of BaPMO induction in a local *Blennius pavo* (Urinj: 35.1; Bakarac: 2.2 pmoles BaPOH/mg/min)[25] and by the content of xenobiotics in HESW as measured by induction of BaPMO after i.p. treatment of experimental carp with equivalents of 1 L of water (Urinj: 52.7; Bakarac: 3.8 pmoles BaPOH/mg/min).[26]

Exposure Experiments

Exposure to 2-aminofluorene (AF) was done in a 5-L flask containing seawater polluted with 3 ppm of AF for 4 hr. AF (500 mg), dissolved in 20 mL of dimethylsulfoxide, was suspended in 1 L of seawater. This suspension was continuously stirred by magnetic stirrer and pumped into a flask with 8 mussels at a flow of 3 mL/min together with a flow of 500 mL/min of seawater.

Exposure to Diesel-2 oil was done in 5-L flask with a group of 8 mussels and a flow of seawater (500 mL/min) saturated with Diesel-2 oil for 24 hr. The saturation was achieved by a continuous flow of seawater through a vertical glass tube (diameter: 24 mm) immersed in the experimental flask and containing a 20-cm column of Diesel-2 oil (about 100 mL). The amount of Diesel-2 oil was renewed every 8 hr.

Exposure to polluted sites was done with a group of control mussels immersed in nylon nets at a depth of 60 cm for 24 hr at the Institute-Pier, Tobacco factory, and Hospital sites.

Isolation of DNA

DNA was isolated from 10% homogenates of mussel digestive glands by removal of protein and RNA by extensive digestions with proteinase K and RNases, respectively, and solvent extractions, essentially as descirbed,[27] except that α-amylase (0.1 mg/mL) was added during the RNase treatment to remove glycogens.

Analysis of DNA

DNA adducts were analyzed by ^{32}P-postlabeling assay,[8] after sensitivity enhancement by butanol extraction,[9] or nuclease P_1 treatment.[10]

RESULTS

Survey of Mussels in Rovinj Area

DNA adducts were analyzed by the ^{32}P-postlabeling assay in the digestive glands of three pooled mussels collected from the unpolluted pristine Lim Channel, moderately contaminated sites of Tobacco factory and Hospital and Institute-Pier site, heavily impacted by untreated wastewaters from a local cannery. Mussels were collected in February 1987. Because at this time mussels in the Lim Channel were spawning, we analyzed adult (>2 year old) males, females, and juveniles (<1 year old) separately. The females revealed the presence of as many as 10 DNA adduct spots (Figure 1a); however, no adducts were detected in the female mussels collected in April and July 1987. DNA of the adult male and juvenile mussels had no adducts (not shown). Out of the three polluted sites, only mussels collected from the moderately polluted Hospital site showed 6 or more adduct spots (Figure 1b). Since the adducts were also present in mussels from the unpolluted site and only one of the three polluted sites, it appears that the latter adducts may be unrelated to the pollution. It is interesting to note that the adducts were observed *only during the spawning phase*.

Survey of Mussels at Oil Refinery Site

Mussels inhabiting a pier in the harbor at Urinj which is heavily impacted by the wastewaters of a large oil refinery were collected in early July 1987. Mussels were young (<1 year; 1–3 cm) and belonged to a new settlement. Digestive glands pooled from about 30 specimens revealed the presence of 1 major and 3 minor DNA adducts (Figure 2b). At the same time, no adducts were detected from sample of the digestive glands pooled from 30 mussels of the same size and age, collected from the reference site (Figure 2a). In mid April 1988, the same pier was settled with a new population of juvenile mussels, with no mussel left from last year's generation. This time, we prepared a sample of whole soft body pooled from about 30 mussels settled on a pier, and a corresponding sample pooled from about 30 mussels, collected from the reference site. Both samples revealed only one minor adduct (not shown). The adducts found in mussels from the Urinj site in July 1987 are so far the only DNA adducts found that can be defined as pollution related.

Exposure to 2-Aminofluorene and Diesel-2 Oil

One day after the 4 hr-exposure to a minimum of 3 ppm concentration of AF, the mussel digestive gland DNA showed exclusively one adduct spot (Figure 3b). A negative control in which DNA hydrolyzing enzymes were omitted did not show any adduct spot (Figure 3c), indicating that the adduct was DNA derived. No spot was detected in the digestive gland DNA of mussels exposed to seawater in the absence of AF (Figure 3a). Likewise, no adduct was detected in the mussel digestive gland DNA exposed to Diesel oil-saturated seawater (not shown). The AF-mussel DNA adduct was identified to be dG-C8-AF as it was chromatographically identical to the N-OH-AF-modified DNA adduct (Figure 3d).

Exposure at Institute-Pier Site

Exposures of mussels collected from the reference site and then exposed in a net to waters of the heavily impacted Institute-Pier site during the winter, spring,

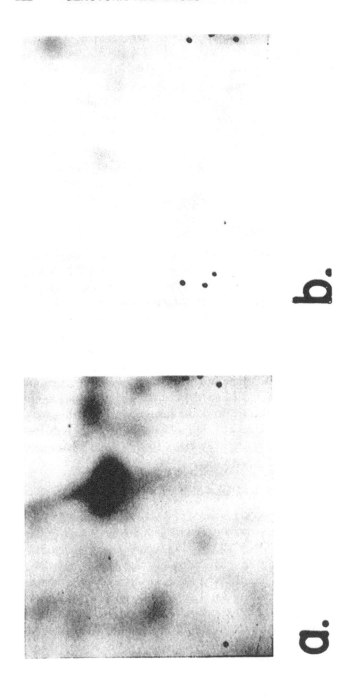

Figure 1. Representative ^{32}P fingerprints of digestive gland DNA adducts in a sample of female mussels collected from an unpolluted Lim Channel site (a) and a sample of mussels collected from a moderately polluted Hospital site (b) near Rovinj, Northern Adriatic. Nuclease P$_1$ enhancement procedure of the ^{32}P-postlabeling assay was employed. The chromatography was performed as described in Reference 9 and adducts were detected by intensifying screen-enhanced autoradiography at −80°C for 15 hr.

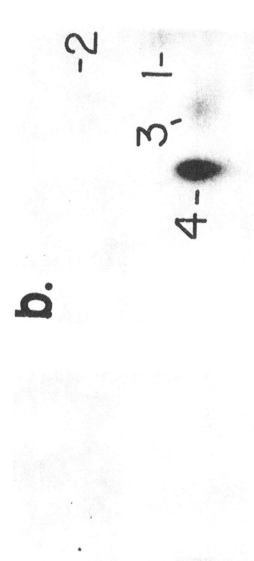

Figure 2. ^{32}P fingerprints of digestive gland DNA adducts in a sample of mussels collected near oil refinery wastewaters outlet at Urinj site (b) and from a reference Bakarac site (a), both in Rijeka Bay, Northern Adriatic. Butanol enhancement version of the ^{32}P-postlabeling assay was employed. See Figure 1 legend for other conditions.

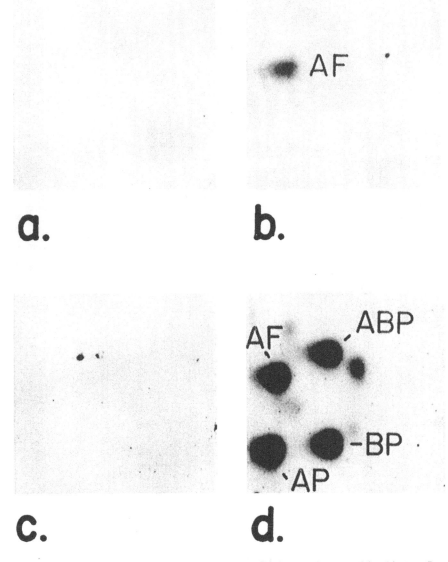

Figure 3. ^{32}P-DNA adduct maps of digestive glands of mussel exposed for 4 hr to a flow of seawater experimentally polluted with 3 ppm of 2-aminofluorene (b) or to a flow of unpolluted seawater (a). Panel c, a negative control in which the AF-exposed DNA was processed after the DNA hydrolyzing enzymes were omitted. Panel d, a positive control of calf thymus DNA modified to a level of 1 adduct/ 10^8 nucleotides in vitro with active metabolites of 2-aminofluorene (AF), 4-aminobiphenyl (ABP), benzo[a]pyrene (BP), and 2-aminophenanthrene (AP) (28), included during the adduct analysis. Adducts were analyzed as described in Figure 2 legend. Panel a, spot AF = dG-C8-2-aminofluorene.

and summer seasons in 1987 and 1988 never induced DNA adducts in the digestive glands of the exposed mussels (not shown).

DISCUSSION

Results of the surveys and experiments described here suggest that mussels can produce pollution-related DNA adducts if they are exposed to a proper pollutant at a proper concentration, like AF at 3 ppm or "mixing zone" of oil refinery wastewaters. The failure to find DNA adducts at the highly polluted Institute-Pier site could be attributed to either types of pollutants or a lower concentration of pollutants. The latter possibility is unlikely because this level of pollution is known to induce significant FCA in mussels[22] or BaPMO activity in fish,[23] and HESW from this site contains inducers of BaPMO[25] and bacterial premutagens.[24] Therefore, the plausible reason for lack of pollution-related DNA adducts should lie in the type of pollutants at this particular site. Based on the adverse biological effects (chromosomal aberrations; induction of BaPMO) induced by these pollutants and on the properties of pollutants from these waters that are hexane-extractable, it infers that they belong to a class of polycyclic aromatic hydrocarbons (PAH). PAHs are known not to be bioactivated in mussels to mutagens[17] or carcinogens.[20] This may explain the lack of induction of pollution-related DNA adducts in mussels living along the spatial gradient of contamination in the Rovinj area. Exposure of mussels to seawater saturated with Diesel-2 oil and the resulting nil-effect with respect to induction of DNA adducts offers an exemplary proof for this assumption. Although concentration of PAHs in this exposure experiment was much higher than was the concentration of AF in a parallel exposure experiment, no PAH-related adducts were detected. Based on this data as well as on the recently described characteristics of the metabolic activation system in mussels,[15,17-20] it can be concluded further that waters at the polluted sites in Rovinj area may not be contaminated with carcinogens such as carcinogenic aromatic amines or other compounds that are substrates for FAD-containing monooxygenase.

ACKNOWLEDGMENTS

This work was performed within the frame of No-obligation Agreement on Scientific Co-operation between our institutions. The work was supported by the U.S. Environmental Protection Agency and the Community for Scientific Work of Croatia through funds made available to the U.S.-Yugoslavia Joint Board on Scientific and Technological Co-operation, J.F. Project 868 (to B. K.), and U.S. Environmental Protection Agency Co-operative Agreement CR-813840 (to R. C. G.). Christine Lochridge is thanked for preparing the thin-layer sheets.

LEGEND OF SYMBOLS

AF	2-aminofluorene
BaPMO	benzo[a]pyrene monooxygenase
BaPOH	benzo[a]pyrene hydroxide
FCA	frequency of chromosomal aberrations
HESW	hexane extract of sea water
N-OH-AF	N-hydroxy-2-aminofluorene
PAH	polycyclic aromatic hydrocarbons

REFERENCES

1. Soileau, S. D. "Carcinogen-DNA Adducts," Environmental Monitoring Systems Laboratory Report, U.S. Environmental Protection Agency (1987), pp. 1–126.
2. Miller, E.C. "Some Current Perspective on Chemical Carcinogenesis in Human and Experimental Animals: Presidential Address," Cancer Res. 38:1479–1496 (1978).
3. Wogan, G. N. and N. J. Gorelick. "Chemical and Biochemical Dosimetry of Exposure to Genotoxic Chemicals," Environ. Health Pers. 62:5–18 (1985).
4. Perera, F. "The Potential Usefulness of Biomarkers in Risk Assessment," Environ. Health Pers. 76:141–145 (1987).
5. Talaska, G., W. W. Au, J. B. Ward, K. Randerath, and M. S. Legator. "The Correlation between DNA Adduct and Chromosomal Aberrations in the Target Organs of Benzidine Exposed Partially-Hepatectomized Mice," Carcinogenesis 8:1899–1905 (1987).
6. Beland, F. A., H. S. Huitfeld, and M. C. Poirier. "DNA Adduct Formation and Removal during Chronic Administration of a Carcinogenic Aromatic Amine," Prog. Exp. Tumor Res. 31:33–41 (1987).
7. Randerath, K., M. V. Reddy, and R. C. Gupta. "^{32}P-labeling Test for DNA Damage," Proc. Natl. Acad. Sci. U.S.A. 78:6126–6129 (1981).
8. Gupta, R. C., M. V. Reddy and K. Randerath. "^{32}P-postlabeling Analysis of Nonradioactive Aromatic Carcinogen DNA Adducts," Carcinogenesis 3:1081–1092 (1982).
9. Gupta, R. C. "Enhanced Sensitivity of ^{32}P-postlabeling Analysis of Aromatic Carcinogen DNA Adducts," Cancer Res. 45:5656–5662 (1985).
10. Reddy, M. V. and K. Randerath. "Nuclease P1-mediated Enhancement of Sensitivity of ^{32}P-postlabeling Test for Structurally Diverse DNA Adducts," Carcinogenesis 7:1543–1551 (1986).
11. Gupta, R. C. and K. Randerath. "Analysis of DNA Adducts by ^{32}P Labeling and Thin Layer Chromatography," in DNA Repair, E. C. Friedberg and P. E. Hanawalt, Eds. (Marcel Dekker Inc., New York, 1988), p. 399.
12. Payne, J. F. "Mixed Function Oxidase in Marine Organisms in Relation to Petroleum Hydrocarbon Metabolism and Detection," Mar. Pol. Bull. 8:112–116 (1977).
13. Vandermeulen, J. H., and W. R. Penrose. "Absence of Aryl Hydrocarbon Hydroxylase (AHH) in Three Marine Bivalves," J. Fish. Res. Board Can. 35:643–647 (1978).
14. Livingstone, D. R., and S. V. Farrar. "Tissue and Subcellular Distribution of Enzyme Activities of Mixed Function Oxigenase and Benzo(a)pyrene Metabolism in the Common Mussel Mytilus edulis L.," Sci. Total Environ. 39:209–235 (1984).

15. Kurelec, B. "Exclusive Activation of Aromatic Amines in the Marine Mussel *Mytilus edulis* by FAD-containing Monooxygenase," *Biochem. Boiophys. Res. Commun.* 127:773–778 (1985).

16. Stegeman, J. J. "Benzo(a)pyrene Oxidation and Microsomal Enzyme Activity in Mussel (*Mytilus edulis*) and other Bivalve Mollusc Species from the Western N. Atlantic," *Mar. Biol.* 89:21–30 (1985).

17. Britvic, S., and B. Kurelec. "Selective Potential for the Activation of Carcinogenic Aromatic Amines to Bacterial Mutagens in the Marine Mussel *Mytilus galloprovincialis*," *Comp. Biochem. Physiol.* 85C:111–114 (1986).

18. Kurelec, B., S. Britvic, S. Krča, and R. K. Zahn. "The Metabolic Fate of Aromatic Amines in the Mussel *Mytilus galloprovincialis*," *Mar. Biol.* 91:523–527 (1986).

19. Kurelec, B., and S. Krča. "Metabolic Activation of 2-Aminofluorene, 2-Acetylaminofluorene and N-hydroxyacetylaminofluorene to Bacterial Mutagen with Mussel (*Mytilus galloprovincialis*) and Carp (*Cyprinus carpio*) Subcellular Preparations," *Comp. Biochem. Physiol.* 88C:171–177 (1987).

20. Kurelec, B., M. Chacko, and R. C. Gupta. "Postlabeling Analysis of Carcinogen-DNA Adducts in Mussel, *Mytilus galloprovincialis*," *Mar. Environ. Res.* 24:317–320 (1988).

21. Johnson, R. A., and T. F. Walseth. "The Enzymatic Preparation of (^{32}P)-ATP, (^{32}P)-GTP, (^{32}P)cAMP and (^{32}P)cGMP, and their Use in Assay of Adenylate and Guanylate and Cyclases, and Cyclic Nucleotide Phosphodiesterases," *Adv. Cyclic Nucleotide Res.* 10:135–167 (1979).

22. Al-Sabti, K., and B. Kurelec. "Induction of Chromosomal Aberrations in the Mussel *Mytilus galloprovincialis* Watch," *Bull. Environ. Contam. Toxicol.* 35:660–665 (1985).

23. Kurelec, B., S. Britvic, M. Rijavec, W. E. Mueller, and R. K. Zahn. "Benzo(a)pyrene Monooxygenase Induction in Marine Fish—Molecular Response to Oil Pollution," *Mar. Biol.* 44:211–216 (1977).

24. Kurelec, B., Z. Matijasevic, M. Rijavec, M. Alacevic, S. Britvic, W. E. G. Mueller, and R. K. Zahn. "Induction of Benzo(a)pyrene Monooxygenase in Fish and the Salmonella Test as a Tool for Detecting Mutagenic/Carcinogenic Xenobiotics in the Aquatic Environment," *Bull. Environ. Contam. Toxicol.* 21:799–807 (1979).

25. Rijavec, M., S. Britvic, M. Protic, and B. Kurelec. "Detection of the Presence of Xenobiotics in Sea Water Samples from the Rijeka Bay Applying Benzo(a)pyrene Monooxygenase Induction," *Thalassia Jugosl.* 7:245–250 (1981).

26. Kurelec, B., M. Protic, M. Rijavec, S. Britvic, W. E. G. Mueller, and R. K. Zahn. "Induction of Benzo(a)pyrene Monooxygenase in Fish after I.P. Application Water Hexane Extract—A Prescreen Tool for Detection of Xenobiotics," in *Carcinogenic Polynuclear Aromatic Hydrocarbons in the Marine Environment*, N. L. Richards and B. L. Jackson, Eds. (Environmental Research Laboratory, Gulf Breeze, FL, 1982), p. 124.

27. Gupta, R. C. "Nonrandom Binding of the Carcinogen N-hydroxy-2-acetylaminofluorene to Repetitive Sequences of Rat Liver DNA *in vivo*," *Proc. Natl. Acad. Sci. U.S.A.*, 81:6943–6947 (1984).

28. Gupta, R. C., S. Sharma, K. Earley, N. Mohapatra, S. Nesnow, and M. Chacko. "^{32}P Adduct Assay: Principle and Application to Carcinogen Exposed Animal and Human DNA," in *Application of Short-Term Bioassays in the Analysis of Complex Environmental Mixtures*, S. S. Sandhu et al., Eds. (Plenum Press: New York, 1987), p. 127.

c-K-*ras* Oncogenes: Prevalence in Livers of Winter Flounder from Boston Harbor

Gerald McMahon, L. Julie Huber, Michael J. Moore, John J. Stegeman, and Gerald N. Wogan

ABSTRACT

Livers of a natural population of winter flounder from a contaminated site in Boston Harbor were examined for the presence of oncogenes by transfection of DNA into NIH3T3 mouse fibroblasts. Tissues analyzed contained histopathologic lesions including abnormal vacuolation, biliary proliferation, and, in many cases, hepatocellular and cholangiocellular carcinomas. Fibroblasts transfected with liver DNA samples from 7 of 13 animals were effective in the induction of subcutaneous sarcomas in nude mice. Further analysis revealed the presence of flounder c-K-*ras* oncogenes in all subcutaneous tumors examined. Direct DNA sequencing and allele-specific oligonucleotide hybridization following polymerase chain reaction DNA amplification of the tumor DNA showed mutations in the 12th codon in this gene. Analysis of DNA of all nude mouse tumors as well as all 13 diseased flounder livers showed mutations at this codon. Liver DNA samples from 5 histologically normal livers of animals from a less polluted site were ineffective in the transfection assay and contained only wild-type DNA sequences. The prevalence of mutations in this gene region correlated with the presence of liver lesions and could signify DNA damage resulting from environmental chemical exposure.

INTRODUCTION

A high incidence of tumors in fish occurs in several water systems of coastal North America. Association of such tumors with the presence of environmental chemicals such as polycyclic aromatic hydrocarbons (PAH) or polychlorinated hydrocarbons suggests a possible chemical etiology for their induction.[1-2] Winter flounder (*Pseudopleuronectes americanus*) from areas of Boston Harbor where sediments were heavily contaminated with PAH and other organic compounds show a high incidence of liver lesions that include hepatocellular and cholangiocellular carcinomas,[3-4] in contrast to a low incidence in livers of fish from adjacent, less polluted coastal areas.[5]

Our purposes in this study were to evaluate the prevalence of oncogenes in diseased livers of feral fish, and to determine whether activating mutations could be found in these genes that could reflect genetic damage resulting from chemical-DNA interactions with specific environmental contaminants. Transforming genes (oncogenes) have been shown to be important in those cellular events leading to the formation of the malignant cell. The genetic characterization of some retroviral oncogenes, such as Harvey *ras* (v-Ha-*ras*) and Kirsten *ras* (v-Ki-*ras*), have revealed homologous nontransforming proto-oncogenes which are associated with all eucaryotes examined (from yeast to man). A comparison of cellular oncogenes, derived from spontaneous and chemically induced tumors, with cellular proto-oncogenes, derived from normal cells, have inferred somatic events which result in stable genetic changes in the tumor cell. Such changes have included gene rearrangement, gene translocation, changes in gene expression, and single amino acid changes. All of these DNA alterations may result in changes in the relative level or type of gene product formed. In the case of the c-*ras* proteins (c-Ha-*ras*, c-Ki-*ras*, and N-*ras*) such single nucleotide changes have resulted in single amino acid changes which have resided in the 12th, 13th, or 61st codons of the gene encoding the 21 kD proteins. Such changes have been implicated in modulation of c-*ras*-dependent GTP hydrolysis and may perturb the transduction of extracellular signals which effect cellular metabolism through this well-conserved, membrane-associated protein. Previous work by many others has suggested that the activation of c-*ras* genes by specific single-base mutations may be one of several steps indicating the progression of a normal cell to one which is malignant. In addition, in vitro mutagenesis of c-*ras* proto-oncogenes has shown that many different mutations within the 12th, 13th, 59th, and 61st codons of the c-Ha-*ras* and N-*ras* genes may lead to qualitatively similar oncoproteins when morphological transformation of NIH3T3 cells is assayed.

Importantly, genotoxic chemical carcinogens often form chemical-DNA adducts which may lead to stable mutations. Many reports have shown that cellular proto-oncogenes can be activated by single-base mutations,[6-7] and that the nature of the activating mutation may be a consequence of specific chemical carcinogen-DNA interaction.[8] Moreover, c-*ras* mutations which arise in tumors of animals after exposure to specific chemicals often result in a very restricted set of nucleotide substitutions at these positions even though many amino acid substitutions at these sites have been shown to lead to transforming proteins. Such a paradox can be partly explained if the binding of chemicals to DNA and the mechanisms underlying DNA repair of such lesions are specific to the chemical. For instance, it has been suggested that the differences in the spectra of mutations which result in mammary carcinomas of methylnitrosourea-treated and dimethylbenz[a]anthracene-treated rats may be related to alterations in processes associated with chemical-DNA binding or the repair of these lesions.[8]

Recently, we have demonstrated that a flounder c-K-*ras* oncogene was present in NIH3T3 mouse fibroblasts transfected with DNA derived from a flounder liver tumor.[9] This report summarizes our results to indicate that transforming genes of the c-Ki-*ras* gene family are highly prevalent in liver lesions of winter flounder from Boston Harbor.

RESULTS

Detection of c-K-*ras* Oncogenes in Flounder Liver DNA

Winter flounder were collected in a series of trawls proximal to the Deer Island sewer outfall in Boston Harbor. Of 227 animals greater than 30 cm in length, 32 (14%) bore gross liver lesions. Histological examination of livers of these 32 animals together with an additional 75 animals revealed that 67 (63%) contained abnormally vacuolated cells, 52 (49%) contained biliary proliferation, and 21 (20%) contained a total of 34 neoplasms of the following types: 17 cholangiocellular carcinomas; 13 hepatocellular carcinomas; 3 mixed hepato-biliary neoplasms; and 1 hemangiosarcoma. None of these histological abnormalities were evident in 28 fish greater than 30 cm in length trawled from reference sites including Cape Cod Bay and Vineyard Sound (south of Cape Cod). DNA was extracted from the livers of 13 Boston Harbor animals exhibiting gross liver lesions and containing varying degrees of histological abnormalities which included focal or diffuse vacuolation, biliary proliferation or degeneration, macrophage aggregation, and fibrosis. Eight of these livers contained carcinomas of cholangiocellular or hepatocellular origin. In addition, DNA was extracted from five winter flounder obtained from a site in Vineyard Sound which served as control animals for these studies.

DNA was transfected into NIH3T3 mouse fibroblasts that were then assayed for the potential to induce subcutaneous fibrosarcomas in athymic nude mice according to the procedure of Fasano et al.[10] DNAs from 7 of 13 diseased livers, when subjected to this assay procedure, were effective in eliciting subcutaneous tumor formation. The proportion of samples that gave positive results is high when compared to similar studies on tumors of human origin, in which 10–40% are typically positive.[11] DNA from the remaining six lesion-bearing animals and from five control animals did not transform NIH3T3 mouse fibroblasts.

To identify transforming genes, samples of DNA from secondary transformants (i.e., from subcutaneous tumors derived by retransfection of NIH3T3 cells with DNA from the fibrosarcomas in nude mice) were isolated and analyzed by Southern blot hybridization. Hybridization of transformant DNA to *ras* DNA probes revealed the presence of c-K-*ras* oncogenes in all of the nude mouse tumor DNAs analyzed. Typical autoradiograms showed the presence of multiple restriction fragments which hybridized to the v-K-*ras* DNA probe and corresponded to restriction fragments identified in flounder DNA.

Detection of 12th Codon Mutations in Flounder c-K-*ras* Oncogenes

Previous research has shown that activation of c-*ras* proto-oncogenes to oncogenic forms can result from single-base mutations in the 12th, 13th, 59th, or 61st codons of the gene.[6] To determine whether such mutations were present in flounder liver c-K-*ras* genes, we employed the polymerase chain reaction (PCR) DNA amplification method[12] as we described previously for similar studies in the rat.[13] The original procedure was modified to utilize *Taq* polymerase, as suggested by Saiki et al.,[14] followed by direct DNA sequencing of the purified PCR product using modified T7 polymerase (Sequenase). Direct DNA sequence analysis of PCR-DNA derived from histologically normal flounder livers did not reveal any DNA sequence divergence when compared to the rat gene for codons 7 through 23 (the region delineated by the primers used in this study). However, sequence analysis of PCR-DNA derived from livers of winter flounder bearing lesions revealed single nucleotide substitutions in the 12th codon of the gene. G-C to A-T or G-C to T-A single-base changes were present in the 12th codon in flounder c-K-*ras* genes of all of the original flounder DNA derived from affected liver lesions derived from contaminated sites (see Table 1).

In contrast to the diseased livers, the analysis of samples from five undiseased animals from Vineyard Sound, control rodent DNAs (rat and mouse), and a series of control c-K-*ras* plasmids exhibited inferred or wild-type sequences when subjected to PCR-DNA sequence analysis. However, the yield of PCR derived from flounder DNA was substantially less than that obtained from mice, rodent, or human DNA using the same primer set. In addition, analysis of the second exon region of the c-K-*ras* gene did not indicate the presence of possible contaminating v-K-*ras* sequences containing the characteristic AGT 12th codon present in most of the samples analyzed in our study. In contrast, the presence of the GTT 12th codon seems to be relatively unique and could not have arisen by contamination of the PCR reaction. Nonetheless, it is important to fully characterize the flounder c-K-*ras* gene, to more precisely optimize the PCR procedures, and to give unequivocal results when mutations should be present.

DISCUSSION

The Deer Island site in Boston Harbor has been shown by others to contain high levels of known carcinogens, both genotoxic, such as benzo[a]pyrene, and

Table 1.   Mutations in the 12th Codon of c-K-*ras* Genes of Liver Samples of Winter Flounder from Boston Harbor.

Parameter	Percent Incidence in Samples Derived From:	
	Deer Island Flats	Vineyard Sound
Liver DNAs positive in nude mouse assay	54 (7/13)[a]	0 (0/5)
Nude mouse tumors containing c-K-*ras* oncogenes	100 (7/7)	na[b]
Liver lesions containing mutations in c-K-*ras*[c]		
GGT-AGT GLY-SER	77 (10/13)	0 (0/5)
GGT-TGT GLY-CYS	15 (2/13)	0 (0/5)
GGT-GTT GLY-VAL	8 (1/13)	0 (0/5)
GGT-GAT GLY-ASP	0 (0/13)	0 (0/5)
Total	100 (13/13)	0 (0/5)

[a]Samples positive/total samples analyzed.
[b]Not applicable.
[c]12th codon mutations were detected by direct DNA sequencing analysis and by dot blot hybridization analyses using PCR-amplified DNA.

epigenetic, such as the chlorinated hydrocarbon, chlordane[16] (Dr. T. O'Connor, personal communication). Winter flounder livers efficiently activate benzo[a]pyrene to mutagenic derivatives[17] and DNA adducts resulting from such epoxides of benzo[a]pyrene or other PAH could contribute to somatic mutations of the type we have reported here. Previous studies in both bacterial[18] and mammalian[19] systems have shown that the predominant mutations induced by treatment of cells with r-7,t-8-dihydroxy-t-9,10-epoxy-7,8,9,10-tetrahydrobenzo[a]pyrene (*anti*-BPDE) involve G-C to T-A or G-C to A-T base changes. In addition, Vousden et al.[20] have shown that treatment of a c-H-*ras*-containing plasmid with *anti*-BPDE results in predominantly G-C mutations in both the 12th and 61st codon of the gene.

We cannot distinguish between mutations induced as a direct consequence of chemical-DNA interaction from those resulting from indirect selection mechanisms. Environmental chemicals could play important roles in the neoplastic process by creating oncoproteins through mutations at the DNA level, a mechanism for which we provide suggestive evidence in this report, or by inducing compensatory regeneration and hyperplasia due to cytotoxic effects, or both.

The finding of c-K-*ras* gene activation in most of the diseased flounder livers represents a strong correlation between environmental chemical exposure and a deleterious disease process which may ultimately result in cancer. Further technical and systematic studies are needed to optimize the detection of mutations in these genes in order to study the possible association of specific liver pathologies with specific genetic changes. Furthermore, if mutations identified in the c-K-*ras* genes can be shown to result from chemically induced DNA damage, then such mutations in winter flounder, and possibly other fish, could be important as indicators for chemical exposure associated with significant environmental disease. Linking of particular mutated alleles with exposure to specific chemicals could complement existing tests in assessing specific chemical hazard in sentinel species. Such associations would also aid in the design of systematic field studies to assess the potential impact of complex genotoxic mixtures in the environment.

ACKNOWLEDGMENTS

We gratefully acknowledge the assistance of Dr. R. A. Murchelano, Mr. L. Bridges, and the crew of the R/V Wilbour, who assisted in obtaining some of these fish. We also acknowledge the advice of Drs. C. Dawe and A. Walsh on the histological diagnosis of liver lesions. Supported by USPHS Grant CA44306.

REFERENCES

1. Couch, J. A., and J. C. Harshbarger. *Environ. Carcinogen. Rev.* 3:63–105 (1985).
2. Malins, D. C., M. M. Krahn, D. W. Brown, L. D. Rhodes, M. S. Myers, B. B. McCain, and S.-L. Chan. *J. Nat. Cancer Inst.* 74:487–494 (1985).
3. Murchelano, R. A., and R. E. Wolke. *Science* 228:587–589 (1985).
4. Boehm, P. NOAA Contract Report, NA-83-FA-C-00022, 1985.
5. Sass, S. L., and R. A. Murchelano. *Aquat. Toxicol.* 11:420–421 (1988).
6. Barbacid, M. *Ann. Rev. Biochem.* 56:779–827 (1987).
7. Balmain, A., and K. Brown, *Adv. Cancer Res.* 51:147–182 (1988).
8. Zarbl, H., S. Sukumar, A. V. Arthur, D. Martin-Zanca, and M. Barbacid. *Nature* 315:382–385 (1985).
9. McMahon, G., L. J. Huber, G. N. Wogan, and J. Stegeman. *Mar. Environ. Res.* 24:345–350 (1988).
10. Fasano, O., D. Birnbaum, L. Edlund, I. Fogh, and M. Wigler. *Mol. Cell. Biol.* 4:1695–1705 (1984).
11. Bos, H. *Mutat. Res.* 195:255–271 (1988).
12. Saiki, R. K., S. Scharf, F. Faloona, K. B. Mullis, G. T. Horn, H. A. Erlich, and N. Arnheim. *Science* 230:1350–1354 (1985).
13. McMahon, G., E. Davis, and G. N. Wogan. *Proc. Natl. Acad. Sci. U.S.A.* 84:4974–4978 (1987).
14. Saiki, R. K., D. H. Gelfand, S. Stoffel, S. J. Scharf, R. Higuchi, G. T. Horn, K. B. Mullis, and H. A. Erlich. *Science* 239:487–491 (1988).
15. Verlaan-de Vries, M., M. E. Bogaard, H. van den Elst, J. H. van Boom, A. J. van der Eb, and J. L. Bos. *Gene* 50:313–320 (1986).
16. *National Status and Trends Program for Marine Environmental Quality*, NOAA, 1987.
17. Stegeman, J. J., T. R. Skopek, and W. G. Thilly, in *Carcinogenic Polynuclear*

Aromatic Hydrocarbons in the Marine Environment, N. Richards, Ed. (EPA-600/9-82-013, 1982), pp. 201–211.

18. Eisenstadt, E., A. J. Warren, J. Porter, D. Atkins, and J. H. Miller. *Proc. Natl. Acad. Sci. U.S.A.* 79:1945–1949 (1982).

19. Yang, J.-L., V. M. Maher, and J. J. McCormick. *Proc. Natl. Acad. Sci. U.S.A.* 84:3787–3791 (1987).

20. Vousden, K. H., J. L. Bos, C. J. Marshall, and D. H. Phillips. *Proc. Natl. Acad. Sci. U.S.A.* 83:1222–1226 (1986).

SECTION FIVE

Metal Metabolism

CHAPTER **14**

Metal-Binding Proteins and Peptides for the Detection of Heavy Metals in Aquatic Organisms

David H. Petering, Mark Goodrich, William Hodgman, Susan Krezoski,
Daniel Weber, C. F. Shaw III, Richard Spieler, and Leslie Zettergren

ABSTRACT

Metallothionein is a key, well characterized metal binding protein in mammalian systems. Its chemical properties, proposed functions, and regulation of transcription by diverse agents are critically reviewed with the objective to relate this information to the proposed use of metallothionein as a biomarker. Careful examination is given to several alternative methods for quantitation of this protein. Then, specific results from the authors' laboratories are used to consider in more detail questions involved in understanding the meaning of measurement of metallothionein or other metal binding proteins or peptides in non-mammalian, aquatic systems including fish, anurans, and microorganisms. It is shown that before such determinations gain meaning, one must know whether measured levels of binding protein represent basal concentrations or induction of protein after exposure to metals, host response to stress, or a complex combination of them. Finally, it is concluded that if metallothionein-like structures are to serve as biomarkers in aquatic systems, much more direct study of specific aquatic organisms will be needed.

INTRODUCTION

Mammalian organisms exposed to elevated levels of a number of metals including Cd, Hg, Zn, and Cu respond with the de novo synthesis of hepatic and renal metallothionein (MT), which then chelates most of the extra metal entering the tissue.[1,2] Other metals such as Pt do not induce MT but avidly bind to preexistent metallothionein.[3] Because of the prominence of metallothionein in studies of heavy metal toxicity, attention has been drawn to its possible use as a biomarker for heavy metal exposure in the environment. Evaluation of this proposal will require field work, which is guided by an understanding of the complexities of metallothionein biochemistry and chemistry and by the recognition that other metal binding structures may be of interest as well. The objective of this article is to provide a selective summary and analysis of the information in the literature on mammalian metallothionein and results from the authors' laboratories on aquatic systems, which address the question of the feasibility of metal-binding structures as biomarkers.

Metallothionein Structure and Metal-Binding Properties

Mammalian metallothionein is a remarkable, 61 amino acid protein, in which 20 cysteinyl thiol (SH) groups serve as ligands for metal chelation.[4] Cadmium, zinc, and copper bind in two metal clusters located in separate structural domains of the protein.[5-7] The first two metals (M) have stoichiometries in the two clusters of M_3S_9 and M_4S_{11}; those for copper appear to be Cu_6S_9 and Cu_6S_{11}. Other metallic species such as Au(I) and Ag(I) bind with ratios of metal to sulfhydryl of or approaching 1 to 1.[8,9] Importantly, depending on the history of the organism from which metallothionein is isolated, the protein may contain multiple metals in mixed-metal clusters. Thus, for example, native rat kidney metallothionein has both Zn and Cu; after exposure of the animals to $CdCl_2$, Cd, Zn, and Cu are found in the protein.[10]

In vitro addition of metal ions to apoMT results in the formation of complexes between the protein and virtually all heavy metals tested.[11] In contrast, in cells metallothionein complexes involving only the following metals have been documented: Zn, Cd, Cu, Hg, Ag, Pt, Au, and Bi.[10,12-16] As described below, Ni and Co also induce the synthesis of MT m-RNA, but their binding to the protein has not been documented. Thus, despite the enormous concentration of sulfhydryl groups in MT, the protein competes effectively in vivo for only some metal ions.

Qualitative orders of equilibrium affinity among these metals include Hg > Cu > Cd > Zn;[17] Pt > Cu (unpublished); Au ~ Cd.[8] The only quantitative, log apparent stability constants for metals bound to metallothionein at pH 7.4 are 11.2 per Zn for all of the metals in Zn_7-MT and 14.7 per Cd for Cd_7-MT.[18] Importantly, this information is directly useful in assessing metal distribution in metallothionein, for the kinetics of metal exchange reactions of the protein except ones involving Pt are rapid and do not control the metal content of the protein.[19]

Table 1.   Possible Functions of Mammalian Metallothionein.

Protection against Toxicity
 Detoxication of heavy metals through chelation[2,3,10,12,20]
 Detoxication of heavy metals (Pt, Au), organic electrophiles, and free radicals through
  reaction with MT thiols[13–15,21–25]
Normal Cellular Functions
 Regulation of Zn metabolism
  Induced storage form (buffer) for Zn entering cells as during the generalized stress
   response of the host and after tissue exposure to hormones and other factors[26–31]
  Intracellular donor of Zn to apoZn metalloproteins[32]
 Role in Cu metabolism
  Intermediate in metabolism of Cu[33]

This observation stands in sharp contrast with the behavior of a number of metalloenzymes, in which rates of metal exchange are typically very slow.[19]

Metallothionein Function

Metallothionein has been studied from a variety of viewpoints for 30 years. The literature on its chemistry, biochemistry, molecular genetics, and participation in normal and toxicological processes in mammalian organisms is immense. Yet, at present there is no unified understanding of its place or significance in biological systems. Table 1 lists a number of documented or potential roles for metallothionein.

The inquiry into the use of metallothionein as a biomarker has its genesis in the central role MT plays in the response of mammalian organisms to Cd exposure. Studies with a number of biological systems show that when Cd enters cells (Figure 1A), it either binds to preexistent metallothionein or induces the synthesis of new protein, which then sequesters (Figure 1B) most of the Cd distributed throughout the cell.[34,35] This binding process protects against acute toxicity.[35] Similar results have been obtained using other toxic metals which also interact with this structure.[36] In the case of hepatic metal exposure, mixed-metal metallothioneins are always formed.[36] In steady-state cycles of biodegradation (Figure 1C) and synthesis of MT, Cd remains bound to the protein.[37] It is the sensitive response of cells to Cd—de novo MT m-RNA and protein synthesis—which has suggested the utilization of MT as a measure of heavy metal exposure.[38]

Recently, attention has been directed toward the nucleophilic reactivity of the multiple sulfhydryl groups in metallothionein. Not only are they reactive toward such metallic compounds as *cis*-dichlorodiammine Pt(II) and some Au(I)-complexes, but they display substantial reactivity toward organic electrophiles such as N-chloroamines, activated double bonds, and activated disulfides.[8,13,39] In addition, they are thought to react with radical species such as ˙CCl_3, organoperoxides or their breakdown products, and oxyradicals, generated by Fenton chemistry.[24,25,40]

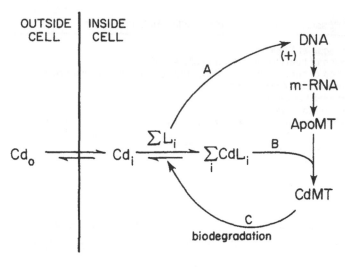

Figure 1. Cadmium distribution in cells: induction (A), binding of Cd (B), and biodegradation (C) of metallothionein.

In other research not closely connected to work in toxicology, it is now evident that metallothionein is centrally involved in biologically essential metabolism of zinc and possibly copper. Although its exact activities remain elusive, induced MT protein, for example, serves as the hepatic acceptor of zinc during its redistribution from plasma into liver in response to a number of stresses such as infection, injection of chemicals, heat, cold, exercise, starvation, and the presence of transplanted tumor cells.[26–28,41] Besides processes involving induction of MT, basal level protein may also be functional in other aspects of zinc metabolism such as the intracellular distribution of zinc to metal-requiring sites.[32] Indeed, basal metallothionein-zinc levels are rapidly responsive to the zinc-status of the host.[10] It is evident in mammals, therefore, that simple measurement of metallothionein protein does not immediately indicate the degree of exposure of the tissue to Cd or other toxic metals.

Control of Metallothionein Gene Expression

Metallothionein synthesis can be induced by a large number of diverse agents (Table 2). Direct inducers are able to induce MT synthesis when placed in contact with cells in culture. In contrast, indirect inducers require some mediation by the organism such as the stress response to cause induction of the protein. Among the inducers there is a large range of inducer strength. Thus, in one study the qualitative order of efficacy of induction of MT m-RNA in mouse hepatoma cells was Cd ~ Ag ~ Hg > Zn > Cu > Ni ~ Co ~ Bi.[38] Furthermore, the rates of biodegradation of MT protein vary with the metal bound. Metallothionein induced in rat liver by Cu has a half-life of 12-16 hr, that stimulated by Zn, 20

Table 2.   Inducers of Metallothionein.

Direct Inducers
  Heavy metals: Cd^{2+}, Zn^{2+}, Cu^{2+}, Hg^{2+},
              Co^{2+}, Ni^{2+}, Bi^{2+}, Ag^{1+} [42]
  Glucocorticoids[43]
  Progesterone[44,45]
  Catecholamines[29,46]
  Glucagon[47]
  Interleukin I[30]
  Interferon[48]
  Phorbol esters[29]
  Iodoacetate[38]
Indirect Inducers
  Carbon tetrachloride[26]
  Ethanol, urethane, lead, manganese[42]

hr, and the Cd-protein, 70–90 hr.[37,50,51] More importantly, whereas Cd, Zn, and Cu remain bound to MT in the steady state of biosynthesis and biodegradation of protein induced by Cd, both Zn and Cu and protein are lost when metallothionein, induced by the latter two metals, biodegrades.

Methods of Measurement of Metallothionein

The determination of metallothionein concentration is also a complex subject. Because the protein has no known catalytic function, its measurement must be based on the quantitative assay of the protein itself. Table 3 lists several approaches to this problem.

In the first, protein purification is carried out prior to metal analysis to ensure that it is the metal content of MT protein that is being determined. Sephadex G-75 chromatography nicely separates out metal bound to macromolecules of about 10,000 Da from the bulk of the larger proteins in solution. By itself, the finding of metal in this MT fraction of the column profile is indicative but not sufficient to define the source of metal as metallothionein. Thus, for example, the testicular Cd-binding protein, once thought to be metallothionein appears to be a different structure.[58]

DEAE ion exchange chromatography on the 10,000-Da material, revealing one or two bands of metal eluting at conductivities of about 4 and 6 mS, provides

Table 3.   Determination of Metallothionein.

1. Isolation of MT by Sephadex G-75 and DEAE chromatography followed by metal analysis[52]
2. Polarographic thiol analysis of cytosol[53,54]
3. Cd-heme method applied to the cytosol[55]
4. Immunochemical assay of MT protein[56]
5. Northern blot analysis for MT messenger RNA[57]

strong supporting evidence that the metal-binding substances are metallothionein. The virtue of this procedure is that the actual metals bound to the protein are determined by atomic absorption spectrophotometry. Knowledge of the metal distribution is a prerequisite to assessment of the exposure of the organism to particular substances. It is also important to repeat this procedure after treating the supernatant with Cd^{2+} to form Cd-MT from any apoMT which may be present. One then can quantitate the metals bound to metallothionein as well as the possibility that some metal-free protein also exists in solution. Naturally, as described this is a slow, material intensive method.

Polarography and Cd-Heme Methods

Both methods 2 and 3 measure metallothionein protein concentrations in situ without isolation. Neither determines the metal composition of the protein. Unless one knows the history of the organism and what this implies about the metal content of MT, measurement of protein concentration alone leaves one without the key information needed to assess the meaning of the protein determination.

Each method pretreats the cytosol to remove proteins, which might interfere with the analysis. Then, in the polarographic method metallothionein is quantitated by measuring electrochemically its sulfhydryl content. In the Cd-heme method Cd^{2+} is mixed with treated supernatant, all non-MT Cd is assumed to be bound by added hemoglobin and precipitated from solution by heating, and the metallothionein concentration determined by measurement of Cd in the supernatant (it is taken to be $1/7$ of the gram equivalent Cd content). Neither method provides information on the metal composition of metallothionein nor do they permit any direct confirmation that the measured quantities relate solely to metallothionein. As such they are best used in well defined systems. The Cd-heme method suffers particularly because Cd^{2+} does not displace Cu or Hg and perhaps other metals from metallothionein. Hence, without any information about the metals actually bound to an unknown sample, even the calculated MT protein content may be misleading.

Immunochemical Assay

Determination of immunological cross reactivity with authentic metallothionein is an important, specific method for quantitating MT protein. However, not all metallothioneins cross react with MT antibodies raised to Cd- or Zn-metallothionein. For example, $\{Au(I)(thioglucose)\}_{20}$-MT displays little of the antigenic determinant to the antibody.[8] As with methods 2, 3, and 5, this quantitation provides no information about the metals bound to the protein. Nevertheless, when protein is in limited supply, it is the method of choice to assay samples for the presence of metallothionein.

M-RNA Analysis

Like methods 2 through 4, this is a very sensitive method. It is also specific for MT m-RNA. It suffers in that it reveals nothing about the metal content of metallothionein protein which may be present and, in fact, only indicates that message is present. One cannot necessarily infer that MT protein is proportional to m-RNA, for at least the steady state level of protein depends on the relative rates of protein biosynthesis and biodegradation.

It is evident from this brief examination of methods used to quantitate MT that each presents problems and limitations. Nevertheless, it is strongly argued that measurements only of protein are inadequate for they leave unresolved the identity of the metals, which are bound to protein. Since in large measure the significance of finding the protein is tied to the particular metals bound to it, without this information it will be difficult to determine what sort of biomarker it is.

Non-mammalian Metal Binding Proteins

There is much support for the view that the principal binding protein for toxic metals in mammalian liver and kidney is metallothionein. Once outside of this class of organisms, however, the generalization no longer holds. Thus, plants and some microorganisms respond to Cd or Cu exposure by synthesizing small peptides called phytochelatins as discussed below.[59] A variety of other organisms make different non-metallothionein binding proteins for Cd and Hg.[20] Hence, care must be taken to identify carefully the nature of the structures to which heavy metals are binding in non-mammalian systems. In the sections that follow, our experience with the responses of aquatic organisms to metals will be described, compared with results in mammals, and used to illustrate the issues involved in the use of metallothionein or other metal-binding structures as biomarkers.

Metal Binding Proteins in Fish

Injection of fish with Cd^{2+} leads to the rapid localization of Cd in a hepatic protein thought to be metallothionein (Figure 2). A first interpretation of this result is that fish, like mammals, respond to acute Cd exposure with the de novo synthesis of new metallothionein protein and that the presence of this protein may be a good indicator of exposure to heavy metals. With additional experiments, the analysis of this experiment becomes more complicated.

Basal Zn,Cu-Binding Protein (Zn,Cu-BP)

Liver or hepatopancreas in many freshwater fishes contains large, sometimes enormous, amounts of a MT-like Zn,Cu-binding protein in the absence of exposure of the host to Cd (Figure 3).[60] The ratio of zinc to copper in the protein varies with the species. This material chromatographs over DEAE Sephadex like mammalian Zn-MT, revealing only one isoprotein in most cases. That this is metallothionein is suggested by the presence of metallothionein genes in rainbow trout.[61] Interestingly, neither the rainbow trout protein nor other Zn,Cu-binding proteins cross react with mammalian metallothionein despite the strong homology between mammalian and trout protein sequences (unpublished information).

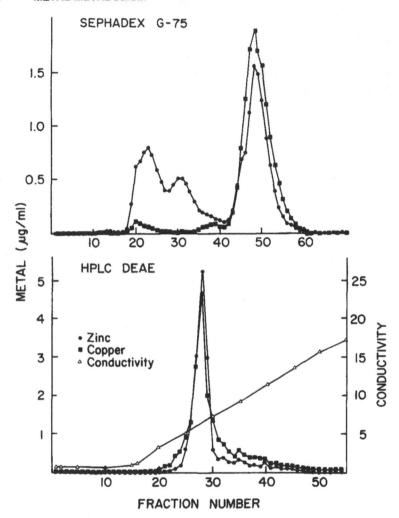

Figure 2. Metal distribution in hepatic cytosol from rainbow trout injected with 200 µg Cd/kg body weight at 0 and 24 hr and sacrificed at 48 hr. Sephadex G-75 profile of cytosol and DEAE anion exchange profile of the lower molecular weight band from Sephadex.

Because environmental zinc or copper can be toxic to fish, it is important to understand whether the hepatic levels of Zn,Cu-BP reflect a normal physiological situation or the response of the organism to toxic concentrations of these metals. Thus, a life-time study of liver metal distribution is underway using rainbow trout raised from eggs entirely in a clean laboratory environment. From the earliest time that liver can be isolated and examined, about 60 days after hatching (16 mg liver per fish), a Zn,Cu-BP can be detected. The concentrations of metals in this fraction undergo little change as the trout grow during the next 5 months.

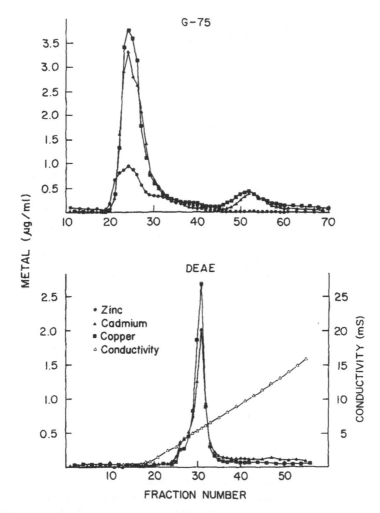

Figure 3. Metal distribution in control liver of Burbot. Sephadex G-75 profile of cytosol and DEAE anion exchange chromatography of the lower molecular weight Zn,Cu-binding protein from Sephadex.

There is nothing to indicate, therefore, that BP is preferentially acquiring potentially toxic zinc or copper to prevent their distribution within this organ.

In summary, freshwater fishes contain large, basal levels of a metallothionein-like BP. Survey of such organisms for elevated metallothionein by methods 2, 3, or 5 in Table 3 without an appreciation that they bear large, uninduced concentrations of BP would lead to the erroneous conclusion that they have been exposed to toxic heavy metals or some other stress. Indeed, had the Cd-heme method been applied, the amount of BP would have been underestimated because the copper sites are unreactive with Cd as they are in authentic metallothionein.

Furthermore, low levels of Cd might simply displace Zn from preexistent Zn,Cu-BP and not change the basal concentration of binding protein. In this case, protein determination alone would not suggest exposure to a toxic metal.

Binding Protein and the Stress Response

A second property of BP confounds the interpretation of its presence in liver. This is its role in the response of fish to stress. It is well known that fish are very sensitive to handling and upon restraint release large amounts of glucocorticoid into plasma.[62] We inquired whether restraint stress leads to the same pattern of metal redistribution seen in mammals, in which Zn leaves the plasma and enters hepatic metallothionein and Cu is released as ceruloplasmin from liver into plasma.[27,28] After 20-min restraint once a day for 3 days during which the animals refuse to feed, largemouth bass show no increase in Zn and 20% increase in Cu in the Cu,Zn-binding protein. This is accompanied by a doubling in plasma copper and an increase of 150% in plasma Zn. Thus, there is a strong metal-related component of this stress response to restraint in fish. Therefore, in the Cd injection experiment mentioned at the outset of this section, handling of the fish during administration of Cd might conceivably affect the amount of BP detected later, which is attributed to induction of BP by cadmium. Indeed, the mere act of capturing a fish for examination might be sufficient to elevate basal binding protein concentrations. Furthermore, given the finding in rodents that fasting elevates liver MT, the results in largemouth bass might be a secondary effect due to fasting brought on by restraint stress.[41]

In another study, elevation in hepatic binding protein metal content was not seen after injection of plaice with endotoxin or turpentine, both of which induce metallothionein synthesis in rodents.[27,63,64] Hence, more study will be needed to assess the extent of similarity between the stress response in mammals and fish.

The fact that fish respond to some stresses with enormous increases in plasma glucocorticoid hormone and plasma Zn suggests that measurement of these parameters can help sort out the role that stress may play in the level of hepatic binding protein that is measured. The ease with which plasma Zn can be determined by atomic absorption spectrophotometry makes it a strong candidate for routine examination in environmental surveys.

Binding Protein in Animals Exposed to Cadmium in Water

Although fish bind substantial amounts of Cd with hepatic BP after injection of the metal, natural routes of intake are not nearly as effective in accumulating Cd in hepatic binding protein as injection.[60,65] Thus, after exposure of rainbow trout to 9 μg Cd/L for 12 months, little of the Cd acquired by the organism is found in liver; much more is in kidney.[60] In reports of Cd accumulation in feral fish, similar concentrations of Cd are found in liver as seen in these experiments.[66]

What metal there is in this organ is bound to a MT-like binding protein, which chromatographs on DEAE Sephadex as one band like the Cd-BP from cadmium-injected animals and basal Zn,Cu-BP. This result is different than that reported by Kay and co-workers, which indicated that Cd binds to a protein different from the one that chelates Zn and that both proteins separate into two isoforms during ion-exchange chromatography.[65]

After exposure to water-borne Cd, gill, like liver, contains very little Cd. Most of the body burden resides in the kidney as a Zn,Cu,Cd-BP.[61,66] In comparison with controls, zinc and copper as well as cadmium have accumulated in this protein. Thus, the total concentration of the protein is related only indirectly to its Cd content, for as in mammalian systems Cd binding in fish is accompanied by additional complexation of some mix of Zn and Cu. As a consequence, it is necessary to assay the nature of the metals bound to metallothionein in order to assess the meaning of a measurement of protein content.

Binding Protein in Animals Exposed to Cadmium in Food

To compare the inefficient uptake of Cd from water with intake from a food source, largemouth bass were fed trout fry, injected with 50 mg Cd/kg, every 3 days for 30 days. At the end of this time their digestive tracts had been exposed to approximately as much Cd as rats which drink 100 μg Cd/mL ad libitum for 30 days. Analysis of the gut and the liver shows that high levels of Cd are bound to intestinal mucosa, mostly in a MT-like species, but that little Cd has reached the liver. As in the case of water exposure, it is bound principally as a Cd-binding protein. Notably, however, even high level, sustained exposure of fish to dietary Cd does not produce much liver Cd-BP.

One needs to be aware that diet may affect Cd uptake as it does in mammals.[67] If this is a factor which can modulate Cd uptake in fish, then interpretation of surveys of binding protein must take account of this confounder, which may vary from organism to organism.

Cadmium Binding Protein in Anurans

Anurans are more tolerant to metal stress than some freshwater fish. For example, while 30 μg Cd/L kills rainbow trout, tadpoles can be immersed in water containing 800 μg Cd/L for 30 days without gross toxicity (unpublished information). Analysis of liver concentration and distribution of Cd after this period shows that most of the metal is sequestered in a metallothionein-like metal binding protein.[68] Interestingly, the amount of Cd in the liver after exposure to 0.1 μg Cd/mL, 16 μg/g wet weight, is comparable to that seen in rats provided drinking water containing 100 μg Cd/mL for 30 days. Considering that each tadpole is maintained in 1 L of water, which is changed 11 times over the 1-month exposure and that rats drink about 30 mL of water a day, the total, integrated Cd placed in the environment of the anuran and rodent during the experiments is 1.1 mg and 90 mg, respectively. Clearly, the tadpole much more efficiently localizes the metal in its liver than does the rat.

Anurans are similarly effective in accumulating hepatic zinc and lead. Hence, this organism may be a particularly sensitive biomonitor for metals in the environment. Based on dose response data for Cd uptake into liver, it is evident that 30 days of incubation in 10 μg Cd/L can readily be detected. One can observe liver cadmium that results from this exposure whether or not Cd-binding protein is measured.

Cadmium Binding in *Euglena gracilis*

The photosynthetic microorganism *Euglena gracilis* responds like higher plants to Cd and synthesizes small glutathione-like peptides which bind to the metal.[69,70] *Euglena* make a complex of Cd, inorganic sulfide, and peptide with the approximate stoichiometry, Cd_2^{2+}, S_2^{2-}, [(Glu-cys)$_3$-gly], which chromatographs on Sephadex G-75 much like Cd-metallothionein.[71,72] Without further anion-exchange chromatography, one may misconstrue this structure for metallothionein.

An intriguing feature of the interaction of Cd with *Euglena gracilis* is that at lower exposure levels (1–3 μg Cd/mL), which do not inhibit proliferation of the cells, Cd is associated with high molecular weight fractions of cytosol chromatographed over Sephadex G-75.[71] It is only at 5 μg Cd/mL, at which cell division is reduced, that the Cd-phytochelatin structure is observed. This is exactly the opposite from mammalian cells, which initially react to the presence of metal by inducing and then forming Cd-metallothionein.[34,35] Only at higher concentrations of Cd does metal begin to appear in high molecular weight species.

The reason for this may be that the underlying detoxification mechanism of *E. gracilis* does not initially involve formation of the phytochelatin structure but the generation of sulfide ion, which binds Cd^{2+} tightly as CdS and removes it from other, deleterious sites of interaction with the cell. Thus, unpublished results suggest that the high molecular weight form alluded to above may be cadmium sulfide. The identification of this material as CdS has also been made in *Candida glabrata*.[73]

When *Euglena gracilis* is grown in the dark as a heterotroph, it does not synthesize phytochelatin peptides to bind Cd and sulfide. Instead, it makes a metallothionein-like protein.[74] This is also true of non-mammalian organisms like the yeast strain *Saccharomyces cerevisiae*, and *Neurospora crassa* fungi, which elaborate distinct, MT-like proteins in response to exposure to certain heavy metals.[75,76] Interestingly, in yeast, copper but not cadmium or zinc induces synthesis of the binding protein.[77] In the fungus, *Candida glabrata*, Cd induces a MT-like protein whereas Cu induces phytochelatin peptide synthesis.[73] It is evident that the metabolism of toxic metals is variable and complicated among microorganisms. Without a detailed analysis of the responses of particular species to an array of metals, it is risky to apply the methods and knowledge derived from mammals for metallothionein to lower organisms.

Use of Metallothionein and Other Metal Binding Proteins/Peptides as Biomarkers

The potential use of metallothionein or related proteins as biomarkers focuses on their involvement in binding toxic metals and in the general host stress response. To evaluate their possible utility, one needs to ask what is gained by measuring tissue MT levels instead of environmental or tissue concentrations of metals. With present techniques such as flameless atomic absorption spectrophotometry or new ones like inductively coupled plasma mass spectroscopy, nanomolar concentrations of most metals can be readily determined. Because there is a limited number of metals in comparison to organic chemicals, which might contaminate the environment, it is, in principle, much easier to measure the metal profile of a sample than its composition of organics. It is also easier to measure total tissue concentrations of metals than to determine their distribution among binding sites such as metallothionein. Therefore, significant information about environmental contamination and bioavailability of metals can be obtained about metal exposure without considering what fraction is bound to particular metal binding proteins.

Where more refined analyses of metal distribution may be important are in studies centering on whether particular degrees of metal exposure or stress are significant for the organism. For example, if one believes that potentially toxic levels of Cd necessarily elicit metallothionein synthesis in a given tissue or cell, then measurement of MT levels may be able to define whether environmental or cellular concentrations of Cd are biologically important or whether they are simply low background levels. Clearly, with the possible involvement of metallothionein in metal exposure, stress, and the response of organisms to organic electrophiles and free radicals, the potential exists to learn much about environmental conditions from examining the behavior of metallothionein in various field situations.

The difficulties associated with the interpretation of the meaning of metallothionein levels in cells have been illustrated in a number of ways. Not even considered are suggestions that out in the environment away from constant conditions maintained in animal facilities there may be seasonal variations in tissue binding protein concentrations.[20] Indeed, in one particularly elegant study it has been shown that Cu and Zn binding to metallothionein-like proteins in blue crab vary inversely during the molt cycle of this creature.[78] Thus, it is our view that if metallothionein and related binding proteins are to be exploited for environmental monitoring purposes, an extensive knowledge base derived from the direct study of particular non-mammalian organisms will need to be established and then applied by investigators familiar with the intricacies of the biology of these structures.

ACKNOWLEDGMENTS

The authors have been supported by the following grants from NIH: ES-04026 and ES-04184.

REFERENCES

1. Webb, M. *The Chemistry, Biochemistry and Biology of Cadmium*, M. Webb, Ed. (Amsterdam: Elsevier/North-Holland, 1979), Chapter 6.
2. Webb, M., and K. Cain. *Biochem. Pharmacol.* 31:137–142 (1982).
3. Bakka, A. B., L. Endresen, A. B. S. Johnsen, P. D. Edminson, and H. E. Rugstad. *Toxicol. Appl. Pharmacol.* 61:215–226 (1981).
4. Kägi, J. H. R., and Y. Kojima. *Metallothionein II*, J. H. R. Kägi, and Y. Kojima, Eds. (Basel: Birkhäuser Verlag, 1985), p. 25.
5. Winge, D. R., and K.-A. Miklossy. *J. Biol. Chem.* 257:3471–3476 (1982).
6. Nielson, K. B., and D. R. Winge. *J. Biol. Chem.* 258:13063–13069 (1983).
7. Nielson, K. B., and Winge, D. R. *J. Biol. Chem.* 259:4941–4946 (1984).
8. Laib, J. E., C. F. Shaw III, D. H. Petering, M. K. Eidness, R. C. Elder, and J. S. Garvey. *Biochemistry* 24:1977–1986 (1985).
9. Johnson, B. A., and I. M. Armitage. *Inorg. Chem.* 26:3139–3144 (1987).
10. Petering, D. H., J. J. Loftsgaarden, J. Schneider, and B. Fowler. *Environ. Health Pers.* 54:73–81 (1984).
11. Nielson, K. B., C. L. Atkin, and D. R. Winge. *J. Biol. Chem.* 260:5342–5350 (1985).
12. Piotrowski, P., B. Trojanowska, Wisniewska-Knypl, and W. Bolanowska. *Toxicol. Appl. Pharmacol.*, 27:11–19 (1974).
13. Zelazowski, A. J., J. S. Garvey, and J. E. Hoeschele. Arch. *Biochem. Biophys.* 229:246–252 (1984).
14. Kraker, A., J. Schmidt, S. Krezoski, and D. H. Petering. *Biochem. Biophys. Res. Commun.*, 130:786–792 (1985).
15. Schmitz, G., D. T. Minkel, D. Gingrich, and C. F. Shaw III. *J. Inorg. Biochem.* 12:293–306 (1980).
16. Webb, M. *Metallothionein II*, J. H. R. Kägi, and Y. Kojima, Eds. (Basel: Birkhäuser Verlag, 1985), p. 109.
17. Holt, D., L. Magos, and M. Webb. *Chem. Biol. Interact.*, 32:125–135 (1980).
18. Bachowski, G., C. F. Shaw III, and D. H. Petering, submitted.
19. Otvos, J. D., D. H. Petering, and C. F. Shaw III. *Comments Inorg. Chem.* 1989, in press.
20. Petering, D. H., and B. A. Fowler. *Environ. Health Pers.* 65:217–224 (1986).
21. Li. T.-Y., D. T. Minkel, C. F. Shaw III, and D. H. Petering. *Biochem. J.* 193:441–46 (1981).
22. Glennas, A., and H. E. Rugstad. *Ann. Rheum. Dis.* 45:101–109 (1986).
23. Endresen, L., A. Bakka, and H. E. Rugstad. *Cancer Res.* 43:2918–2926 (1983).
24. Cagen, S. Z., and C. D. Klaassen. *Toxicol. Appl. Pharmacol.* 54:229–237 (1980).
25. Coppen, D. E., D. E. Richardson, and R. J. Cousins. *Proc. Soc. Exp. Biol. Med.* 189:100–109 (1988).
26. Oh, S. H., J. T. Deagen, P. D. Whanger, and P. H. Weswig. *Am. J. Physiol.* 234:E282–85 (1978).
27. Sobocinski, P. Z., W. J. Canterbury Jr., C. A. Mapes, and R. E. Dinterman. *Am. J. Physiol.* 234:E399–E406 (1978).
28. Ujjani, B., G. Krakower, G. Bachowski, S. Krezoski, C. F. Shaw III, and D. H. Petering. *Biochem. J.* 233:99–105 (1986).

29. Brady, F. O., B. S. Helvig, A. E. Funk, and S. H. Garrett. *Metallothionein II*, J. H. R. Kägi, and Y. Kojima, Eds. (Basel: Birkhäuser Verlag, 1985), p. 555.
30. Cousins, R. J., and A. S. Leinart. *FASEB J.* 2:2884–2890 (1988).
31. Imbra, R. J., and M. Karin. *Mol. Cell Biol.* 7:1358–1363 (1987).
32. Krezoski, S. K., J. Villalobos, C. F. Shaw III, and D. H. Petering. *Biochem. J.* 255:483–491 (1988).
33. Bremner, I. *J. Nutr.* 117:19–29 (1987).
34. Bryan, S.E., and H. A. Hidalgo. *Biochem. Biophys. Res. Commun.* 68:858–866 (1976).
35. Enger, M. D., L. T. Ferzoco, R. A. Tobey, and C. E. Hildebrand. *J. Toxicol. Environ. Health* 7:675–690 (1981).
36. Winge, D. R., R. Premakumar, and K. V. Rajagopalan. *Arch. Biochem. Biophys.* 170:242–252 (1975).
37. Chen, R. W., P. D. Whanger, and P. H. Weswig. *Biochem. Med.* 12:95–105 (1975).
38. Durnam, D. M., and R. D. Palmiter. *Mol. Cell. Biol.* 4:484–491 (1984).
39. Li, T.-Y., D. T. Minkel, C. F. Shaw III, and D. H. Petering. *Biochem. J.* 193:441–446 (1981).
40. Thomas, J. P., G. J. Bachowski, and A. W. Girotti. *Biochim. Biophys. Acta* 884:448–461 (1986).
41. McCormick, C. C., M. P. Menard, and R. J. Cousins. *Am. J. Physiol.* 240:E414–421 (1981).
42. Bracken, W. M., and C. D. Klaassen. *J. Toxicol. Environ. Health* 22:163–74 (1987).
43. Failla, M. L., and R. J. Cousins. *Biochim. Biophys. Acta* 538:435–444 (1978).
44. Bremner, I., R. B. Williams, and B. W. Young. *J. Inorg. Biochem.* 14:135–146 (1981).
45. Slater, E. P., A. C. Cato, M. Karin, J. D. Baxter, and M. Beato. *Mol. Endocrinol.* 2:485–491 (1988).
46. Brady, F. O., and B. Helvig. *Am. J. Physiol.* 247:E318–322 (1984).
47. DiSilvestro, R. A., and R. J. Cousins. *Am. J. Physiol.* 247:E436–441 (1984).
48. Friedman, R. L., and K. T. Stark. *Nature* 314:637–639 (1985).
49. Winge, D. R., R. Premakumar, and K. V. Rajagopalan. *Arch. Biochem. Biophys.* 186:466–475 (1978).
50. Feldman, S. L., and R. J. Cousins. *Biochem. J.* 160:583–590 (1976).
51. Bremner, I., W. G. Hoekstra, N. T. Davies, and N. T. Young. *Biochem. J.* 174:883–892 (1978).
52. Minkel, D. T., K. Poulsen, S. Wielgus, C. F. Shaw III, and D. H. Petering. *Biochem. J.* 191:475–485 (1980).
53. Olafson, R. W., and R. G. Sim. *Anal. Biochem.* 100:343–351 (1979).
54. Olafson, R. W. *J. Biol. Chem.* 256:1263–1268 (1981).
55. Eaton, D. L., and B. F. Toal. *Toxicol. Appl. Pharmacol.* 66:134–142 (1982).
56. Vander Mallie, R. J., and J. S. Garvey. *J. Biol. Chem.* 254:8416–8421 (1979).
57. Swerdel, M. R., and R. J. Cousins. *J. Nutr.* 112:801–809 (1982).
58. Deagen, J. T., and P. D. Whanger. *Biochem. J.* 231:279–283 (1985).
59. Grill, E., E.-L. Winnacker, and M. H. Zenk. *Proc. Natl. Acad. Sci. U.S.A.* 84:439–443 (1987).
60. Krezoski, S., J. Laib, P. Onana, T. Hartmann, P. Chen, C. F. Shaw III, and D. H. Petering. *Mar. Environ. Res.* 24:147–150 (1988).
61. Bonham, K., M. Zafarullah, and L. Gedamu. *DNA* 6:519–28 (1987).

62. Spieler, R.E. *J. Fish. Res. Bd. Can.* 31:1240–1242 (1974).
63. Overnell, J., R. McIntosh, and T. C. Fletcher. *Experientia* 43:178–181 (1987).
64. Searle, P. F., B. L. Davison, G. W. Stuart, T. M. Wiikie, G. Norstedt, and R. D. Palmiter. *Mol. Cell. Biol.* 4:1221–1230 (1984).
65. Kay, J., D. C. Thomas, M. W. Brown, A. Cryer, D. Shurben, J. F. deL. Solbe, and J. S. Garvey. *Environ. Health Pers.* 65:133–140 (1986).
66. Bohn, A., and R. O. McElroy. *J. Fish. Res. Bd. Can.* 33:2836–2840 (1976).
67. Waalkes, M. *J. Toxicol. Environ. Health* 18:301–314 (1986).
68. Suzuki, K. T., and H. Akitomi. *Comp. Biochem. Physiol.* 75C:211–215 (1983).
69. Taylor, P., D. Weber, D. Gingrich, C. F. Shaw III, and D. H. Petering. *Heavy Metals in the Environment* (1987), p. 250–252.
70. Shaw III, C. F., D. H. Petering, D. Weber, and D. J. Gingrich. *Metal Ion Homeostasis: Molecular Biology and Chemistry*, D. Winge, and D. Hamer, Eds. (New York: Alan R. Liss, in press).
71. Gingrich, D. J., D. N. Weber, C. F. Shaw III, J. S. Garvey, and D. H. Petering. *Environ. Health Pers.* 65:77–85 (1986).
72. Weber, D. N., C. F. Shaw III, and D. H. Petering. *J. Biol. Chem.* 262:6962–6964 (1987).
73. Mehra, R. K., E. B. Tarbet, W. R. Gray, and D. R. Winge. *Proc. Natl. Acad. Sci. U.S.A.* 85:8815–8819 (1988).
74. Mazus, B., K. H. Falchuk, and B. L. Vallee. *Fed. Proc.* Abst. 2737 (1988).
75. Winge, D. R., K. B. Nielson, W. R. Gray, and D. H. Hamer. *J. Biol. Chem.* 260:14464–14470 (1985).
76. Lerch, K. *Nature* 284:368-370 (1980).
77. Butt, T. R., E. J. Sternberg, J. A. Gorman, P. Clark, D. H. Hamer, M. Rosenberg, and S. T. Crooke. *Proc. Natl. Acad. Sci. U.S.A.* 81:3332–3336 (1984).
78. Engel, D. W. and M. Brouwer. *Biol. Bull.* 173:239–251 (1987).

CHAPTER 15

Metallothionein as a Biomarker of Environmental Metal Contamination: Species-Dependent Effects

William H. Benson, Kevin N. Baer, and Carl F. Watson

ABSTRACT

The metal-binding protein, metallothionein (MT), is induced by selected heavy metals and has appeared promising as a biological marker of metal exposure. Mammalian MT, however, is induced by additional factors such as environmental stresses, and therefore, the use of MT as a biomarker for metal exposure might be premature. Because of the paucity of data concerning the induction of MT following stress in teleosts, the present comparative investigation focused on the influence of cadmium and environmental stresses on MT induction in mice and bluegill sunfish. Following exposure to environmental stresses, significant decreases in zinc and copper were observed in the gill MT-like fraction, as compared to control. In teleost liver, no significant alterations in metal content were observed in the MT-like fraction, as compared to control. These results indicate that the induction of teleost hepatic MT-like proteins appears to be more metal-specific.

INTRODUCTION

Increasing environmental exposure to heavy metals creates the need for development of biological markers useful in contaminant effects surveys in aquatic and terrestrial ecosystems. An advantage to the use of biomarkers in such surveys would be the ability to measure a contaminant or its metabolite at the critical cellular target. Moreover, biomarkers might be used in a quantitative manner to predict early detrimental effects of a contaminant. Metallothionein (MT), an intracellular, low molecular weight metal-binding protein appears promising as a biomarker of metal exposure. For example, in rats and mice, a dose-related increase in MT has been demonstrated with exposure to cadmium.[1,2] In addition, an increase in MT excretion appears to be related to cadmium concentration in kidney and renal dysfunction.[3,4] Such a relation was demonstrated in the plasma and urine of cadmium-exposed workers by use of a radioimmunoassay for MT.[4,5]

It has further been demonstrated that MT induction in mammals is influenced by a variety of factors in addition to metal exposure. For example, environmental stresses, such as cold, have been observed to increase hepatic MT levels.[2,6] The induction of MT during periods of stress may be regulated by hormonal control at the cellular level. The synthetic glucocorticoid dexamethasone has been demonstrated to enhance zinc uptake and induce MT in vitro.[7,8] In addition, glucagon[9] as well as epinephrine and norepinephrine increase MT-bound zinc.[10] Results from a recent investigation by Hidalgo et al.,[11] however, suggest that glucocorticoids have an inhibitory role on the maintenance of MT concentration in the liver and that these hormones have a permissive role on the flux of MT from organs into the blood, at least in some circumstances. Although the physiological mechanism is unclear, these results indicate a function of MT in zinc regulation during periods of stress. From these considerations, the induction of MT during the stress phenomenon may confound the use of this protein as a biomarker of metal exposure in mammals.

In teleost tissues, numerous investigators have demonstrated a correlation with MT concentration and metal exposure.[12-14] Regarding the influence of stress factors, stress (due to catching), endotoxin, dexamethasone, cortisol, and turpentine failed to induce MT in plaice.[15] In addition, Thomas and Wofford[16] did not observe any fluctuations in hepatic acid-soluble thiol concentrations in mullet after feeding or during short-term starvation, or after acute or chronic physical trauma. Such findings have led investigators to suggest the use of MT as a specific biomarker of metal exposure in teleosts. However, an investigation by Olsson et al.[17] demonstrated variations in MT levels during the annual reproductive cycle in rainbow trout. These limited findings suggest that, at least in teleosts, the influence of stress on MT induction is not clear. This necessitates a reevaluation of the use of MT as a specific biomarker of metal exposure in teleost species. In view of this consideration, the present comparative investigation focused on the influence of cadmium and environmental stresses on MT induction in mammals and teleosts.

EXPERIMENTAL

Selection of Animal Models

The mammalian model used was adult, male Swiss-Webster mice (25 to 35 g). Unless otherwise indicated, animals were housed in a temperature-controlled environment (22 to 25°C) and maintained on a 12-hr light:12-hr dark photoperiod. All animals were permitted food and water ad libitum.

The teleost model was juvenile bluegill sunfish, *Lepomis macrochirus* (approximately 40 to 70 mm in length; 5 to 13 g body weight). Unless otherwise indicated, bluegill were maintained in a 300-L tank. The holding tank was aerated and supplied with a continuous flow of dechlorinated tap water (20 ± 1°C) having the following mean water quality characteristics: pH 7.1, hardness 50 mg/L, alkalinity 57 mg/L, dissolved oxygen 8.0 mg/L. A photoperiod of 16-hr light:8-hr dark was maintained and bluegill were fed a commercial diet.

Exposures

To examine the influence of metal pretreatment on the acute toxicity of cadmium, mice received a pretreatment dose of 20 mg Cd/kg 24 hr prior to administration of a challenge dose (100 mg Cd/kg). Following administration of the challenge, mortality was observed for a 96-hr period. Cadmium solutions were prepared by dissolving $CdCl_2 \times 2^1/_2 H_2O$ in deionized water. All doses were administered by oral gavage, and the maximum dose volume did not exceed 20 mL/kg. Animals exposed to cold pretreatment were housed in individual metal cages at 4°C for 12 hr and challenged 12 hr after removal from the 4°C environmentally controlled room. Following administration of the challenge dose, mortality again was observed for a 96-hr period.

Studies were conducted to determine the influence of cadmium exposure on MT-like protein induction in bluegill sunfish. Fish were exposed in 30-L tanks to cadmium using 48-hr static, renewal bioassays (20 ± 1°C). Bluegill were exposed to 0, 1, 10, and 100 μg Cd/L. A stock solution of cadmium ($CdCl_2 \times 2^1/_2 H_2O$) was added to each exposure chamber prior to introduction of the test fish. Water parameters and metal concentrations were monitored in both the initial exposure and renewal chambers. Bluegill exposed to cold stress were maintained in 30-L tanks which were cooled at a rate of 1 to 2°C per hour, reaching a final temperature of 4°C at the end of a 12-hr period. In separate experiments, bluegill were exposed to hypoxic stress defined as 2 mg O_2/L for 12 hr.

Analysis of Metal-binding Proteins

Analysis of metal-binding proteins was achieved by modification of a procedure described by Benson and Birge.[13] A Sephadex G-75 column (2.5 × 50 cm) was calibrated for molecular weight determinations using bovine serum albumin (1 mg), carbonic anhydrase (0.5 mg), myoglobin (1 mg), MT rabbit standard (1 mg), cytochrome c (1 mg), insulin chain B (0.5 mg), and reduced glutathione (0.5 mg), which have respective molecular weights of 66,000, 29,000, 16,900, 15,000, 12,400, 5,700, and 307 Da. All preparative procedures were conducted at 4°C. Tissues were removed and homogenized in four volumes of buffer (20 mM Tris-HCl), pH 7.2; 10 mM 2-mercaptoethanol; 250 mM sucrose) and centrifuged for 10 min at 10,000 g. For mammalian preparations, two livers were pooled to obtain one sample (3.5 g); for teleost preparations, approximately ten animals for gills and five animals for livers were pooled to obtain one sample (1.5 g). The resulting supernatant was centrifuged for 60 min at 100,000 g to obtain the cytosolic fraction. A 2.5-mL sample was then loaded on a Sephadex G-75 column equilibrated with buffer B (50 mM Tris-HCl, pH 8.2: 5 mM 2-mercaptoethanol) and eluted with the same buffer at a flow rate of 18 mL/hr (3.66 mL/cm^2/hr). Fractions (3.0 mL for mammalian tissues and 2.4 mL for teleost tissues) were collected and directly analyzed for cadmium, zinc, and copper. Mammalian and teleost MT-like protein was determined indirectly by summing metal concentrations of protein fraction (10% or greater of the peak concentration) eluted in the 10,000 molecular weight range (Ve/Vo = 1.8 to 2.1) from the Sephadex G-75 column.

In mammals, quantification of MT-like reserve capacity was conducted by the method of Probst et al.[18] The final supernatant obtained from the scheme for isolation of metal-binding proteins described above was divided into two fractions and either distilled-deionized water or exogenous cadmium was added. These fractions were then termed the "cadmium-unsaturated" and "cadmium-saturated" supernatants, respectively. In these preparations, 2 μmol of exogenous cadmium was demonstrated to saturate the mammalian MT-like protein fraction. The samples were vortexed and centrifuged at 100,000 g for 60 min and loaded on the gel-filtration column, as previously described. However, during this procedure 2-mercaptoethanol was omitted in the elution buffer. As described above, the hepatic MT-like protein concentration was determined indirectly by summing cadmium concentrations of fractions eluted in the 10,000 molecular weight range (Ve/Vo = 1.8 to 2.1) from the Sephadex G-75 column. The MT-like reserve capacity was defined as the difference between the "cadmium-saturated" and "cadmium-unsaturated" values.

Hematology and Blood Chemistry

Blood was obtained from mice and bluegill by decapitation and dorsal gill incision,[19] respectively. For hematocrit determination, blood was withdrawn into 75 mm heparinized microhematocrit capillary tubes. Hematocrit tubes were centrifuged at 4000 rpm for 15 min using a clinical centrifuge and hematocrit was analyzed using a microcapillary reader. The plasma obtained by centrifugation was then analyzed for glucose (YSI Model 23A). In mammals, determination of white blood cell (WBC) count was determined using a Model S + 5 Coulter Counter. Teleost WBC count was estimated by a manual procedure.[20]

Data Analysis

Mortality data were analyzed by binomial distribution.[21] Treatment effects were evaluated using one-way analysis of variance. Mean separation of data was achieved by Waller-Duncan k-ratio t test.[22] For heterogeneous data, a Proc Rank

Table 1.   Influence of Stress and Metal Pretreatments on Hematological and Blood Chemistry Parameters in Mammals.[a]

Pretreatments	Hematocrit (%)	Glucose (mg/dL)	WBC (10^3/mm^3)
Control	40 ± 0.6	190 ± 5	11 ± 1.0
Cold (4°C, 12 hr)	39 ± 0.4	216 ± 5*	6 ± 0.2*
20 mg Cd/kg	41 ± 0.6	206 ± 9	10 ± 1.0

[a]Values are expressed as mean ± standard error (n = 6).
*Significantly different from control at $p < 0.05$.

nonparametric procedure was used for mean separation.[22] Differences were considered significant at $p < 0.05$.

RESULTS AND DISCUSSION

Mammals

The influence of stress and metal pretreatments on hematological and blood chemistry parameters in mammals is presented in Table 1. Parameters were chosen that would have indicated whether animals had exhibited a significant stress response. Exposure to cold resulted in a significant increase in glucose and WBC, as compared to control. The observed hyperglycemia and lymphopenia demonstrated that animals were under sufficient stress during cold exposure. In animals pretreated with cadmium, however, no significant differences were observed in any of the selected parameters.

The influence of stress and metal pretreatments on the acute toxicity of cadmium to mammals is presented in Table 2. A significant degree of tolerance to the cadmium challenge dose was observed following cold stress (4°C, 12 hr). Likewise, pretreatment with 20 mg Cd/kg provided a significant degree of tolerance. When administered the challenge dose of 100 mg/kg, mortalities of 40%

Table 2.   Influence of Stress and Metal Pretreatments on the Acute Toxicity of Cadmium to Mammals.

Pretreatments	Mortality 96 hr after Challenge[a] 100 mg Cd/kg
Control	8/10
Cold (4°C, 12 hr)	4/10*
20 mg Cd/kg	0/10*

[a]Dead/total.
*Significantly different from control at $p < 0.05$.

Table 3. Influence of Stress and Metal Pretreatments on MT-like Reserve Capacity in Mammals.[a]

Pretreatments	Unsaturated Cytosol (nmol Cd)	Saturated Cytosol (nmol Cd)	MT-like Reserve Capacity (nmol Cd)
Control	ND[b]	12 ± 1	12 ± 1
Cold (4°C, 12 hr)	ND[b]	24 ± 5*	24 ± 5*
20 mg Cd/kg	16 ± 3	72 ± 6*	56 ± 5*

[a]Values are expressed as mean ± standard error (n = 3).
[b]Not detected.
*Significantly different from control at $p < 0.05$.

and 0% were observed in the respective cold- and cadmium-pretreated animals, compared with control mortality of 80%.

Restriction of food intake has been observed to increase MT synthesis in rats;[23] therefore, food consumption, body weight, and body temperature were monitored prior to administration of the cadmium challenge. In the present investigation, however, food intake was not significantly altered from control during cold exposure.[20] Therefore, the significant increase in MT-like reserve capacity was attributed to the cold pretreatment.

Table 3 summarizes the influence of stress and metal pretreatments on MT-like reserve capacity in mammals. The observed tolerance to acute cadmium toxicity following cold and metal pretreatments was associated with significant induction of hepatic MT-like reserve capacity, which is the amount of protein available to bind cadmium. Following cold and metal pretreatments, values for MT-like reserve capacity were 24 and 56 nmol cadmium, respectively, compared with 12 nmol cadmium for control. Several investigators have suggested that the appearance of MT in plasma and urine could be used as a specific index of cadmium toxicity and that assay of the protein in these fluids could be used as a biomarker of cadmium exposure.[24-26] The source of urinary MT remains unknown, but it has been postulated that the MT detected in urine may be of hepatic and/or renal origin.[25] Results from the present investigation, indicate that exposure to environmental stresses, such as cold, may contribute to the level of MT already present in the extracellular fluid of mammals following metal exposure. In view of this, the levels of MT resulting from environmental stresses may confound such a simplistic approach to biological monitoring in mammals.

Teleosts

To examine the influence of cadmium exposure on MT-like protein induction in teleost species, bluegill were exposed 0, 1, 10, and 100 μg Cd/L. Figure 1 presents the trace metal analysis of hepatic MT-like protein following metal exposure. Exposure to 10 and 100 μg Cd/L for 48 hr resulted in significant increases in the cadmium and zinc content of the MT-like protein fraction. Due

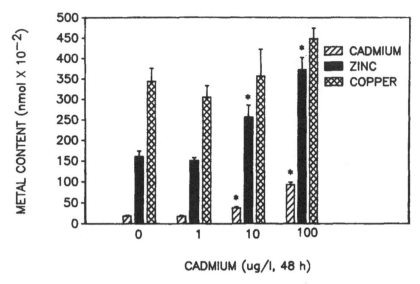

Figure 1. Trace metal analysis of teleost hepatic MT-like protein following metal exposure. Values presented as means with standard error (n = 3). *Significantly different from control at $p < 0.05$.

to the large content of copper associated with bluegill liver, large deviations were observed in copper content of the MT-like protein fraction. This phenomenon has been observed in other teleost species (D. H. Petering, personal communication (1989)), and can render evaluation of copper-binding proteins following environmental and/or chemical stresses difficult.

As with mammals, studies were conducted to determine the influence of environmental stresses on MT-like protein induction in bluegill sunfish. Selected hematological and blood chemistry parameters were examined to assure that fish were stressed under the experimental conditions of cold, hypoxia, and metal exposure (Table 4). Significant secondary responses (hematocrit and lymphopenia) were observed with cold stress. Significant hyperglycemia was observed in both cold and hypoxia stressed animals. In metal-exposed bluegill, a significant increase in WBC was observed. This may be due to direct stimulation of the immune defense by cadmium.[27] In agreement with Larsson et al.,[28] WBC measurement appears to be a sensitive indicator of metal exposure in teleosts.

Figure 2 graphically presents the trace metal analysis of hepatic MT-like protein from bluegill sunfish following environmental stress exposure. The cadmium, zinc, and copper content of hepatic MT-like proteins were not significantly altered, compared with control. On the other hand, exposure to environmental stresses had a profound effect on metal content of gill MT-like protein (Figure 3). Following cold exposure, zinc content of the MT-like fraction significantly decreased, while copper content significantly decreased under conditions of cold as well as hypoxia.

Table 4.   Influence of Stress and Metal Exposure on Hematogical and
Blood Chemistry Parameters in Bluegill Sunfish.[a]

Stresses	Hematocrit (%)	Glucose (mg/dL)	WBC (10^3/mm³)
Cold[b]			
Control	40 ± 0.8	82 ± 7	12 ± 0.8
Stress	36 ± 0.8*	150 ± 14*	8 ± 0.6*
Hypoxia[c]			
Control	38 ± 0.6	60 ± 1	12 ± 0.9
Stress	41 ± 2.0	103 ± 17*	14 ± 0.4
Cadmium			
Control	39 ± 0.3	69 ± 5	11 ± 0.4
10 µg/L	37 ± 1.0	71 ± 2	19 ± 0.4*

[a]Values are expressed as mean ± standard error (n = 6).
[b]Gradual cooling (1–2°C/hr) to 4°C for 12 hr.
[c]2 mg O_2/L for 12 hr.
*Significantly different from control at $p < 0.05$.

CONCLUSIONS

In mammals, a variety of stresses have been demonstrated to induce MT-like protein synthesis in the liver.[2,6] Exposure to a variety of stresses in teleosts, however, has failed to induce hepatic MT-like protein synthesis.[15,16] In the present investigation, significant alterations in zinc and copper content were observed in the gill MT-like protein fraction as a result of environmental stress exposure. This response appears to be unique to the gill, since the hepatic MT-like protein metal content was not significantly altered by environmental stresses.

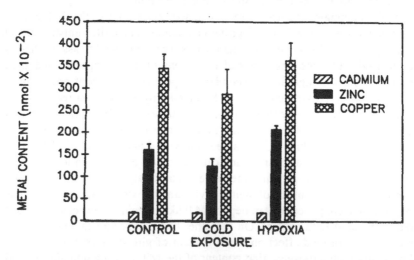

Figure 2.   Trace metal analysis of teleost hepatic MT-like protein following environmental stress exposure. Values presented as means with standard error (n = 3). *Significantly different from control at $p < 0.05$.

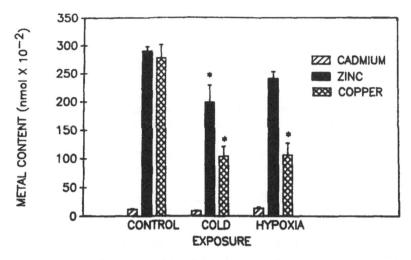

Figure 3.   Trace metal analysis of teleost gill MT-like protein following environmental stress exposure. Values presented as means with standard error (n = 3). *Significantly different from control at p < 0.05.

The significance of the decreased zinc MT-like protein content in gill during stress exposure is not apparent. Zinc may be released from this fraction as a result of increased metabolic demand during stress. Zinc deficiency, zinc, as well as copper have been demonstrated to decrease in the MT-like protein fraction of mammals.[29] The hepatic zinc content of MT-like proteins in teleosts has been demonstrated to be released during periods of high metabolic demand such as sexual maturation.[17] Results from the present investigation indicate that the metal content of gill MT-like protein may be involved in an adaptive response to stress. Furthermore, gill MT-like proteins appear to be influenced by environmental stresses to a greater extent than hepatic MT-like proteins.

Findings from the present investigation indicate environmental stresses may confound the use of gill MT-like proteins for monitoring metal exposure. The liver MT-like protein appears to be more metal specific and may be promising as a biomarker of environmental metal contamination. As mentioned above, physiological conditions such as sexual maturation and nutritional status may, however, alter the metal content of hepatic MT-like proteins. Therefore, caution should be taken when evaluating hepatic MT-like proteins in relation to metal exposure.

REFERENCES

1. Kotsonis, F. N., and C. D. Klaassen. "The Relationship of Metallothionein to the Toxicity of Cadmium after Prolonged Oral Administration to Rats," *Toxicol. Appl. Pharmacol.* 46:39–54 (1978).

2. Baer, K. N., and W. H. Benson. "Influence of Chemical and Environmental Stressors on Acute Cadmium Toxicity," *J. Toxicol. Environ. Health* 22:35–44 (1987).

3. Sugihara, N., C. Tohyama, M. Murakami, and H. Saito. "Significance of Increase in Urinary Metallothionein of Rats Repeatly Exposed to Cadmium," *Toxicology* 41:1–9 (1986).

4. Tohyama, C., Y. Mitane, E. Kobayashi, N. Sugihira, A. Nakano, and H. Saito. "The Relationship of Urinary Metallothionein with Other Indicators of Renal Dysfunction in People Living in a Cadmium-polluted Area in Japan," *J. Appl. Toxicol.* 8:15–21 (1988).

5. Shaikh, Z. A., C. Tohyama, and C. V. Nolan. "Occupational Exposure to Cadmium: Effect on Metallothionein and Other Biological Indices of Exposure and Renal Function," *Arch. Toxicol.* 59:360–364 (1987).

6. Oh, S. H., J. T. Deagen, P. D. Whanger, and P. H. Weswig. "Biological Function of Metallothionein. V. Its Induction in Rats by Various Stresses," *Am. J. Physiol.* 234:E282–E285 (1978).

7. Failla, M. L., and R. J. Cousins. "Zinc Accumulation and Metabolism in Primary Cultures of Adult Rat Liver Cells," *Biochem. Biophys. Acta* 543:293–304 (1978).

8. Etzel, K. R., S. G. Hapiro, and R. J. Cousins. "Regulation of Liver Metallothionein and Plasma Zinc by the Glucocorticoid Dexamethasone," *Biochem. Biophys. Res. Commun.* 89:1120–1126 (1979).

9. Etzel, K. R. and R. J. Cousins. "Hormonal Regulation of Liver Metallothionein Zinc: Independent and Synergistic Action of Glucagon and Glucocorticoids," *Proc. Soc. Exp. Biol. Med.* 167:233–236 (1981).

10. Brady, F. O. and B. Helvig. "Effects of Epinephrine and Norepinephrine on Zinc Thionein Levels and Induction in the Rat Liver," *Am. J. Physiol.* 247:E319–E322 (1984).

11. Hidalgo, J., M. Giralt, J. S. Garvey, and A. Armario. "Physiological Role of Glucocorticoids on Rat Serum and Liver Metallothionein in Basal and Stress Conditions," *Am. J. Physiol.* 254:E71–E78 (1988).

12. Kito, H., T. Tazawa, Y. Ose, T. Sato, and T. Ishikawa. "Protection by Metallothionein against Cadmium Toxicity," *Comp. Biochem. Physiol.* 73C:135–139 (1982).

13. Benson, W. H. and W. J. Birge. "Heavy Metal Tolerance and Metallothionein Induction in Fathead Minnows: Results from Field and Laboratory Investigations," *Toxicol. Environ. Chem.* 4:209–217 (1985).

14. Hamilton, S. J., and P. M. Mehrle. "Metallothionein in Fish: Review of Its Importance in Assessing Stress from Metal Contaminants," *Trans. Am. Fish.* Soc. 115:596–609 (1986).

15. Overnell, J., R. McIntosh, and T. C. Fletcher. "The Enhanced Induction of Metallothionein by Zinc, Its Half-life in the Marine Fish *Pleuronectes platessa*, and the Influence of Stress Factors on Metallothionein Levels," *Experientia* 43:178–181 (1987).

16. Thomas, P., and H. W. Wofford. "Effects of Metals and Organic Compounds on Hepatic Glutathione, Cysteine, and Acid-soluble Thiol Levels in Mullet (*Mugil cephalus* L.)," *Toxicol. Appl. Pharmacol.* 76:172–182 (1984).

17. Olsson, P.-E., C. Haux, and L. Forlin. "Variations in Hepatic Metallothionein, Zinc and Copper Levels during an Annual Reproductive Cycle in Rainbow Trout, *Salmo gairdneri*," *Fish Physiol. Biochem.* 3:39–47 (1987).

18. Probst, G. S., W. F. Bousquet, and T. S. Miya. "Correlation of Hepatic Metallo-

thionein Concentrations with Acute Cadmium Toxicity in the Mouse," *Toxicol. Appl. Pharmacol.* 39:61–69 (1977).

19. Watson, C. F., K. N. Baer, and W. H. Benson. "Dorsal Gill Incision: A Simple Method for Obtaining Blood Samples in Small Fish," *Environ. Toxicol. Chem.* 8:457–461 (1989).

20. Baer, K. N. "Influence of Chemical and Environmental Stresses on Metal-binding Proteins: Species-dependent Effects," Ph.D. Dissertation, Northeast Louisiana University, Monroe, LA (1988).

21. Sokal, R. R., and F. J. Rolf. *Biometry: The Principles and Practice of Statistics in Biological Research* (New York: W.H. Freeman and Company, 1981), pp. 70–82.

22. *SAS Procedures Guide for Personal Computers, Version 6 Edition* (Cary, NC: SAS Institute Inc., 1985), p. 373.

23. Bremner, I., and N. T. Davies. "The Induction of Metallothionein in Rat Liver by Zinc Injection and Restriction of Food Intake," *Biochem. J.* 149:733–738 (1975).

24. Chang, C. C., R. Lauwerys, A. Bernard, H. Roels, J. P. Buchet, and J. S. Garvey. "Metallothionein in Cadmium-Exposed Workers," *Environ. Res.* 23:422–428 (1980).

25. Lee, Y. H., Z. A. Shaikh, and C. Tohyama. "Urinary Metallothionein and Tissue Metal Levels of Rats Injected with Cadmium, Mercury, Lead, Copper or Zinc," *Toxicology* 27:337–345 (1983).

26. Shaikh, Z. A., and L. M. Smith. "Biological Indicators of Cadmium Exposure and Toxicity," *Experientia* 40:36–42 (1984).

27. Wedemeyer, G. A., and D. J. McLeoy. "Methods for Determining the Tolerance of Fishes to Environmental Stressors," in *Stress and Fish*, A.D. Pickering, Ed. (New York: Academic Press, 1981), pp. 247–275.

28. Larsson, A., C. Haux, and M. L. Sjobeck. Fish Physiology and Metal Pollution: Results and Experiences from Laboratory and Field Studies," *Ecotoxicol. Environ. Saf.* 9:250–281 (1985).

29. Petering, D. H., J. Loftsgaarden, J. Schneider, and B. Fowler. "Metabolism of Cadmium, Zinc, and Copper in Rat Kidney: The Role of Metallothionein and Other Binding Sites," *Environ. Health Pers.* 54:73–81 (1984).

CHAPTER 16

METALLOTHIONEIN: A POTENTIAL BIOMONITOR OF EXPOSURE TO ENVIRONMENTAL TOXINS

Justine S. Garvey

Evidence is presented in support of the thesis that the ubiquitous metal-binding protein, metallothionein (MT), may serve as a biomonitor of exposure to environmental toxins such as heavy metals and to environmental stress. MT has been demonstrated to be induced by exposure of vertebrates, invertebrates, and microorganisms to elements of toxic potential such as cadmium, platinum, gold and mercury, and by the stresses of restraint and food and water deprivation. Methods have been developed to permit accurate and sensitive detection and quantitation of MT in mammals, including radioimmunoassay (RIA) and enzyme-linked immunosorbent assay (ELISA), both assays based on the use of both polyclonal and monoclonal antibodies to the protein. Panels of specific antibodies may be readily produced to monitor MT levels in a selected series of non-mammalian sentinel species.

Metallothionein (MT) is a low molecular weight (6000–8000 in mammals) metal-binding protein found in vertebrates, invertebrates, and microorganisms (References 1 and 2 are excellent summaries of MT research from 1957 to 1978 (1) and to 1985 (2)). In most mammals MT isoforms contain 61 amino acids of which 20 are cysteine, the metal-binding agent; the protein in the ascomycetes, *Neurospora crassa*, contains 25 amino acids of which 7 are cysteine.[3,4] The principal role of MT appears to be the maintenance of homeostasis of zinc and copper (Reference 5 is an excellent summary of research on MT and metal metabolism). MT is induced directly by zinc, copper, and cadmium and is induced indirectly by various heavy metals of toxicological interest, e.g., Hg, Au, Pt.[6] Zn-induced synthesis and secretion of MT was shown in primary cultures of rat hepatocytes, indicating a useful methodology to study the cellular metabolism of metals and MT.[7] The elevation of MT levels in cells and physiological fluids following exposure to toxic metals has led to numerous investigations over the last 30 years designed to establish the potential of MT as a biomonitor of such exposure.[8,9] And interest in the biological function of the protein has led to studies involving MT induction by stresses associated with environmental conditions.[10–12]

The interest of my laboratory in MT dates from 1976 with the successful development of an antibody to mammalian MT using rabbits injected with an isoform of MT isolated from rat liver.[13] Using this antibody a radioimmunoassay (RIA) for MT was developed[14] and subsequently an enzyme-linked immunosorbent assay (ELISA) was developed as a supplement to the RIA.[15] The antibody recognized with similar affinity MT isoforms from human, rat, equine, and hamster species. This affinity has since been shown for MT isoforms from mouse (isolates prepared by M. Bhattacharyya, Argonne National Laboratories and D. Solaiman, Duquesne University), sheep (isolates prepared by J. Apgar, USDA, Ithaca, NY and C. McCormick, Cornell University), calf (in collaboration with D. Winge, University of Utah[16]), moose and reindeer (isolates prepared by G. Nordberg and colleagues, University of Umea, Sweden[17]). The cross-reactivity exhibited by these various mammalian MTs when responding to an antibody raised against a MT isoform of one species (the rat) is strong support for the thesis that an antibody raised against a MT isoform of one member of a closely related family of species of ecological interest (avian, aquatic) will prove to have similar affinity for MTs from the other members of that family. The polyclonal antibody developed for detection and quantitation of mammalian MTs is specific for the two principal determinants of mammalian isoforms, the invariant sequence M-D-P-N-C- at the amino terminus and the sequence of residues 20–25 (-K-C-K-E-C-K- in human MT-2). The latter region exhibits some substitutions in other isoforms (e.g., R may replace K22 or K25, Q or N may replace E23, G may replace K20). The experimental determination of the principal determinants[18] is supported by theoretical analysis[19] based on the protocols of Hopp and Woods for relative hydrophilicity[20] and the protocols of Chou and Fasman for prediction of secondary structure.[21,22] The predicted secondary structure is one of random coils and reverse turns, the latter characterizing the principal antigenic determinants.[19]

Other investigators have provided evidence that mammalian MTs exist as two clusters or domains (alpha, residues 30–61; beta, residues 1–29), the alpha cluster typically binding four divalent metals and the beta cluster typically binding three divalent metals.[23–28] Although the theoretical analysis[19] does not rule out the possibility of significant antigenic determinants on the alpha cluster, experiments

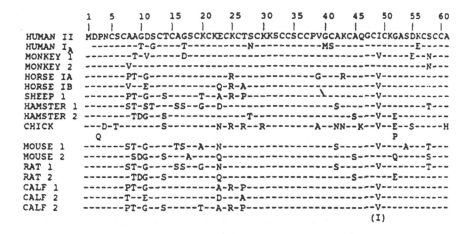

Figure 1. Amino acid sequences of MTs from various mammalian species and chicken. Standard one-letter amino acid symbols are shown. Chicken MT sequence aligned to accommodate two insertions. *Note:* The substitutions for Human I_A should be shifted one residue to the left. (Reproduced with permission from McCormick, C. C., C. S. Fullmer, and J. S. Garvey. "Amino Acid Sequence and Comparative Antigenicity of Chicken Metallothionein," *Proc. Natl. Acad. Sci., U.S.A.* 85:309–313 (1988). Copyright 1988, National Academy of Sciences U.S.A.)

on the isolated beta and alpha segments of both rat and calf MTs confirm that the alpha cluster is minimally reactive in the RIA whereas the beta cluster exhibits the reactivity of the complete molecule.[16,18] Figure 1 indicates the sequences of various MT isoforms, including chicken MT. Notable is the absence of the aromatic amino acids and histidine in the mammalian MTs. Chicken MT exhibits a negligible response in the RIA, a consequence of major changes from mammalian MTs in the two principal antigenic determinants of mammalian MTs.[29] This directs attention to the primary sequences of MTs from related species (avian, aquatic) when selecting a sentinel species for a reference MT for antibody production.

MT AS A BIOMONITOR: MAMMALIAN SPECIES

Monitoring of mammalian species for evidence of exposure to toxic metals and/or to stress is readily achieved by immunochemical means using the mentioned assays (RIA and ELISA) for the quantitation of MT in physiological fluids and cytosols, supplemented when appropriate by immunocytochemical localization assays for cells or tissue sections and by physical-chemical methods such as atomic absorption spectrophotometry when specific metals are to be identified. With respect to the choice of assay, alternative methods exist for cytosols of relatively high MT content;[30–37] other investigators have also developed

RIAs.[38–40] Experience indicates that the RIA is a superior method for assay of physiological fluids of relatively low MT content.[41,42]

Assay Protocol

The RIA developed in my laboratory is a fluid phase, competitive binding, double antibody assay.[43] A series of varying amounts of a known reference MT is added to a set of tubes for subsequent development of a standard curve for quantitation of unknowns; samples to be quantitated are added in known amount to another set of tubes. To all tubes there is added the primary antibody followed by a constant aliquot of ^{125}I-labeled MT to compete with the unlabeled competitor MT. Then a secondary antibody (goat anti-rabbit IgG in the case of the rabbit-induced antibody) is added to precipitate the antigen-antibody complexes. The ^{125}I content of the collected precipitates is determined by gamma scintillation counting. The fraction of labeled antigen bound (Y) is plotted as a function of the competitor MT concentration (Q), where $Q = \log(\text{pg MT})$ or $\log(100(\text{pmol MT}))$. The characteristic response is sigmoid, being linear over 2 to 3 orders of magnitude in competitor MT concentration. An improvement in representation is obtained by transforming Y to logit form $(Z = \log(100\ Y/(1\text{-}Y))$ and developing an inverse variance weighted logit-log regression (Figure 2). This regression (standard curve) typically has a correlation coefficient in excess of 0.99 permitting quantitation of unknowns over the range 100–20,000 pg MT with 5 percent accuracy. It is customarily possible to develop extensions of the central regression over the range 1–100 pg MT and 20,000–100,000 pg MT, thereby permitting quantitation in these ranges with 10–20% accuracy. The RIA has been a preferred assay in my laboratory because of its accuracy. The ELISA was developed as a supplementary assay, permitting more rapid processing of samples with but slight reduction in accuracy. The format of the ELISA[15] differs from that of the RIA in being a solid phase assay with the reference MT coating the wells of a 96-well microtiter plate after an incubation and washing step. Competitor MT and primary antibody are added as in the RIA; following an incubation and washing step the secondary antibody (specific for immunoglobulin of the species in which the primary antibody is produced) is added. The secondary antibody is conjugated with an enzyme (e.g., alkaline phosphatase). Following another incubation a substrate (4-methylumbelliferyl phosphate) is added and the fluorescent cleaved substrate (umbelliferone) is detected in a fluorimeter. The typical response (Figure 3) is similar to that of the RIA. As in the RIA the developed standard curve is used for the quantitation of unknowns.

In selecting either the RIA or the ELISA for a given analysis there are several factors to consider. The RIA is more work intensive; the ELISA is more rapid. The RIA has proven superior in quantitating serums (typically of low MT content); the ELISA is an acceptable alternative for urines and cytosols. Table 1 shows the results of an RIA and an ELISA in quantitation of MT in lyophilized cytosols from rat liver (cytosols prepared by R. Flos, University of Barcelona, Spain).

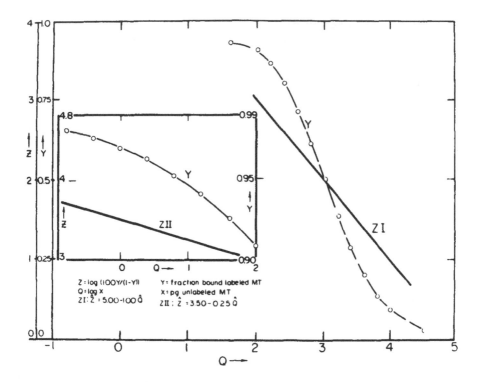

Figure 2. Typical standard curve for MT: radioimmunoassy. The sigmoid response (Y vs Q) is linearized in logit form (Z vs Q). Accuracy of quantitation of unknown is usually ±5 percent over the central region. The characteristic correlation coefficient (in excess of 0.99) of the regression decreases to 0.80–0.85 for the extension (insert) developed over the concentration range 100–1 pg MT.

Each sample contained 1/2 μL of reconstituted cytosol (MT content of cytosols varies widely depending on treatment and metal exposure; typically, concentrations are in the range 10^{-3} to 10^{-6} g MT/g wet weight tissue). The standard curve in this experiment was developed using a rat MT-1 isoform as reference MT and a 50/50 mix of isoforms 1 and 2 as competitor MT. There is no statistical significance in the difference between the respective values in the two assays; standard errors of 5 to 10 percent (RIA) and 10 to 15 percent (ELISA) characterized the two assays.

The sensitivity of the RIA for MT is evidenced in assays of normal serums and urines. Samples from young non-smokers with no history of unusual exposure to heavy metals showed the range of MT concentrations in normal serum to be 0.01–1 pg/μL and the range in normal urines to be 1–10 pg/μL.[19] The sources were diverse (Syracuse, NY; Farmington, CT; Brussels, Belgium). Typical aliquots of samples are 100 μL (serum) and 25 μL (urine); reconstituted cytosols from liver, kidney, or lung are accurately assayed in aliquots of less than 1 μL.

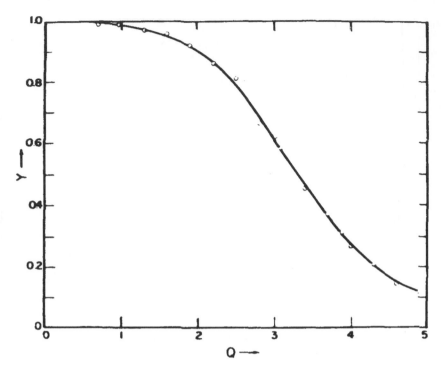

Figure 3. Typical standard curve for MT: ELISA. The sigmoid response (Y vs Q) may be linearized as in the RIA when expressed in logit form (Z vs Q). Symbolism as in Figure 2. In logit form accuracy is comparable to that of RIA. (Reproduced with permission from Thomas, D. G., H. J. Linton, and J. S. Garvey. "Fluorometric ELISA for the Detection and Quantitation of Metallothionein," *J. Immunol. Meth.* 89:239–247 (1986). Copyright 1986, Elsevier Science Publishers.)

Cadmium Exposure

As mentioned previously,[6] MT is inducible by exposure to various metals. Particular attention has been given to the effects of cadmium in the environment and the role of MT in response to such exposure.[44,45] A study performed in my laboratory investigated MT concentrations in the serum of rats exposed to sublethal i.p. injections of either $ZnSO_4$ or $CdCl_2$.[46] A sham group was included to obtain normal levels of MT: these values were in the range 1–1.5 pg/μL. As indicated in Figure 4, the responses to Cd in terms of serum MT greatly exceeded the responses to Zn. After each injection (on days 0, 3, and 6) the response to Zn increased briefly, then declined toward control values; the response to Cd remained at elevated levels, an indication of the long biological half-life of Cd in the body.[46–48] A series of analyses were performed on samples of physiological fluids from workers in this country and abroad who were exposed to Cd in their working environments, the exposure times varying from 8 to 40 years.[49–52] In extreme cases Cd and MT concentrations in urine were 100-fold the concentra-

Table 1. Comparison of RIA and ELISA in Quantitation of Cytosols.[a]

| X | pg MT/μL | | X | pg MT/μL | |
	ELISA	RIA		ELISA	RIA
1	275 (258)	199 (133)	11	1516 (324)	1317 (134)
2	420 (174)	324 (125)	12	1551 (330)	1511 (134)
3	635 (184)	897 (130)	13	1802 (352)	2193 (147)
4	664 (212)	816 (128)	14	1806 (354)	2199 (148)
5	911 (244)	1081 (127)	15	2487 (430)	2835 (166)
6	994 (256)	1033 (128)	16	2768 (494)	2330 (162)
7	1140 (292)	943 (132)	17	3098 (492)	2887 (222)
8	1205 (296)	1015 (126)	18	3155 (502)	3535 (287)
9	1306 (298)	1390 (133)	19	5775 (799)	5995 (626)
10	1365 (306)	1419 (137)	20	6682 (869)	6477 (704)

[a]Cytosols from liver tissues of stressed rats (isolated by Dr. R. Flos, University of Barcelona, Spain); samples were 1/2 μL of reconstituted lyophilized cytosols. Responses in pg MT/μL of the 1 mL reconstituted samples. The standard curves were developed using rat MTs 1 and 2 as the competitor antigen. In logit-log form they were [Z = log(100 Y/(1-Y), Y = fraction ^{125}I labeled rat MT bound, Q = log(pg MT)]. RIA: Z = 4.5897 − 0.8360 Q; r^2 = 0.9895. ELISA: Z = 4.5059 − 0.7068 Q; r^2 = 0.9655. There is no significance between RIA and ELISA values for any of the 20 dilutions (X). (Reproduced with permission from Thomas, D. G., H. J. Linton, and J. S. Garvey. "Fluorometric ELISA for the Detection and Quantitation of Metallothionein," *J. Immunol. Meth.* 89:239–247 (1986). Copyright 1986, Elsevier Science Publishers.)

tions in urines of nonexposed workers. Of particular interest is the relation between MT and Cd concentrations in urine (Figure 5) which indicates that MT levels are reasonably well-correlated with Cd levels and suggests the utility of MT as a biomonitor of Cd exposure.[53,54] A related analysis of liver tissue from a patient with lung cancer (fatal) after 20 years of working in a Cd environment demonstrated MT levels 20 times those of normal liver tissue.[55] An experiment to determine by RIA the MT content in lung cells and lavage fluid of rats exposed to airborne Cd for varying periods of time showed that MT concentrations in alveolar cells increased by a factor of 60 as exposures increased from 0 to 24 (Table 2). Analysis of MT concentrations at 24, 48, and 72 hr after the last of 17 exposures demonstrated the presence of the protein in alveolar macrophages and the relative absence of MT in polymorphonuclear leucocytes.[56]

The analyses of MT levels in mammals exposed to Cd provide support for the potential of MT as a biomonitor of exposure to Cd. Cytosols typically exhibit concentrations permitting less than 1 μL of a reconstituted cytosol to be assayed (corresponding to less than 1 mg wet weight tissue); in the case of physiological fluids either the RIA or an ELISA will accurately quantitate MT in aliquots of 25–100 μL.

Exposure to Other Heavy Metals

The RIA has been used to detect and quantitate MT in isolates from mammals exposed to, or treated with, compounds containing potentially toxic metals such

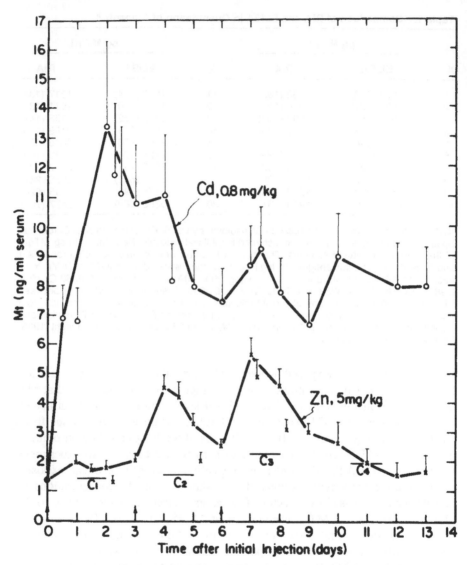

Figure 4. Response of rats to injections of ZnSO₄ or CdCl₂. Concentration of MT in serums of rats given three i.p. injections of either ZnSO₄ (5 mg Zn/kg body weight/injection) or CdCl₂ (0.8 mg Cd/kg body weight/injection). Injections on days 0, 3, and 6. C1, C2, C3, C4 are control values. (Reproduced with permission from Garvey, J. S., and C. C. Chang. "Detection of Circulating Metallothionein in Rats Injected with Zinc or Cadmium," *Science* 214:805–807 (1981). Copyright 1981, Science.)

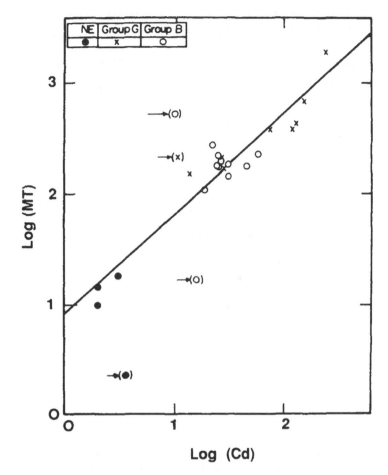

Figure 5. Relation between MT and Cd in urine of Cd-exposed workers. Mt and Cd concentrations in urine of workers exposed for varying periods of time (0 to 40 years)
to a cadmium environment. Symbols: •, nonexposed; ○ and x, two differing
environments. Arrows indicate deviant responses (4 of the total of 24). The
regression (20 responses) has a correlation coefficient of 0.962. (Reproduced
with permission from Garvey, J. S. "The Application of a Radioimmunoassay for
Sensitive Detection of Metallothionein (Thionein) in Physiologic Fluids of Humans
and Rats," in *Nephrotoxic Mechanisms of Drugs and Environmental Toxins*, G.
Porter, Ed. (New York: Plenum Press, 1982), pp. 437–449. Copyright 1982,
Plenum Press.)

as mercury (Hg), gold (Au), platinum (Pt), and lead (Pb). Rats treated with
HgCl$_2$ by B. Albini (University of Buffalo, NY) produced a protein identified
by RIA as an Hg-MT.[53] Rats treated with lead produced Pb-binding proteins
(isolated by B. Fowler and G. Duval, NIEHS) among which were Pb-binding
metallothioneins as shown by their affinity characteristics in the RIA. Au-binding
proteins isolated from cells of normal and rheumatoid arthritis patients, the latter

Table 2. Effect of Cadmium Exposure on Metallothionein (MT) Levels in
          Free Alveolar Cells and Lavage Fluids.

Sample	Number of Exposures	μg MT/mg Protein
Cells[a]	0	0.35 (0.03)[c]
Fluid[b]	0	Not detectable
Cells	12	2.22 (0.10)
Fluid	12	0.42 (0.03)
Cells	18	7.20 (0.22)
Fluid	18	1.12 (0.15)
Cells	24	21.73 (0.65)
Fluid	24	2.67 (0.15)

Note: MT content of free alveolar cells[a] and lavage fluid[b] from rats (5–6 per indicated response) exposed 0 to 24 times to airborne Cd. MT determinations by RIA; protein determined by Lowry procedure. Results[c] expressed with standard deviation in parentheses. (Reproduced with permission from Hart, B. A., and J. S. Garvey. "Detection of Metallothionein in Bronchoalveolar Cells and Lavage Fluid Following Repeated Cadmium Inhalation," *Environ. Res.* 40:391–398 (1986). Copyright 1986, Academic Press.)

treated with the drug auranofin, were determined to contain an increased amount of MT as auranofin exposure increased; the ratios of MT content in epithelial cells and in synovial fibroblasts of the arthritic patients with respect to normal values were 100:1 and 10:1, respectively.[57] This indicated that MT could be used as a biomonitor of exposure to gold-containing compounds. Gold thiomalate, also used in treatment of arthritis, was demonstrated by RIA to lead to replacement of Zn or Cd in existing MTs in equine kidney cells.[58] The series of aurothioneins produced, from the original CdZn-MTs to AuCd-MTs to fully Au-substituted MT, could be distinguished in the RIA by subtle changes in affinity of antibody and antigen due to conformational changes (Figure 6). The compound dichlorodiammineplatinum (DDP) exists in two forms, *cis*-DDP and *trans*-DDP. The former is useful in treatment of tumors, the latter is not; either can prove toxic. The RIA has demonstrated that Pt in these compounds replaces Zn or Cd in existing CdZn-MTs to yield Pt-MTs, the process inducing additional MT production.[59] In all these cases detection of enhanced MT levels provides evidence of heavy metal exposure; identification of the element involved, when this is uncertain, is provided by complementary methods such as atomic absorption spectrophotometry. Additional details may be provided by immunocytochemical localization of MT as shown in Figure 7 (an experiment performed by M. Elmes, University College, Cardiff, Wales). The protocol involved use of the polyclonal antibody produced in my laboratory and a peroxidase-labeled second antibody staining method developed and optimized by Dr. Elmes. The results demonstrate that MT is present in normal human adult liver and in normal human fetal liver (22 weeks); the piecemeal necrosis in hepatic tissue characteristic of Wilson's disease (faulty copper metabolism) is also made evident.

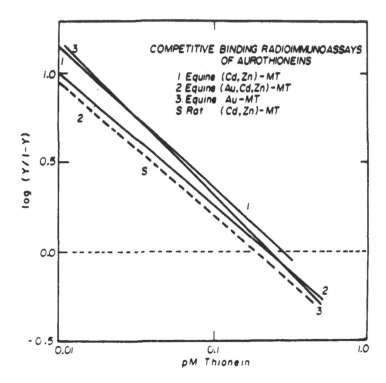

Figure 6. Radioimmunoassay of aurothioneins. Inverse variance weighted logit-log regressions developed for various equine kidney MTs following treatment with aurothiomalate. Symbolism: S, rat liver CdZn-MT; 1, equine kidney CdZn-MT; 2, Equine kidney AuCdZn-MT; 3, fully Au-chelated equine kidney MT. The slope of curve S (correlation coefficient 0.992) differs significantly only from that of curve 3 ($p < 0.05$), indicating the slight change in affinity of antibody and antigen due to conformational changes in the regions of the principal antigenic determinants (curve 3). Abscissa is in pmol MT. (Reproduced with permission from Laib, J., C. F. Shaw III, D. H. Petering, M. K. Eidsness, R. C. Elder, and J. S. Garvey. "Formation and Characterization of Aurothioneins: Au, Zn, Cd-Thionein, Au, Cd-Thionein, and (Thiomalato-Au)$_x$-Thionein," *Biochemistry* 24:1977–1986 (1985). Copyright American Chemical Society).

MT AS A BIOMONITOR OF STRESS

The RIA has demonstrated that MT in liver cells of rats subjected to the stress induced by physical restraint or by food and water deprivation increases with the duration of the state of stress.[10,11] The method is applicable to states of stress induced by environmental conditions in feral animals. Restraint stress is more severe than food and water deprivation, but either condition produces MT levels readily distinguished from control levels (Figure 8). The increase in serum MT in such cases is detectable,[60] and differences between control and stressed animals are significant; however, levels in cytosols are preferred for accurate monitoring.

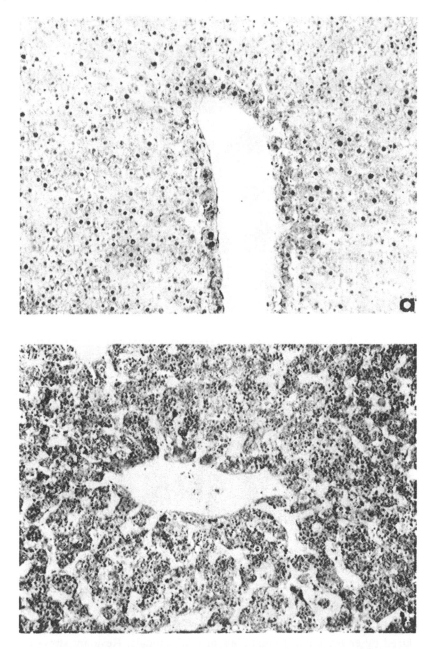

Figure 7. Immunocytochemical localization of MT in human liver. Metallothionein immunoreactivity demonstrated with the polyclonal specific anti-MT antibody[13] as primary antibody and a second antibody staining method (immunoperoxidase, haematoxylin counterstain) developed by J. P. Clarkson, M. E. Elmes, and B. Jasani (University of Wales College of Medicine, Cardiff, Wales). (a) Normal human liver. Cytoplasmic and nuclear immunostaining of MT is present in liver cells with a perivenular accentuation of the cytoplasmic staining. (b) Fetal human liver, gestation 22 weeks. The cytoplasm of all liver cells is uniformly immunostained for MT. Primitive blood cell nuclei stained with haematoxylin are prominent.

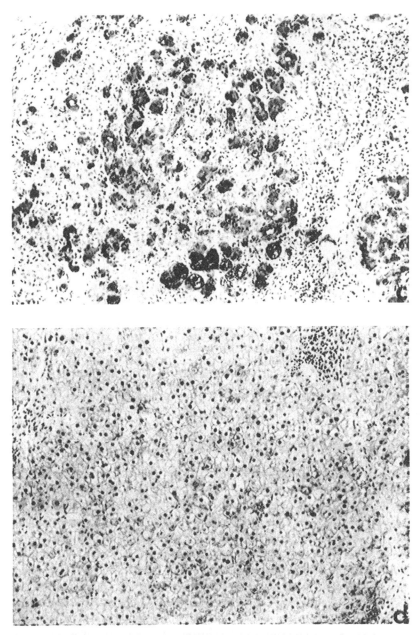

Figure 7. (c) Diagnostic liver biopsy from a case of untreated Wilson's disease. Strong immunostaining of clusters of surviving liver cells is prominent against a background of chronic inflammation. (d) Diagnostic liver biopsy from the same case of Wilson's disease after 1 year of penicillamine therapy. Regenerated liver cells showing near normal MT immunostaining are seen with focal areas of chronic inflammation. [Photomicrographs (original magnification × 125) and histology/pathology assessment courtesy of M. E. Elmes and B. Jasani.]

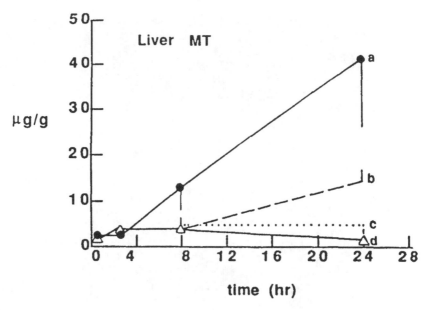

Figure 8.   Effect of restraint stress and food and water deprivation on MT induction. Response of liver MT in rats subjected to the stresses of restraint and of food and water deprivation. Indicated are the mean values and standard error for responses from at least 5 rats. Symbols: control, △-----△; restraint, ●—●; deprivation, -----; Restraint + resting, ... The 24-hr values are each significantly different from the other 24-hr values ($p < 0.05$). (Reproduced with permission from Hidalgo, J., A. Armario, R. Flos, and J. S. Garvey. "Restraint Stress Induced Changes in Rat Liver and Serum Metallothionein and in Zn Metabolism," *Experientia* 42:1006–1010 (1986). Copyright 1986, Birkhäuser-Verlag.)

MT AS A BIOMONITOR: NON-MAMMALIAN SPECIES

In recent years, stimulated by advances in molecular biological techniques, there has been notable attention to the development of metabolic information and of the sequences of MT isoforms from both mammalian and non-mammalian species.[61-71] The partial list of sequences (Figure 1) is indicative of such recent attention. Figure 9 is an abbreviated list to illustrate significant variations from human MT isoforms of aquatic and microorganism isoforms. The specificity of the antibody produced against mammalian MT has been established in numerous experiments. Of the various MTs or candidate MTs (metal-binding proteins) from non-mammalian species all have proven minimally responsive in the RIA or ELISA. This does not indicate that the proteins are not MTs but rather that they do not possess the principal antigenic determinants of mammalian MTs, discussed earlier. MTs or candidate metal-binding proteins which have been assayed include those from trout (*Salmo gairdneri*[72]), oysters (*Crassotrea glomerata* and *Ostrea lutaria*;[73] *Crassotrea virginica*, isolates prepared by B. Fowler, NIEHS, and G. Roesijadi, University of Maryland), plaice (*Pleuronectes platessa*, isolates prepared by J. Overnell, NERC Institute of Marine Biochemistry,

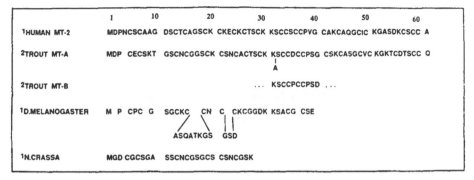

Figure 9.   Amino acid sequences of typical Non-mammalian MTs. Sequences of MTs from trout (*Salmo gairdneri*), *D. melanogaster*, and *N. crassa* compared to the sequence of human MT-2. Trout MT-B differs from MT-A only as indicated. 1-Sequence adapted from Table 3 of Reference 3. 2—Sequences adapted from Bonham, K., M. Zafarullah, and L. Gedamu. *DNA* 6:519–528 (1987).

Aberdeen, Scotland), scallops (*Placopecten magellanicus*, isolates prepared by B. Fowler, NIEHS), mussels (*Mytilus edulis*, isolates prepared by G. Roesijadi, University of Maryland), crabs (*Scylla serrata*, isolates prepared by R. W. Olafson, University of Victoria, Canada), insects (*Drosophila melanogaster*, protein synthesized from genetic information by G. Maroni, University of North Carolina; *Daphina pulcaria*, isolates prepared by C. F. Shaw and colleagues, University of Wisconsin), algae (*Euglena gracilis*[74]), and ascomycetes (*Neurospora crassa*, isolates prepared by K. Lerch, University of Zurich, Switzerland). In all these cases of non-mammalian MTs one can readily exploit the biomonitor character of MT by developing an antibody specific for the species of interest. By careful selection of a prototypic species the antibody can be useful in detection and quantitation of isoforms from closely related species.

In this regard, marine species are of particular interest in view of the widespread pollution of aquatic environments.[75] The induction of MT or MT-like proteins following exposure to metals has been reported in trout,[76,77] plaice,[78] mussels,[79,80] clams,[81] and scallops.[82] Other investigators have identified certain 60-70 kDa proteins, labeled "stress proteins", associated with the response to heat and/or Cd (SPR, "stress protein response") in molluscs[83,84] and relate them to similar proteins (50 percent homology) found in vertebrates, invertebrates, plants, and microorganisms. It is suggested that these proteins may be used for biomonitoring following development of appropriate immunochemical assays.[83] The mussel, *M. edulis*, has been suggested as a sentinel species for biomonitoring of environmental pollution because of its wide distribution and its sensitivity toward toxicants.[80] An aspect to be further pursued is that of non-MT proteins (such as SPR species) which have been associated with environmental stress. Progress has also been reported in developing antibodies to cell surface antigens in molluscs and clams for pathobiological purposes.[85,86] Immunochemical methods might be supplemented by molecular biological protocols as developed by

L. Shugart for assessing genetic damage caused by pollutants in aquatic environments.[87]

BIOMONITORS: PLANTS AND MICROORGANISMS

Although the role of MT in the response to vertebrates and invertebrates to metals and to stress is well-documented, there are other molecules which participate in these responses, particularly in plants and microorganisms. Recent experiments have demonstrated the existence in plants of metal-binding proteins for which the terms "phytochelatins,"[88] "phytometallothioneins,"[89] or "Class III Metallothioneins"[90] have been suggested. The phytochelatins have been identified as peptides of repeating units of Glu-Cys terminating in Gly; they are not primary gene products[88] and are genotypically unrelated to known forms of MT.[91] In response to treatment with Cd, cell cultures from more than a dozen plants, including *Agrostis gigantea* and *Brassica capitata*,[88] have produced phytochelatins. The same peptide, complexing to 776 Da, has been identified in *Datura innoxia*[89] and complexes to 4000–8000 Da have been identified in *Zea mays*.[91] In most cases the inducing metal has been Cd; in some cases the inducing metal has been Cu, Hg, Zn, or Pb.[88] A CdBP of 3000 Da has also been isolated from *Lycopersicon esculentum*.[89] It is expected that immunochemical methods will prove useful in utilizing certain of these agents as supplementary biomonitors.

SUMMARY

The metal-binding protein, metallothionein (MT), is a potential biomonitor for assessment of exposure of animals, including feral species, to environmental toxins and to stress. MT is ubiquitous and is induced not only by zinc and copper but by numerous metals which may be associated with environmental pollution. The RIA and the ELISA developed for detection and quantitation of MT have a well-documented utility in performing this assessment in the case of all mammalian isoforms studied to date. To extend the field of investigation to non-mammalian species, including microorganisms, it is necessary to develop antibodies specific for MTs from species of interest. A first step in this regard is to identify sentinel species, and particularly feral species of both aquatic and terrestrial habitat. A second step is to select from sentinel species a prototypic species that will allow limiting the number of antibody reagents produced by virtue of isoform homology in principal antigenic determinants. MT can serve as the biomonitor for all mammals, for perhaps all other vertebrates and invertebrates, and for some microorganisms. For plants it is conceivable that phytochelatins (phytometallothioneins) may serve as the biomonitor.

ACKNOWLEDGMENT

Reported research by Justine S. Garvey was supported by NIH Grant ESO1629.

REFERENCES

1. Kägi, J. H. R., and M. Nordberg, Eds. *Metallothionein* (Basel, Switzerland: Birkhäuser-Verlag, 1979), passim.
2. Kägi, J. H. R., and Y. Kojima, Eds. *Metallothionein II* (Basel, Switzerland: Birkhäuser-Verlag, 1987), passim.
3. Kägi, J. H. R., and Y. Kojima. "Chemistry and Biochemistry of Metallothionein," in *Metallothionein II* (Basel, Switzerland: Birkhäuser-Verlag, 1987), pp. 25–61.
4. Lerch, K. "Amino-Acid Sequence of Copper-Metallothionein from *Neurospora Crassa*," in *Metallothionein* (Basel, Switzerland: Birkhäuser-Verlag, 1979), pp. 173–179.
5. Bremner, I. "Nutritional and Physiological Significance of Metallothionein," *Metallothionein II* (Basel, Switzerland: Birkhäuser-Verlag, 1987), pp. 81–107.
6. Nielson, K. B., C. J. Atkin, and D. R. Winge. "Distinct Metal-binding Configurations in Metallothionein," *J. Biol. Chem.* 260:5342–5350 (1985).
7. Thomas, D. G., A. D. Dingman, and J. S. Garvey. "The Function of Metallothionein in Cell Metabolism," in *Metallothionein II* (Basel, Switzerland: Birkhäuser-Verlag, 1987), pp. 539–543.
8. Hennig, H. F.-K. O. "Metal-binding Proteins as Metal Pollution Indicators," *Environ. Health Pers.* 65:175–187 (1986).
9. Fowler, B. A. "Structure, Mechanism, and Toxicity," in *Changing Metal Cycles and Human Health*, J. O. Nriagu, Ed. (Berlin, West Germany: Springer-Verlag, 1984), pp. 391–404.
10. Hidalgo, J., A. Armario, R. Flos, and J. S. Garvey. "Restraint Stress Induced Changes in Rat Liver and Serum Metallothionein and in Zn Metabolism," *Experientia* 42:1006–1010 (1986).
11. Hidalgo, J., L. Campmany, M. Borras, J. S. Garvey, and A. Armario. "Metallothionein Response to Stress in Rats: Role in Free Radical Scavenging," *Am. J. Physiol.* 255:E518–E524 (1988).
12. Whanger, P. D., and J. W. Ridlington. "Role of Metallothionein in Zinc Metabolism," in *Biological Roles of Metallothionein*, E. C. Foulkes, Ed. (New York: Elsevier North Holland, 1982), pp. 263–277.
13. Vander Mallie, R. J., and J. S. Garvey. "Production and Study of Antibody Produced Against Rat Cadmium Thionein," *Immunochemistry*, 15:857–868 (1978).
14. Vander Mallie, R. J., and J. S. Garvey. "Radioimmunoassay of Metallothionein," *J. Biol. Chem.* 254:8416–8421 (1979).
15. Thomas, D. G., H. J. Linton, and J. S. Garvey. "Fluorometric ELISA for the Detection and Quantitation of Metallothionein," *J. Immunol. Meth.* 89:239–247 (1986).
16. Winge, D. R., W. R. Gray, A. Żelazowski, and J. S. Garvey. "Sequence and Antigenicity of Calf Metallothionein II," *Arch. Biochem. Biophys.* 245:254–262 (1986).
17. Wikner, M., G. F. Nordberg, M. Nordberg, and J. S. Garvey. "Copper and Cadmium

Binding Proteins from Liver and Kidney of Moose and Reindeer," 7th Int. Congress on Circumpolar Health, Umea, Sweden (1987).

18. Winge, D. R., and J. S. Garvey. "Antigenicity of Metallothionein," *Proc. Natl. Acad. Sci., U.S.A.* 80:2472–2476 (1983).

19. Garvey, J. S. "Metallothionein: Structure/Antigenicity and Detection/Quantitation in Normal Physiological Fluids," *Environ. Health Pers.* 54:117–127 (1984).

20. Hopp, P. T., and K. R. Woods. "Prediction of Protein Antigenic Determinants from Amino Acid Sequences," *Proc. Natl. Acad. Sci., U.S.A.* 78:3824–3828 (1981).

21. Chou, P. Y., and G. D. Fasman. "β-Turns in Proteins," *J. Mol. Biol.* 116:135–175 (1977).

22. Chou, P. Y., and G. D. Fasman. "Prediction of the Secondary Structure of Proteins from Their Amino Acid Sequence," *Adv. Enzymol.* 47:45–148 (1978).

23. Otvos, J. S., and I. M. Armitage. "Structure of the Metal Clusters in Rabbit Liver Metallothionein," *Proc. Natl. Acad. Sci., U.S.A.* 77:7094–7098 (1980).

24. Boulanger, Y., I. M. Armitage, K.-A. Miklossy, and D. R. Winge. "^{113}Cd NMR Study of a Metallothionein Fragment," *J. Biol. Chem.* 257:13717–13719 (1982).

25. Winge, D. R., and K.-A. Miklossy. "Domain Nature of Metallothionein," *J. Biol. Chem.* 257:3471–3476 (1982).

26. Pande, J., M. Vašák, and J. H. R. Kägi. "Interaction of Lysine Residues with the Metal Thiolate Clusters in Metallothionein," *Biochemistry* 24:6717–6722 (1985).

27. Furey, W. F., A. H. Robbins, L. L. Clancy, D. R. Winge, B. C. Wang, and C. D. Stout. "Crystal Structure of Cd, Zn Metallothionein," *Science* 231:704–710 (1986).

28. Avdeef, A., A. J. Żelazowski, and J. S. Garvey. "Cadmium Binding by Biological Ligands. III. Five- and Seven-Cadmium Binding in Metallothionein: A Detailed Thermodynamic Study," *Inorg. Chem.* 24:1928–1933 (1985).

29. McCormick, C. C., C. S. Fullmer, and J. S. Garvey. "Amino Acid Sequence and Comparative Antigenicity of Chicken Metallothionein," *Proc. Natl. Acad. Sci., U.S.A.* 85:309–313 (1988).

30. Chen, R. N., and H. E. Ganther. "Relative Cadmium Binding Capacity of Metallothionein and other Cytosolic Fractions in Various Tissues of the Rat," *Environ. Phys. Biochem.* 5:378–388 (1975).

31. Onosaka, S., and M. G. Cherian. "The Induced Synthesis of Metallothionein in Various Tissues of Rats in Response to Metals," *Toxicology* 22:91–101 (1981).

32. Żelazowski, A. J. and J. K. Piotrowski. "A Modified Procedure for Determination of Metallothionein-like Protein in Animal Tissues," *Acta Biochim. Polon.* 24:97–103 (1977).

33. Kotsonis, F. N., and C. D. Klaassen. "Comparison of Methods for Estimating Hepatic Metallothionein in Rats," *Toxicol. Appl. Pharmacol.* 42:583–588 (1977).

34. Wong, K.-L., and C. D. Klaassen. "Relationship Between Liver and Kidney Levels of Glutathione and Metallothionein in Rats," *Toxicology* 19:39–47 (1981).

35. Olafson, R. W., and R. G. Sim. "An Electrochemical Approach to Quantitation and Characterization of Metallothionein," *Anal. Biochem.* 100:343–351 (1979).

36. Olafson, R. W., "Differential Pulse Polarographic Determination of Murine Metallothionein Induction Kinetics," *J. Biol. Chem.* 256:1263–1268 (1981).

37. Eaton, D. L., and B. F. Toal. "Evaluation of the Cd/hemoglobin Affinity Assay for the Rapid Determination of Metallothionein in Biological Tissues," *Toxicol. Appl. Pharmacol.* 66:134–142 (1982).

38. Brady, F. O., and R. L. Kafka. "Radioimmunoassay of Rat Liver Metallothionein," *Anal. Biochem.* 98:89–94 (1979).
39. Tohyama, C., and Z. A. Shaikh. "Metallothionein in Plasma and Urine of Cadmium-exposed Rats Determined by a Single-antibody Radioimmunoassay," *Fund. Appl. Toxicol.* 1:1–7 (1981).
40. Mehra, R. K., and I. Bremner. "Development of a Radioimmunoassay for Rat Liver Metallothionein-1 and its Application to the Analysis of Rat Plasma and Kidneys," *Biochem. J.* 213:459–465 (1983).
41. Waalkes, M. P., J. S. Garvey, and C. D. Klaassen. "Comparison of Methods of Metallothionein Quantification: Cadmium Radioassay, Mercury Radioassay, and Radioimmunoassay," *Toxicol. Appl. Pharmacol.* 79:524–527 (1985).
42. Dieter, H. H., L. Muller, J. Abel, and K. H. Summer. "Metallothionein-Determination in Biological Materials: Interlaboratory Comparison of 5 Current Methods," in *Metallothionein II* (Basel, Switzerland: Birkhäuser-Verlag, 1987), pp. 351–358.
43. Garvey, J. S., R. J. Vander Mallie, and C. C. Chang. "Radioimmunoassay of Metallothioneins," *Methods Enzymol.* 84:121–138 (1982).
44. Webb, M., Ed. *The Chemistry, Biochemistry and Biology of Cadmium* (Amsterdam: Elsevier North Holland, 1979), passim.
45. Foulkes, E. C., Ed. *Biological Roles of Metallothionein* (Amsterdam: Elsevier North Holland, 1982), passim.
46. Garvey, J. S., and C. C. Chang. "Detection of Circulating Metallothionein in Rats Injected with Zinc or Cadmium," *Science* 214:805–807 (1981).
47. Friberg, L. "Cadmium and the Kidney," *Environ. Health Pers.* 54:1–11 (1984).
48. Suzuki, K. T. "Studies of Cadmium Uptake and Metabolism by the Kidney," *Environ. Health Pers.* 54:21–30 (1984).
49. Chang, C. C., R. Lauwerys, A. Bernard, H. Roels, J. P. Buchet, and J. S. Garvey. "Metallothionein in Cadmium-exposed Workers," *Environ. Res.* 23:422–428 (1980).
50. Roels, H., R. Lauwerys, P. Buchet, A. Bernard, J. S. Garvey, and H. J. Linton. "Significance of Urinary Metallothionein in Workers Exposed to Cadmium," *Int. Arch. Occup. Environ. Health* 52:159–166 (1983).
51. Falck, F. Y., L. J. Fine, R. G. Smith, J. S. Garvey, A. M. Schork, B. G. England, K. D. McClatchey, and H. J. Linton. "Metallothionein and Occupational Exposure to Cadmium: A Potential Biological Monitor," *Brit. J. Ind. Med.* 40:305–313 (1983).
52. Nordberg, G. F., J. S. Garvey, and C. C. Chang. "Metallothionein in Plasma and Urine of Cadmium Workers," *Environ. Res.* 28:179–182 (1982).
53. Garvey, J. S. "The Application of a Radioimmunoassay for Sensitive Detection of Metallothionein (Thionein) in Physiologic Fluids of Humans and Rats," in *Nephrotoxic Mechanisms of Drugs and Environmental Toxins*, G. Porter, Ed. (New York: Plenum Press, 1982), pp. 437–449.
54. Shaikh, Z. A., and K. Hirayama. "Metallothionein in the Extracellular Fluids as an Index of Cadmium Toxicity," *Environ. Health Pers.* 28:267–271 (1979).
55. Garry, V. F., B. L. Pohlman, M. R. Wick, J. S. Garvey, and R. Zeisler. "Chronic Cadmium Intoxication: Tissue Response in an Occupationally Exposed Patient," *Am. J. Med.* 10:153–161 (1986).
56. Hart, B. A. and J. S. Garvey. "Detection of Metallothionein in Bronchoalveolar Cells and Lavage Fluid Following Repeated Cadmium Inhalation," *Environ. Res.* 40:391–398 (1986).

57. Glennås, A., P. E. Hunziker, J. S. Garvey, J. H. R. Kägi, and H. E. Rugstad. "Metallothionein in Cultured Human Epithelial Cells and Synovial Rheumatoid Fibroblasts after *In Vitro* Treatment with Auranofin," *Biochem. Pharmacol.* 35:2033–2040 (1986).
58. Laib, J., C. F. Shaw III, D. H. Petering, M. K. Eidsness, R. C. Elder, and J. S. Garvey. "Formation and Characterization of Aurothioneins: Au, Zn, Cd-Thionein, Au, Cd-Thionein, and (Thiomalato-Au)$_x$-Thionein," *Biochemistry* 24:1977–1986 (1985).
59. Zelazowski, A. J., J. S. Garvey, and J. D. Hoeschele. "*In Vivo* and *In Vitro* Binding of Platinum to Metallothionein," *Arch. Biochem. Biophys.* 229:246–252 (1984).
60. Hidalgo, J., M. Giralt, J. S. Garvey, and A. Armario. "Physiological Role of Glucocorticoids on Rat Serum and Liver Metallothionein in Basal and Stress Conditions," *Am. J. Physiol.* 254:E71–E78 (1988).
61. Durnam, D. M., and R. D. Palmiter. "Transcriptional Regulation of the Mouse Metallothionein-I Gene by Heavy Metals," *J. Biol. Chem.* 256:5712–5716 (1981).
62. Karin, M., and R. I. Richards. "Human Metallothionein Genes—Primary Structure of the Metallothionein-II Gene and a Related Processed Gene," *Nature* 299:796–802 (1982).
63. Karin, M., G. Cathala, and M. C. Nguyen-Huu. "Expression and Regulation of a Human Metallothionein Gene Carried on an Autonomously Replicating Shuttle Vector," *Proc. Natl. Acad. Sci., U.S.A.* 80:4040–4044 (1983).
64. Karin, M., A. Haslinger, H. Holtgreve, R. I. Richards, P. Krauter, H. M. Westphal, and M. Beato. "Characterization of DNA Sequences Through which Cadmium and Glucocorticoid Hormones Induce Human Metallothionein-II$_A$ Gene," *Nature* 308:513–519 (1984).
65. Sequin, C., B. K. Felber, A. D. Carter, and D. H. Hamer. "Competition for Cellular Factors that Activate Metallothionein Gene Transcription," *Nature* 312:781–785 (1984).
66. Schmidt, C. J., M. F. Jubier, and D. H. Hamer. "Structure and Expression of Two Human Metallothionein-I Isoform Genes and a Related Pseudogene," *J. Biol. Chem.* 260:7731–7737 (1985).
67. Wake, S. A., and F. B. Mercer. "Induction of Metallothionein mRNA in Rat Liver and Kidney after Copper Chloride Injection," *Biochem. J.* 228:425–432 (1985).
68. Palmiter, R. D. "Molecular Biology of Metallothionein Gene Expression," in *Metallothionein II* (Basel, Switzerland: Birkhäuser-Verlag, 1987), pp. 63–80.
69. Cousins, R. J. "Synthesis and Degradation of Liver Metallothionein," in *Metallothionein* (Basel, Switzerland: Birkhäuser-Verlag, 1979), pp. 293–301.
70. Mayo, K. E., and R. D. Palmiter. "Glucocorticoid Regulation of Metallothionein Gene Expression," in *Biochemical Actions of Hormones, Vol. XIII*, G. Litwack, Ed. (New York: Academic Press, 1985), pp. 69–88.
71. Butt, T. R., E. J. Sternberg, C. K. Mirabelli, and S. T. Crooke, "Regulation of Metallothionein Gene Expression in Mammalian Cells by Gold Compounds," *Mol. Pharmacol.* 29:204–210 (1986).
72. Kay, J., D. G. Thomas, M. W. Brown, A. Cryer, D. Shurben, J. F. del G. Solbe, and J. S. Garvey. "Cadmium Accumulation and Protein Binding in Tissues of the Rainbow Trout, *Salmo gairdneri*," *Environ. Health Pers.* 65:133–139 (1986).
73. Nordberg, M., I. Nuottaniemi, M. G. Cherian, G. F. Nordberg, T. Kjellström, and J. S. Garvey. "Characterization Studies on the Cadmium-Binding Proteins from Two Species of New Zealand Oysters," *Environ. Health Pers.* 65:57–62 (1986).
74. Gingrich, D. J., D. N. Weber, C. F. Shaw III, J. S. Garvey, and D. H. Petering.

"Characterization of a Highly Negative and Labile Binding Protein Induced in *Euglena gracilis* by Cadmium," *Environ. Health Pers.* 65:77–86 (1986).

75. Roesijadi, B., and R. B. Spies, Eds. *Marine Environmental Research, Vol. 24* (Essex, England: Elsevier Applied Science, 1988), passim.

76. Krezowski, S., J. Laib, P. Onana, T. Hartmann, P. Chen, C. F. Shaw III, and D. H. Petering, "Presence of Zn, Cu-Binding Protein in Liver of Freshwater Fishes in Absence of Elevated Exogenous Metal: Relevance to Toxic Metal Exposure," *Ibid.*, pp. 147–150.

77. Olsson, P.-E., A. Larsson, and C. Haux. "Metallothionein and Heavy Metal Levels in Rainbow Trout (*Salmo gairdneri*) During Exposure to Cadmium in the Water," *Ibid.*, pp. 151–153.

78. Overnell, J., T. C. Fletcher, and R. McIntosh. "Factors Affecting Hepatic Metallothionein Levels in Marine Flatfish," *Ibid.*, pp. 155–158.

79. Viarengo, A., L. Canesi, M. Pertica, G. Mancinelli, M. Orunesu, A. Mazzucotelli, and J. M. Bouquegneau. "Biochemical Characterization of a Copper-Thionein Involved in Cu Accumulation in the Lysosomes of the Digestive Gland of Mussels Exposed to the Metal," *Ibid.*, pp. 163–166.

80. Kluytmans, J. H., F. Brands, and D. I. Zandee. "Interactions of Cadmium with the Reproductive Cycle of *Mytilis edulis* L.," *Ibid.*, pp. 189–192.

81. Roesijadi, G. "Natural Occurrence of High Levels of Cadmium Bound to Metallothionein-like Proteins in the Kidney of the Clam *Protothaca staminea*," *Ibid.*, pp. 172–173 (Abstract).

82. Fowler, B. A., E. Gould, J. S. Garvey, and W. E. Bakewell, "Comparative Studies on the 45000 Dalton Cadmium-binding Protein (45K CdBP) from the Scallop *Placopecten magellanicus*: Immunological Properties and Copper Competition Studies," *Ibid.*, pp. 171–172 (Abstract).

83. Sanders, B. M., "The Role of the Stress Proteins Response in Physiological Adaptation of Marine Molluscs," *Ibid.*, pp. 207–210.

84. Steinert, S. A., and G. V. Pickwell, "Expression of Heat Shock Proteins and Metallothionein in Mussels Exposed to Heat Stress and Metal Ion Challenge," *Ibid.*, pp. 211–214.

85. Reinisch, C. L., D. L. Miosky, and R. Smolowitz. "The Use of Monoclonal Antibodies in Molluscan Pathobiology," *Ibid.*, pp. 354–355 (Abstract).

86. Smolowitz, R., and C. L. Reinisch. "Immunochemical Detection of the Origin of Hematopoietic Neoplasia in the Soft Shell Clam," *Ibid.*, p. 355 (Abstract).

87. Shugart, L. "An Alkaline Unwinding Assay for the Detection of DNA Damage in Aquatic Organisms," *Ibid.*, pp. 321–325.

88. Grill, E. "Phytochelatins, The Heavy Metal Binding Peptides of Plants: Characterization and Sequence Determination," in *Metallothionein II* (Basel, Switzerland: Birkhäuser-Verlag, 1987), pp. 317–322.

89. Rauser, W. E., "The Cd-binding Protein from Tomato Compared to Those of Other Vascular Plants," *Metallothionein II* (Basel, Switzerland: Birkhäuser-Verlag, 1987), pp. 301–308.

90. Kägi, J. H. R., and Y. Kojima. "Chemistry and Biochemistry of Metallothionein," in *Metallothionein II* (Basel, Switzerland: Birkhäuser-Verlag, 1987), p. 38.

91. Bernhard, W., and J. H. R. Kägi. "Purification and Characterization of Atypical Cadmium-binding Polypeptides from *Zea mays*," in *Metallothionein II* (Basel, Switzerland: Birkhäuser-Verlag, 1987), p. 309–315.

CHAPTER 17

Effect of Cadmium on Protein Synthesis in Gill Tissue of the Sea Mussel *Mytilus edulis*

M. B. Veldhuizen-Tsoerkan, D. A. Holwerda, C. A. van der Mast
and D. I. Zandee

ABSTRACT

Cellular toxicity of cadmium was studied in gill tissue of the sea mussel, *Mytilus edulis*. Mussels were exposed to cadmium chloride at 50 or 250 μg Cd/ L for short periods. Then the gills were excised and incubated with [^{35}S]-methionine or cysteine for 4 hr. Uptake of radiolabeled amino acids by the isolated gills was not affected by cadmium, whereas the incorporation of label was significantly decreased after Cd exposure. Two dimensional gel electrophoresis was used to analyze the de novo synthesized gill proteins. It revealed that the expression of particular proteins was differentially altered by cadmium. One dimensional gel analysis by [^{35}S]-cysteine labeled gill proteins demonstrated that cadmium induced, in a concentration-dependent manner, a cysteine-rich protein with a molecular weight of approximately 13 kDa, consisting of two isomers with low isoelectric points.

INTRODUCTION

Sea mussels, *Mytilus edulis,* have a high capacity to accumulate cadmium and other heavy metals without notable toxic effects.[1] Induction of metal-binding, metallothionein-like proteins by cadmium exposure has been demonstrated in

M. edulis.[2-5] Metallothioneins are specific metal-binding proteins of low molecular weight, having a high cysteine content. They play a crucial role in the cellular pathways and detoxification of heavy metals. Recent data indicate that the capacity of *M. edulis* to produce metal-binding proteins is limited, thereby restricting the organism's ability to tolerate a further increase in metal concentration.[6] We have recently found that cadmium is toxic to *M. edulis* at a relatively low concentration; the survival time during environmental anoxia was markedly diminished after only 2 weeks of exposure to 50 μg Cd/L (results to be published). Based on this finding, a study was started on the toxic effects of cadmium at the macromolecular level.

Cadmium is a ubiquitous environmental pollutant and may be expected to exert its toxic effects either by complexing directly with cellular macromolecules or indirectly by interfering with the metabolism of essential metals. In a variety of organisms and cells in culture, cadmium was found to interfere with DNA, RNA, and protein metabolism. Cadmium induced DNA single-strand breaks[7,8] and inhibited the processes of DNA repair and replication.[8-11] At higher concentrations, cadmium inhibited the incorporation of radioactive precursors into RNA and proteins.[9,10,12-15]

With marine invertebrates, the effects of cadmium at the molecular level are poorly known. Recently, Sanders[16] found that both heat shock and cadmium can induce heat shock proteins in *M. edulis*. At the same time, the expression of "normal" cellular proteins was decreased. However, according to Steinert and Pickwell[17] cadmium appeared to be a poor inducer of heat shock proteins in *M. edulis* and did not inhibit the incorporation of amino acids into cellular proteins. In fish hepatocytes, cadmium was found to affect the pattern of protein synthesis.[18,19]

The present study was designed to assess possible effects of cadmium on protein synthesis in isolated gill tissue of *M. edulis*. The gills were chosen for these experiments, since they are ideally constructed for an absorptive role, in having a high surface area, a cell monolayer and a rich vascularization, and because they are known as cadmium-accumulating tissue.[20,21] Moreover, morphological cadmium-induced changes have been found in molluscan gill tissue[22-24] that are apparently linked to physiological and biochemical alterations.

The experiments were carried out with gills isolated from mussels that had been exposed to cadmium chloride for short periods of time. The excised gills were incubated with [^{35}S]-methionine or -cysteine in order to estimate the rate of amino acid incorporations into cellular proteins. The *de novo* synthesized gill proteins were analyzed by means of one and two dimensional gel electrophoresis.

MATERIALS AND METHODS

Animals

Mussels, *Mytilus edulis*, were collected in the Eastern Scheldt in October 1987 and February 1988. Mean shell length was 6.0 ± 0.5 cm. After transportation to Utrecht, the animals were kept in aquaria with recirculating sea water at 12°C and were not fed.

Exposure System

Mussels were exposed to approximately 50 or 250 µg Cd/L in glass aquaria of 80 L to which sea water and metal solution were supplied with pumps at rates of 1 L/hr and 10 mL/hr, respectively. The animals were kept under natural light regime. The sea water temperature was 12°C. Cadmium was added as $CdCl_2 \cdot H_2O$ (Merck No. 2011). Prior to the further experimental procedure, the animals were kept overnight in unspiked sea water to eliminate adherent cadmium.

Metal Analysis

Cadmium in gill tissue was assayed by decomposing lyophylized tissue in 65% (w/v) nitric acid (Merck No. 456) at 80°C for 1.5 hr, using teflon (PTFE) bombs placed in a sandbath.[25] Cd concentration was measured by atomic absorption spectrophotometry, using a Varian SpectrAA-10, equipped with a deuterium lamp for background correction.

Incubation of Isolated Gills with Radiolabeled Amino Acid

Gills were isolated from mussels exposed to cadmium for 4, 7, or 15 days. The middle parts of the outer gill lamella were incubated individually in 0.5 mL of standard medium (0.45 µL filtered sea water, 32 mM imidazol, pH 7.6, 25 µg/mL chloroamphenicol, and 1 µM each of 19 unlabeled amino acids) with 12.5 or 25 µCi/mL [^{35}S]-methionine or [^{35}S]-cysteine (Amersham International). Incubations were carried out in the wells of multidashes (24 wells, Nucleon) for 4 hr at room temperature, using a rotation shaker plate (Brouwer Scientific). Incubations were stopped by placing on ice for 5 min. Gill tissue was removed from the incubation medium, rinsed, and washed twice with distilled water. After centrifugation for 2 min at 16,000 g at room temperature the supernatants were discarded. Pellets were weighed and resuspended 1:5 (w/v) in buffer (5 mM Tris.HCl, pH 7.1, 0.1 mM PMSF and 1 mM DTT). Pellets were disrupted by sonication for 2 × 30 sec. The homogenate was centrifuged at 16,000 g for 60 min at 4°C. A 10-µL aliquot of supernatant (SN) from each sample was used for protein determination according to Bradford.[26] SN aliquots of 5 µL were transferred to Whatman 3MM filters to measure incorporation of the label by hot 10% trichloroacetic acid precipitation. Label uptake was determined as total radioactivity present in the SN aliquots of 5 µL. Radioactivity was counted in a Beckman LS 75 liquid scintillation counter.

Label incorporation is expressed as counts per minute (cpm) per microgram cellular protein or as percentage of total radioactivity. Label uptake is expressed as cpm per microliter of supernatant. Statistical analysis of the data was performed by Student *t* test.

Two Dimensional Gel Electrophoresis

Two dimensional isoelectric focusing in polyacrylamide gels (IEF/SDS-PAGE) was employed to resolve the [^{35}S]-methionine or -cysteine labeled proteins, as described by O'Farrel et al.[27] and Garrels.[28] SN-samples, containing 75 μg of protein, were mixed with an equal volume of lysis buffer: 9.95 M urea, 4% Nonidet P-40, 2% Bio-Rad ampholytes (1.3% pH range 5–7 and 0.7% pH range 3–10), 0.1 M DTT and 0.3% SDS and were brought into the cathodic end of the gel for isoelectric focusing. The first dimension gels were cast from the mixture: 8.5 M urea, 2% Nonidet P-40, 3.5% acrylamide/bisacrylamide (ratio 30:1.5%) and 2% ampholytes (1.3% pH range 5–7 and 0.3% pH range 3–10). The gels were polymerized with 2 μL 10% ammonium persulfate and 1.5 μL TEMED per milliliter of gel mixture. The gels were loaded in glass tubes (12 mm × 2.5 mm inside diameter) and allowed to polymerize for 2 hr. They were prerun for $\frac{1}{2}$ hr at 200 V, $\frac{1}{2}$ hr at 400 V, and 1 hr at 600 V in a Bio-Rad model-175 tube cell. The lower reservoir was filled with 25 mM H_3PO_4 (anode electrode solution) and the upper reservoir with 50 nM NaOH (cathode electrode solution). After the samples were loaded and the reservoirs refilled, the gels were run for 20 hr at 800 V. The gels were then equilibrated for 2 × 15 min with gentle shaking in 125 mM Tris.HCl, pH 6.8, 3% SDS, 50 mM DTT and 0.001% bromophenol blue and loaded on the second dimension. Analysis in the second dimension was performed by SDS-PAGE on 10–20% polyacrylamide gradient gels (see below).

Following electrophoresis, the gels were stained in 40% methanol and 10% acetic acid with 0.25% Coomassie Brilliant blue R and treated with Enhance (New England Nuclear), according to the instructions of the manufacturer. The gels were dried and fluorographed with Hyper-film-MP (Amersham) at −70°C.

Computer Analysis

The fluorographed protein spots were analyzed with the IBAS image analysis system (Zeiss/Kontron, Eching, FRG). This system is organized around a fast pipeline-structured array processor (cycle time 100 ns). Images were digitized 10 times and averaged to improve signal to noise ratio (frame size 512 × 512 pixels, 8 bits = 256 grey levels). With the digital filter operation TRACKG (filter size 30 × 30 pix) the grey values of the spots are replaced by the locally most likely background values. The resulting image served as shading correction image for the original. Measurements were carried out on the corrected image. To delimit the individual spots, the procedure DISDYN was applied. This dynamic discrimination method operates with a local threshold which is dependent on the local neighborhood region. The local thresholds are derived from the local background grey level. Before determination of the integrated optical density as a measure for the amount of protein in the spots of interest, the input grey range is calibrated according to the corresponding optical density values.

One Dimensional Gel Electrophoresis

[35S]-cysteine labeled gill proteins were carboxymethylated by incubation in 0.2 M iodoacetate for 1 hr at 37°C in the dark, and resolved either by SDS-PAGE or by one dimensional gel isoelectric focusing.

SDS-PAGE was performed by the method of Laemmli.[29] Samples containing 70 μg of protein were mixed 2:1 with loading buffer (0.125 M Tris.HCl, pH 6.8, 5% SDS, 25% glycerol, 25% β-mercaptoethanol and 20 μM bromophenol blue), incubated at 95°C for 10 min and applied to 10–20% gradient gel. Gels were run at 80 V for 16 hr in a Bio-Rad Proteon II slab cell. Phosphorylase B (M_r = 92 kDa), serum albumin (M_r = 67 kDa), ovalbumin (M_r = 45 kDa), catalase (4 subunits, M_r = 40 kDa) and cytochrome c (M_r = 12.5 kDa) were used as standard markers.

One dimensional denaturing isoelectric focusing in a vertical mini gel system (Bio-Rad mini Proteon II slab cell) was performed essentially after Robertson et al.[30] Samples of 20 μg protein were mixed with an equal volume of lysis buffer. Lysis buffer and gel mixture were composed as described for IEF/SDS-PAGE, except that ampholytes were used of pH range 3–10. Samples were applied to the cathodic part of the gel. Electrophoresis was performed at room temperature for 3 hr at 200 V. After electrophoresis was completed, gels were fixed with 10% trichloroacetic acid for 10 min and then transferred to 1% trichloroacetic acid for 16 hr to remove ampholines.

Staining and fluorography of SDS-PAGE and IEF-gels were carried out as described above.

RESULTS

Cadmium Accumulation

The in vivo exposure resulted in accumulation of cadmium in the gills of *M. edulis*. Cadmium concentration both increased with dosage (Figure 1A) and time (Figure 1B).

[35S]-Methionine Incorporation

In preliminary experiments the optimal conditions for incorporation of label by isolated gills were assessed. Addition of chloramphenicol, an inhibitor of bacterial protein synthesis, and of 19 unlabeled amino acids to the incubation medium had a positive effect on the incorporation of label. Maximal incorporation in the gills was reached after 4 hr of incubation in standard medium at room temperature (Figure 2).

Exposure to 250 μg Cd/L for 7 days caused a significant ($p < 0.05$) decrease of 30% in the rate of methionine incorporation (mussels from October 1987). When the exposure time was extended to 15 days, the incorporation rate in gills of the exposed group continued to decline, attaining a significant ($p < 0.001$)

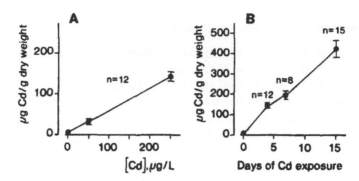

Figure 1. Cadmium accumulation in gills of mussels exposed (A) to 50 and 250 μg Cd/L for 4 days, (B) to 250 μg Cd/L. Mean ± SEM.

decrease of 40% (Figure 3). The experiment was duplicated on mussels collected in February 1988. Uptake of [^{35}S]-methionine by the isolated gills was not affected by cadmium (Figure 4A). Inhibition (34%) of the protein synthesis rate was already significant ($p < 0.01$) after 4 days of exposure to 250 μg Cd/L. The previously observed significant decrease in methionine incorporation rate following 7 and 15 days of exposure to 250 μg Cd/L was confirmed but, in this experiment, extension of the exposure time to 7 and 15 days did not result in an additional inhibition of the incorporation rate that remained at a level of approximately 76% of the control (Figure 4A, B).

Two dimensional gel electrophoresis of de novo synthesized gill proteins, followed by image analysis, revealed that protein synthesis was differentially affected by 7 days of exposure to 250 μg Cd/L (Figure 5, Table 1). The majority of protein spots showed a decreased integrated optical density (Table 1). Syn-

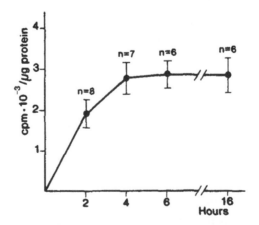

Figure 2. Time course of [^{35}S]-methionine incorporation. Isolated gills were incubated with 12.5 μCi [^{35}S]-methionine at 22.5°C. Mean ± SEM.

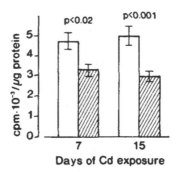

Figure 3. Effect of cadmium on the [^{35}S]-methionine incorporation. Isolated gills were incubated with 25 μCi [^{35}S]-methionine for 4 hr at 20°C. □, gills from control
mussels; ■, gills from mussels exposed to 250 μg Cd/L. Mean of 10 animals
± SEM.

thesis of several proteins was almost completely inhibited (numbers 7–9, 30–
32, 40–42; Table 1, Figure 5), whereas synthesis of others was increased (numbers 3, 5, 6, 11, 22, and 38; Table 1, Figure 5).

[^{35}S]-Cysteine Incorporation

Experiments with [^{35}S]-cysteine were carried out on mussels from February
1988. Uptake of [^{35}S]-cysteine by the isolated gills appeared to be stimulated
by cadmium (Figure 6A). A significant decrease of label incorporation rate of
34% was found after 4 days of exposure to 50 μg Cd/L. The inhibition of protein
synthesis rate increased to 45% when the cadmium concentration was raised to
250 mg Cd/L. The effect of cadmium became more conspicuous when the
incorporated radioactivity was expressed as percentage of the total radioactivity
(Figure 6C). In a duplicate experiment, no significant effect of cadmium on label
uptake was found (Figure 7A). Exposure for 4 days to 50 and 250 μg Cd/L
resulted in a significant inhibition of 32% and 40%, respectively, of the protein
synthesis rate (Figure 7B). In the next experiment, the exposure to 250 μg Cd/
L was extended to 15 days. Uptake of cysteine was not affected by cadmium
(Figure 8A), whereas the rate of label incorporation was significantly decreased
(Figure 8B). The observed decrease of 33% became less conspicuous than that
after 4 days of exposure to 250 μg Cd/L.

Two dimensional gel electrophoresis of [^{35}S]-cysteine labeled gill proteins,
followed by image analysis, indicated that 4 days of exposure to 50 μg Cd/L
differentially changed the protein synthesis pattern (Figure 9, Table 2). The
majority of protein spots had a decreased integrated optical density, while the
synthesis of several proteins was almost completely inhibited (numbers 3–5, 13,
20–21, 24, 27, and 29; Figure 9, Table 2).

One dimensional SDS-PAGE of the carboxymethylated, [^{35}S]-cysteine labeled
gill proteins revealed that cadmium induced the synthesis of a cysteine-rich

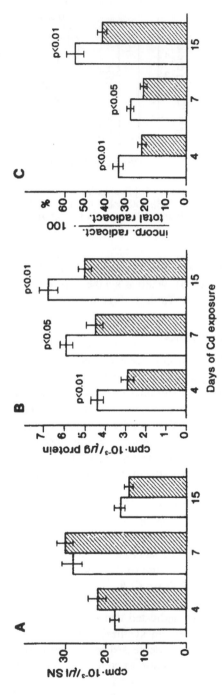

Figure 4.   Effect of cadmium on uptake (A) and incorporation rate (B, C) of [^{35}S]-methionine in gill tissue. Isolated gills were incubated with 25 μCi [^{35}S]-methionine for 4 hr at 20°C. □, gills from control mussels; ▧, gills from mussels exposed to 250 μg Cd/L. Mean of 10 mussels ± SEM.

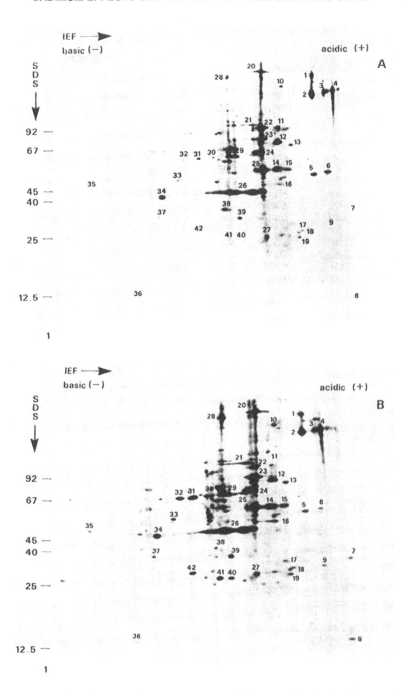

Figure 5. Two dimensional IEF/PAGE patterns of [^{35}S]-methionine labeled gill proteins from mussels exposed to 250 μg Cd/L for 7 days. A—exposed group; B—control group. Fluorography was for 30 days. Lane 1: molecular weight markers. Numbers indicate the most conspicuous spots. Integrated optical density of these spots is given in Table 1.

Table 1.   Differential Effect of Cd Exposure on the Synthesis of Gill Proteins.

Protein Spot No.	Integrated Optical Density[a]			Protein Spot No.	Integrated Optical Density[a]		
	Cd Exposed Group	Control Group	Ratio[b]		Cd Exposed Group	Control Group	Ratio[b]
1	1.01	1.19	85	22	2.36	1.87	126
2	3.64	3.70	98	23	1.65	4.23	39
3	1.06	0.88	121	24	3.58	5.33	67
4	1.43	1.68	85	25	2.87	6.13	47
5	0.78	0.69	113	26	17.30	29.45	59
6	1.09	0.24	454	27	0.41	1.07	38
7	0.00	0.40	0	28	0.56	1.65	34
8	0.00	0.45	0	29	1.57	3.93	40
9	0.02	0.23	9	30	0.13	1.26	10
10	0.36	0.98	40	31	0.33	3.37	10
11	0.66	0.34	194	32	0.05	2.01	3
12	2.07	1.99	104	33	0.16	0.60	27
13	0.37	0.83	45	34	1.20	2.40	50
14	2.79	5.07	55	35	0.08	0.26	31
15	0.59	1.34	44	36	0.10	0.41	24
16	0.15	0.77	20	37	0.13	0.40	33
17	0.04	0.11	36	38	0.71	0.28	254
18	0.07	0.25	28	39	0.37	1.19	31
19	0.11	0.32	34	40	0.05	0.71	7
20	1.44	2.08	69	41	0.07	1.39	5
21	0.86	2.23	39	42	0.07	1.05	7

[a]Integrated optical density of IEF/SDS-PAGE patterns (Figure 5) of [35S]-methionine labeled gill proteins.
[b]Integrated optical density of Cd exposed group as a percentage of integrated optical density of control group.

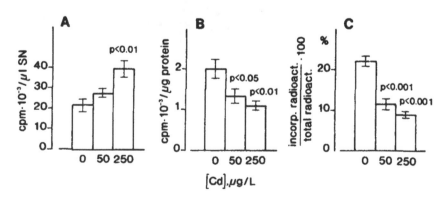

Figure 6.   Effect of cadmium on uptake (A) and incorporation rate (B, C) of [35S]-cysteine in gill tissue from mussels exposed to 0, 50, and 250 μg Cd/L for 4 days. Isolated gills were incubated with 25 μCi [35S]-cysteine for 4 hr at 20°C. Mean of 5 animals ± SEM.

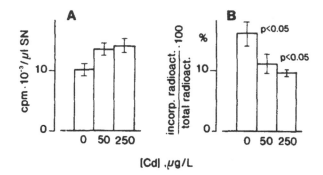

Figure 7. Effect of cadmium on uptake (A) and incorporation rate (B) of [^{35}S]-cysteine in gill tissue from mussels exposed to 0, 50, and 250 µg Cd/L for 4 days. Gills were incubated with 12.5 µCi [^{35}S]-cysteine for 4 hr at 20°C. Mean of 5 animals ± SEM.

protein with molecular weight of approximately 13–14 kDa (Figure 10). The induction of this protein increased with dosage (Figure 10, lanes 3 and 4) and time (Figure 10, lanes 4 and 5) of Cd exposure.

One dimensional isoelectric focusing of the same [^{35}S]-cysteine labeled samples showed that this newly induced, cysteine-rich protein consists of two isomers with isoelectric point between pH 3 and 5 (Figure 11). Induction of these proteins was strongly dependent on the Cd concentration: the higher the Cd concentration, the more pronounced the expression of these proteins (Figure 10, lanes 3, 4, and 5).

DISCUSSION

Cadmium-binding proteins in the sea mussel, *Mytilus edulis,* have been reported to contain a small amount of methionine residues and a high cysteine

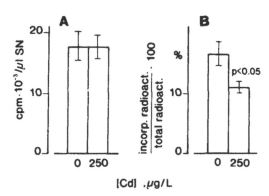

Figure 8. Effect of cadmium on uptake (A) and incorporation rate (B) of [^{35}S]-cysteine in gill tissue from mussels exposed to 0 and 250 µg Cd/L for 15 days. Gills were incubated with 25 µCi [^{35}S]-cysteine for 4 hr at 22°C. Mean of 5 animals ± SEM.

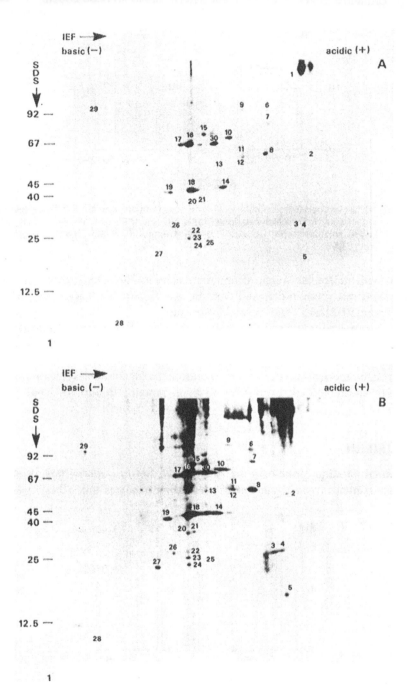

Figure 9. Two dimensional IEF/PAGE patterns of [^{35}S]-cysteine labeled gill proteins from mussels exposed to 50 μg Cd/L for 4 days. A—exposed group; B—control group. Fluorography was for 45 days. Lane 1: molecular weight markers. Numbers indicate the most conspicuous spots. Integrated optical density of these spots is given in Table 2.

Table 2.   Differential Effect of Cd Exposure on the Synthesis of Gill Proteins.

Protein Spot No.	Integrated Optical Density[a]			Protein Spot No.	Integrated Optical Density[a]		
	Cd Exposed Group	Control Group	Ratio[b]		Cd Exposed Group	Control Group	Ratio[b]
1	2.81	3.84	73	16	2.78	3.35	83
2	0.05	0.27	19	17	0.54	0.86	63
3	0.00	0.56	0	18	2.23	5.53	40
4	0.01	0.42	2	19	0.50	1.17	43
5	0.03	0.76	4	20	0.01	0.31	3
6	0.07	0.46	15	21	0.01	0.19	5
7	0.14	0.51	27	22	0.00	0.29	0
8	0.55	2.23	25	23	0.24	0.37	65
9	0.09	0.16	56	24	0.00	0.38	0
10	0.87	1.65	53	25	0.10	0.18	56
11	0.07	0.28	25	26	0.09	0.30	30
12	0.19	0.66	29	27	0.00	0.79	0
13	0.01	0.22	5	28	0.25	0.17	147
14	0.88	3.36	26	29	0.01	0.42	2
15	0.38	0.54	70	30	1.79	3.31	54

[a]Integrated optical density of IEF/SDS-PAGE patterns (Figure 9) of [^{35}S]-methionine labeled gill proteins.
[b]Integrated optical density of Cd exposed group as a percentage of integrated optical density of control group.

content, up to about 25 residue-%.[5] Therefore, [^{35}S]-cysteine labeling of gill proteins was applied to specifically trace the induction of thioneins by Cd exposure, whereas [^{35}S]-methionine labeling was used to study the general effect of Cd on total protein synthesis. Amino acid incorporation is commonly used to indicate cytotoxicity[31] and it reflects both the cellular uptake of amino acid and the protein synthesis activity.

The present data show that the rate of protein synthesis in gills is significantly decreased by short-term in vivo Cd exposure. The effect of cadmium was most pronounced after 4 days and remained significant after 15 days of exposure. Inhibition of cysteine incorporation (32–34%) was observed at a Cd concentration of 50 µg/L (Figures 6B, C–7B), while an equal decrease (34%) of methionine incorporation was found at a Cd concentration 5 times higher after 4 days of exposure (Figure 4B, C). Prolonged exposure to cadmium (250 µg/L, 15 days) did not result in a further decrease of incorporation rate, that amounted to 33% for [^{35}S]-cysteine and 24% for [^{35}S]-methionine (Figures 4B, C and 8B). Stabilization of the cadmium effect at a certain level may indicate that the inhibition of label incorporation is an acute response to Cd exposure which is counteracted by, e.g., synthesis of metal-binding proteins after longer exposure times.

Two dimensional gel analysis revealed that expression of gill proteins is differentially affected by cadmium. Synthesis of methionine-labeled proteins was generally decreased, however, to a different extent and the expression of specific

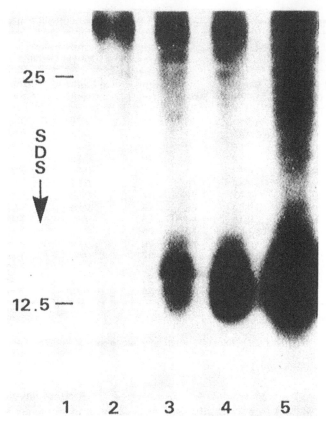

25 —

S
D
S
↓

12.5 —

1   2   3   4   5

Figure 10. *Induction of low molecular weight, cysteine-rich protein in the gills of Cd-exposed mussels. SDS-PAGE of [³⁵S]-cysteine labeled gill proteins. Fluorography was for 14 days. Only the low molecular weight part of the gel fluorograph is shown. Lanes: 1—molecular protein markers; 2—control group; 3—group exposed to 50 μg Cd/L for 4 days; 4—group exposed to 250 μg Cd/L for 4 days; 5—group exposed to 250 μg Cd/L for 15 days.*

proteins was even enhanced (Figure 5, Table 1). The majority of cysteine-labeled proteins had a decreased integrated optical density and the expression of several proteins was beyond the detection level (Figure 9, Table 2). Induction of thioneins, as cysteine-rich proteins, by cadmium was not apparent on a two dimensional gel, as the protein samples were not subjected to prior carboxymethylation. The latter resulted in poor two dimensional resolution (results not shown). Carboxymethylation is usually applied to block SH-residues of thioneins, that can be smeared all over the gel, presumably through formation of S-S-linkages with other proteins during electrophoresis.[32] Carboxymethylation followed by one dimensional gel analysis of [³⁵S]-cysteine labeled gill proteins showed that cadmium induced, in a concentration-dependent manner, cysteine-rich proteins with low molecular weight (13–14 kDa), consisting of two isomers with low

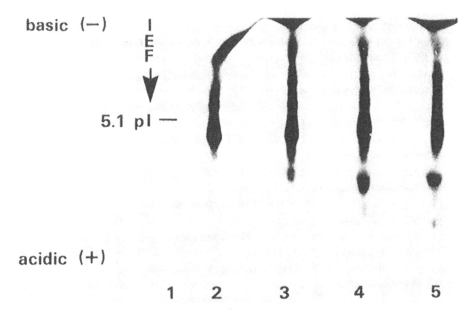

basic (−)

I
E
F
↓

5.1 pl —

acidic (+)

1   2   3   4   5

Figure 11. Induction of two cysteine-rich proteins with low isoelectric point in the gills of Cd exposed mussels. One dimensional IEF of [^{35}S]-cysteine labeled gill proteins. Fluorography was for 7 days. Only the acidic part of the gel fluorograph is shown. Lanes: 1—IEF standard, β-lactoglobuline B, pl 5.1; 2—control group; 3—group exposed to 50 μg Cd/L for 4 days; 4—group exposed to 250 μg Cd/L for 4 days; 5—group exposed to 250 μg Cd/L for 15 days.

isoelectric points (Figures 10 and 11). These proteins are presumably metallothioneins, but proof of their cadmium-binding capacity is lacking in this study.

The observed differences between cysteine and methionine incorporation into the gill proteins may reflect different cellular mechanisms of Cd toxicity. The decrease of cysteine incorporation is probably related to the cellular metabolism of cysteine. The synthesis of metallothioneins by metal stress may cause a decrease in the intracellular level of cysteine available for the synthesis of other molecules, e.g., high molecular weight proteins and glutathione. In mammalian cells, the glutathione level was shown to be markedly decreased by Cd treatment.[11] In fishes, cadmium depressed the [^{35}S]-cysteine incorporation into high molecular weight proteins, whereas synthesis of cysteine-rich, low molecular weight, metal-binding proteins was changed.[19]

Depression of the [^{35}S]-methionine incorporation rate may represent a general cellular stress response. A recent study with *M. edulis* showed that both heat-shock stress and cadmium caused a repression of "normal" cellular proteins, although 72 hr of exposure to cadmium at 1 ng/L or 1 μg/L inhibited the incorporation of [^{35}S]-methionine into gill proteins to a lesser degree than did heat shock.[16] On the other hand, in the experiments of Steinert and Pickwell[17] mussels were subjected to Cd exposure at 200 and 600 μg/L for 4 or 28 hr, but no inhibitory effect on methionine incorporation into gill proteins was observed.[17] In both studies, however, label incorporation was measured by densitometry of gel fluorographs, while in our experiments the incorporation rate was assessed in individual gill pieces by hot trichloroacetic acid precipitation.

Cadmium was also shown to inhibit protein synthesis, as measured by a decreased incorporation of labeled amino acids other than cysteine, in mammalian cells.[9,10,13-15] In these experiments, an effect of cadmium on protein synthesis was observed at Cd concentration of 200–220 μg/L after several hours of exposure.

To make sure that inhibition of amino acid incorporation was not due to a decreased membrane permeability, uptake of amino acids by the isolated gills was also measured. The uptake of [^{35}S]-methionine appeared not be be affected by in vivo Cd exposure (Figure 4A), while a tendency to an enhanced uptake of [^{35}S]-cysteine was observed in some experiments (Figures 6A and 7A).

Possible mechanisms for a decrease of protein synthesis may include disfunction at the level of replication, transcription, or translation. In mammalian cells, cadmium is known to cause DNA damage,[7,8] to inhibit the processes of DNA replication and repair,[8-11] and to depress RNA synthesis.[9,10] It has also been reported that cadmium inhibits protein synthesis in vitro by phosphorylation of the α-subunit of the eucaryotic initiation factor 2, the loss of reversing factor activity, and the disaggregation of polyribosomes.[33] It is worthwhile to note that the initial phase of the stress response is characterized by a shutdown of normal protein synthesis due to inhibition of initiation.[34,35]

In order to get more insight into the level and the mechanism of the inhibitory action of cadmium on protein synthesis in the sea mussel, *M. edulis*, additional research is required.

ACKNOWLEDGMENTS

The authors would like to thank Dr. M. Terlou for developing the computer program to analyze two dimensional gels, Mr. F. Kindt and his colleagues for photography, the Department for Image Processing and Design for preparation of the graphics, and Miss M. H. van Hattum for typing the manuscript. This work was supported by the Ministry of Transport and Public Works, Tidal Waters Division.

REFERENCES

1. Poulsen, E., H. U. Rijsgard, and F. Molenberg. "Accumulation of Cadmium and Bioenergetica in the Mussel *Mytilus edulis*," *Mar. Biol.* 68:25–29 (1982).
2. Noël-Lambot, F. "Distribution of Cadmium, Zinc and Copper in the Mussel *Mytilus edulis*. Existence of Cadmium-Binding Proteins Similar to Metallothioneins," *Specialia Exp.* 32(3):324–326 (1975).

3. George, S. G., E. Carpene, T. L. Coombs, J. Overnell, and A. Youngson. "Characterization of Cadmium Binding Protein from *Mytilus edulis* (L.), Exposed to Cadmium," *Biochim. Biophys. Acta* 580:225–233 (1979).

4. Nolan, C. V., and E. J. Duke. "Cadmium Accumulation and Toxicity in *Mytilus edulis:* Involvement of Metallothioneins and Heavy-Molecular Weight Protein," *Aquat. Toxicol.* 4:153–163 (1983).

5. Frazier, J. M. "Cadmium-Binding Proteins in the Mussel, *Mytilus edulis,*" *Environ. Health Pers.* 65:39–43 (1986).

6. Harrison, F. L., J. R. Lam, and J. Novacek. "Partitioning of Metals among Metal-Binding Proteins in the Bay Mussel, *Mytilus edulis,*" *Mar. Environ. Res.* 24(1-4):167–170 (1987).

7. Ochi, T., and M. Ohsawa. "Induction of 6-thioguanine-Resistent Mutants and Single-Strand Scission of DNA by Cadmium Chloride in Cultural Chinese Hamster Cells," *Mutation Res.* 111:69–78 (1983).

8. Burkart, W., and B. Ogorek. "Genotoxic Action of Cadmium and Mercury in Cell Cultures and Modulation of Radiation Effects," *Toxicol. Environ. Chem.* 12:173–183 (1986).

9. Nocentini, S. "Inhibition of DNA Replication and Repair by Cadmium in Mammalian Cells. Protective Interaction of Zinc," *Nucleic Acids Res.* 15:4211–4225 (1987).

10. Ochi, T., M. Mogi, M. Watanabe, and M. Ohsawa. "Induction of Chromosomal Aberrations in Cultured Chinese Hamster Cells by Short-Term Treatment with Cadmium Chloride," *Mutation Res.* 137:103–109 (1984).

11. Ochi, T., R. Takahashi, and M. Ohsawa. "Indirect Evidence for the Induction of a Prooxidant State by Cadmium Chloride in Cultured Mammalian Cells and a Possible Mechanism for the Induction," *Mutation Res.* 180:257–266 (1987).

12. Hidalgo, H. A., V. Koppa, and S. E. Bryan. "Effect of Cadmium on RNA-polymerase and Protein Synthesis in Rat Liver," *FEBS Lett.*, 64:159–162 (1976).

13. Beattie, J. H., M. Marion, and F. Denizeau. "The Modulation by Metallothionein of Cadmium-Induced Cytotoxicity in Primary Rat Hepatocyte Cultures," *Toxicology* 44:329–339 (1987).

14. Mitane, Y., Y. Aoki, and K. T. Suzuki. "Accumulation of Newly Synthesized Serum Proteins by Cadmium in Cultured Rat Liver Parenchymal Cells," *Biochem. Pharmacol.* 36:3657–3663 (1987).

15. Din, W. S., and J. M. Frazier. "Protective Effect of Metallothionein on Cadmium Toxicity in Isolated Rat Hepatocytes," *Biochem. J.* 230:395–402 (1985).

16. Sanders, B. M. "The Role of the Stress Proteins Response in Physiological Adaptation of Marine Molluscs," *Mar. Environ. Res.* 24:207–210 (1987).

17. Steinert, S. A., and G. V. Pickwell. "Expression of Heat Shock Proteins and Metallothionein in Mussels Exposed to Heat Stress and Metal Ion Challenge," *Mar. Environ. Res.* 24:211–214 (1987).

18. Baksi, S. M., and J. M. Frazier. "A Fish Hepatocyte Model for the Investigation of the Effects of Environmental Contamination," *Mar. Environ. Res.* 24(1-4):141–145 (1987).

19. Baksi, S. M., N. Libbus, and J. M. Frazier. "Induction of Metal Binding Proteins in Striped Bass, *Morone saxatilis,* following Cadmium Treatment," *Comp. Biochem. Physiol.* 91C(2):355–363 (1988).

20. George, S. G., and T. L. Coombs. "The Effects of Chelating Agents on the Uptake and Accumulation of Cadmium by *Mytilus edulis,*" *Mar. Biol.* 39:261–268 (1977).

21. Carpene, E., and S. G. George. "Absorption of Cadmium by Gills of *Mytilus edulis*," *Mol. Physiol.* 1:23–34 (1981).
22. Engel, D. W., and B. A. Fowler. "Copper and Cadmium Induced Changes in the Metabolism and Structure of Molluscan Gill Tissue," in *Marine Pollution: Functional Responses,* F.P. Vernberg, A. Calabrese, F.P. Thurberg, and F.J. Vernberg, Eds. (New York: Academic Press, 1979), pp. 239–256.
23. Sunila, I. "Toxicity of Copper and Cadmium to *Mytilus edulis* L. (Bivalvia) in Brackish Water," *Ann. Zool. Fennici* 18:213–223 (1981).
24. Sunila, I., and R. Lindström. "The Structure of the Interfilamentar Junction of the Mussel (*Mytilus edulis* L.) Gill and its Uncoupling by Copper and Cadmium Exposures," *Comp. Biochem. Physiol.* 81C(2):267–272 (1985).
25. Hemelraad, J., D. A. Holwerda, and D. I. Zandee. "Cadmium Kinetics in Freshwater Clams. I. The Pattern of Cadmium Accumulation in *Anodonta cygnea*," *Arch. Environ. Contam. Toxicol.* 15:1–7 (1986).
26. Bradford, M. M. "A Rapid and Sensitive Method for Quantitation of Micro-gram Quantities of Protein Utilizing the Principle of Protein-Dye Binding," *Anal. Biochem.* 72:248–252 (1976).
27. O'Farrel, P. Z., H. M. Goodman, and P. H. O'Farrel. "High Resolution Two-Dimensional Electrophoresis of Basic as well as Acidic Proteins," *Cell* 12:1133–1142 (1977).
28. Garrels, J. I. "Two-Dimensional Electrophoresis and Computer Analysis of Proteins Synthesized by Clonal Cell Lines," *J. Biol. Chem.* 257:7961–7977 (1979).
29. Laemmli, U. K. "Cleavage of Structural Proteins during the Assembly of the Head of Bacteriophage T-4," *Nature* 227:680–685 (1970).
30. Robertson, E. F., H. K. Dannelly, P. J. Malloy, and H. C. Reeves. "Rapid Electric Focusing in a Vertical Polyacrylamide Minigel System," *Anal. Biochem.* 167:290–294 (1987).
31. Holbrook, D. J., Jr. "Effects of Toxicants on Nucleic Acid and Protein Metabolism," in *Introduction to Biochemical Toxicology,* E. Hodgson, and F. E. Guthrie, Eds. (New York: Elsevier North Holland, 1980), pp. 262–284.
32. Koizumi, S., N. Otaki, and M. Kimura. "Estimation of Thionein Synthesis in Cultured Cells by Slab Gel Electrophoresis," *Ind. Health* 20:101–108 (1982).
33. Hurst, R., J. R. Schatz, and R. L. Matts. "Inhibition of Rabbit Reticulocyte Lysate Protein Synthesis by Heavy Metal Ions Involves the Phosphorylation of the α-Subunit of the Eukaryotic Initiation Factor 2," *J. Biol. Chem.* 262:15939–19545 (1987).
34. Lindquist, S. "The Heat-Shock Response," *Ann. Rev. Biochem.* 55:1151–1191 (1986).
35. Craig, E. A. "The Heat-Shock Response," *Crit. Rev. Biochem.* 18:239–280 (1985).

SECTION SIX

Application of Biomarkers in Field Evaluation

CHAPTER 18

Sentinel Species and Sentinel Bioassay

W. R. Lower and R. J. Kendall

INTRODUCTION

In Webster's *New International Dictionary, Second Edition* (1930), "sentinel" is defined as "One who watches or guards . . . from surprise, . . . observe(s) and give(s) notice of danger." This definition of sentinel is carried over into the two concepts of sentinel species as organisms and sentinel bioassays as measurable changes in biological functions, both of particular relevance as biological indicators of environmental contamination. The use of both sentinel species and sentinel bioassays is germane to laboratory and field analyses and to assessing the human and environmental health effects of toxic substances and hazardous conditions. The two concepts can be applied independently or together, but the combined use of sentinel species and sentinel bioassay is probably the most effective. The ideal application of sentinel is as an early warning system, in keeping with the dictionary definition. In this sense it is a "biological radar" warning of approaching hostile conditions.

A sentinel species can be defined as any domestic or wild microorganism, plant or animal that can be used as an indicator of exposure to and toxicity of a xenobiotic that can be used in assessment of the impact on human and/or environmental health because of the organism's sensitivity, position in a community, likelihood of exposure, and geographic and ecological distribution or abundance. A sentinel species can be any organism or tissue of an organism that fulfills the need as a suitable subject upon which to perform bioassay and chemical analysis. The organism may be common, with broad occurrence geographically and ecologically, or it may occur only in special places. There does not need to be any restriction on what organism can qualify as a sentinel species. It may be commonly used in the laboratory or it may be reserved for special tests. A sentinel species can be from any taxon from bacteria to man.

A sentinel bioassay can be defined as changes in a measurable biological function which can be evaluated in a number of species covering a range of taxa, or a biological function that assesses especially important aspects of environmental and human health, with some of the assays sensitive to specific xenobiotics and others to a broad range of xenobiotics. Sentinel bioassay involves the concept of detection and prevention of disease, common to preventive medicine, with the disease, in this discussion, referring to a disfunction of pollution origin affecting the health of other organisms as well as human health. Sentinel bioassay also involves the use of agreed upon and defined organisms and applied biological bioassay endpoints for assessment and monitoring.

This is an early formulation of the concepts of sentinel species and sentinel bioassay and, although the concepts have been used by others, it is difficult to assign the first coining of the terms. The definition of the terms as first used may also be different from the definition in this paper. The current discussion combines sentinel species and sentinel bioassay and contributes to their use by formalizing these conceptions jointly and by outlining their definitions and utility. Certainly, this formalized drafting of both candidate sentinel species and candidate sentinel bioassays, and their application, will evolve and change.

The utility of sentinel species is based upon the uptake and metabolism of toxic xenobiotics and the occurrence of measurable biological changes as bioassay endpoints caused by xenobiotics. The nomination of a species as a sentinel is determined by a number of factors, including the experience and background of the investigator. The concept and definition of sentinel species requires an enumeration of criteria supplied for each candidate. Criteria might include:

- geographic and ecological distribution—a species with a wide geographic and ecological distribution may be useful because of the ability to compare results in the same species from different monitored locations;
- a restricted home range of individuals;
- the presence of a particular species at a particularly important location, whether or not it is widely distributed;
- the availability of one or more easily measured and/or significant biological endpoint(s) for bioassay in a species;
- the ability of a species to be used as a probe of a particular aspect of the environment, e.g., a particular niche, habitat, trophic position, unique biochemistry, etc.;
- the easy and accurate identification of a species, particularly when the species is to be used by investigators who are not experts in the taxonomy of the organism.

The current common use of a species for any purpose, whether microorganisms or large metazoa or metaphyta, need not necessarily influence its choice as a sentinel species. Too often a species has been used for bioassay purposes because it is available and something is known about it, not necessarily because it is particularly suited as a bioassay organism. All organisms are potentially useful. Organisms high on the food chain may be particularly useful, but it depends on the chemical(s) (does it bioaccumulate?), and on the motivation for monitoring (food chains integrate many routes of exposure, but identification of specific routes of exposure may also be critical). In picking a species or bioassay as a potential sentinel one may want to confer with experts on the subjects for advice and comments to take advantage of their knowledge.

The current discussion will provide only provisional comments on candidate sentinel species and sentinel bioassays. The lists will change as the criticisms of others provide a more comprehensive understanding and examination of the use of the two concepts.

In one sense, all bioassays have been applying the concepts of sentinel species and sentinel bioassay, in that particular species have been chosen and particular biological functions have been used. In many cases, however, both species and endpoints appear to have been developed for bioassay as an afterthought from use or study for other purposes. For example, *Drosophila melanogaster* was used for genetic research for six or more decades before attempts were made to adapt its use for bioassay purposes with limited success. The freshwater alga, *Selenastrum capricornutum*, is commonly and successfully used as a bioassay for testing water. However, upon conferring with an algalogist, Dr. Charles Gowens, about species of algae suited for bioassay, not only was *S. capricornutum* not one of the organisms that the algalogist suggested, but he asked why in the world even consider it when, from the algalogists view, there are many better candidates.

SENTINEL SPECIES

Wildlife toxicology, perhaps more than other areas of toxicology, has used the concepts of sentinel to identify and remedy problems, although in the past the sentinel aspect may have been more retrospective than prospective, e.g., the finding of DDT as the cause of egg shell thinning resulting in the decrease in populations of pelicans. The pelican could be considered to have become a sentinel species and the thinning of the egg shell the sentinel bioassay.

The species used are often a function of habitat and purpose of a study and the species chosen depends upon the trophic level, food chains, and the biological endpoints of interest as well as the geographic and ecological distribution. For example, if groundwater contamination is a major consideration, the species chosen should be directly or indirectly exposed to the aqueous environment such as aquatic invertebrates, aquatic plants, shellfish, fish, waterfowl, herons, and muskrats. Terrestrial plant and animal species would assess soil contamination. Identifying entire food chains for evaluation (such as plants → earthworms → shrews → mice → snakes and hawks) may be particularly valuable in evaluating food chain transfer and bioaccumulation. Airborne contamination is more of a problem to assess biologically; plants appear to be very good candidates, although certain insects, such as colonial bees, and nesting birds could be of use.

The use of indigenous species offers an advantage of having organisms exposed to multiple contaminants potentially over their lifetime while the introduction of species on-site may provide a greater measure of control. The species chosen can be those for which toxicological information is available from previous field and laboratory research, those in which capture and observation is facilitated, or those which have other desirable characteristics such as occurrence in a niche of interest, critical placement in a food web, expression and measurement of particularly relevant biological function, etc. In field pen studies, animals and plants with known histories can be placed on-site and the exposure route and duration can be controlled. Pens of plants or animals placed off the ground result in more direct exposure to airborne contaminants. Pens properly placed can allow for direct substrate contact and potential ingestion of contaminated soil, food and water, while pens placed on the ground in streams allow direct contact with contaminated water.

Plants

Plants, both domestic and wild, are particularly suited as sentinel species. There is a large number of cultivated crop and ornamental plants that may qualify as sentinel species and may be particularly useful for monitoring. In selecting potential sentinel plant species, the following characteristics might be included: monocot and dicot; C_4 and C_3 carbon fixation; nitrogen fixing species (mostly legumes); herbaceous and woody growth; annual and perennial plants. Various plant parts might also be considered: roots or tubers; flowers and fruits; foliage; and stems or stalks.

The listing of domestic plants might include, as a first approximation, *Zea mays* (corn)—an important C_4 crop plant; and *Glycine max* (soybean)—a C_3, nitrogen fixing annual species and one of the dominant world-wide agricultural crops; another nitrogen fixing C_3 plant might be the perennial *Medicago sativa* (alfalfa), the roots of which can reach depths of 10 m or more. Other species might include *Daucus carota* (carrot), *Beta vulgaris saccharata* (sugar beet), and *Solanum tuberosum* (potato), for root related structures, and the fruit of *Lycopersicon esculentum* (tomato), another solanaceous species related to the potato. This list is composed primarily of temperate species, but the list should also include tropical and desert agricultural species and agricultural species of particular regional importance.

Wild plants represent an enormous reservoir of potential sentinel monitoring species. The very number of possible species makes the choice of sentinel species a considerable challenge. When a wild species of plant—or any other organism—is used as a sentinel, great care must be taken to assure that all of the individuals are indeed the same species or, if several species are mixed, the mixture must be noted. In any case, the taxonomy of any organism must be completely defined to ensure an unequivocal identification of a species.

Plant bioassays have been employed in a variety of field studies. The single-gene waxy locus of corn (*Zea mays*) has been used under a variety of conditions. This bioassay system promulgated by M. Plewa involves the synthesis of amylopectin and amylose in pollen cells and detects both forward and reverse mutations.[1] Several species, hybrids, and strains of *Tradescantia* have been used to investigate environmental mutagenicity. Two *Tradescantia* assay systems have been employed: one for detecting single-gene mutation in stamen hair cells and the other for chromosomal aberrations as micronuclei in pollen mother cells. The waxy corn pollen system in combination with the *Tradescantia* stamen hair test and observations of nitrogen fixation of soybean (*Glycine max*), corn height at anthesis, percent pollen abortion of corn, and excretion of urinary δ-aminolevulinic acid in *Peromyscus leucopus/maniculatus* (white-footed/deer mouse) have been used to evaluate toxicity in the vicinity of a lead smelter.[2-6] This test system in concert with *Salmonella typhimurium*, the *Tradescantia* micronucleus test, and the corn yellow/green (yg-2/yg-2) leaf spot assay has also been used to evaluate the mutagenicity of a number of city sewage sludges and with the *Tradescantia* stamen hair test to evaluate environmental mutagenicity near an oil refinery and a petrochemical complex.[7-8]

A series of studies using the *Tradescantia* stamen hair assay to monitor ambient air for genetic hazards at 18 U.S. sites was instituted by Brookhaven National Laboratory from 1976 through 1980. One of the sites monitored was Elizabeth, New Jersey, which has a large number of petrochemical refineries and at which mutation frequency in plants exposed to ambient air was 90.6% above the control level. Two clean sites, the Grand Canyon, Arizona, and Pittsboro, North Carolina, registered no increase in mutations above the control level.[9-10] Part of each air sample was analyzed by GC/MS, and a part was also tested for mutagenicity in *Salmonella typhimurium*. The results obtained in the *S. typhimurium* showed close agreement with those obtained from the *Tradescantia* assay. However, no compound or class of compounds could be associated with the mutagenic responses observed and the mutagenic response at a particular site was dependent upon the ambient temperature and wind direction relative to the monitoring site.

The *Tradescantia* stamen hair assay has also been used to determine the mutagenicity of sewage sludges, polluted water, bottom sediment from a drinking-water reservoir, diesel emission exposures, air pollution from petrochemical plants, soil and air pollution from a lead smelter, chemical smokes at a military installation, effluent from a titanium factory and the environment around nuclear power plants and to evaluate the hazards of peroxyacetyl nitrate and related photo-oxidants in the ambient environment and under laboratory conditions.[2,8,11-18] Most of the efforts to use this system have yielded positive results although one, using clone KU-7 to monitor the area around a nuclear power plant for 3 years in Japan, did not show significant changes in mutation frequencies as compared to the control.

The *Tradescantia* micronucleus assay detects chromosomal aberrations as micronuclei at the tetrad stage in meiotic cells of pollen. This unique assay has been used for testing several sites, as well as sewage sludges, pesticides, diesel exhaust, and obscurant smokes.[9,19-22]

A natural population of *Osmunda regalis* (royal fern) growing in a river heavily contaminated with paper recycling waste was reported to have a high incidence of chromosomal aberrations.[23] Solvent extracts of samples of the paper recycling waste collected from the wastewater treatment facility and tested in the laboratory also showed mutagenicity in the Ames test with S-9 fraction and in the soybean meitotic crossing-over assay.

Other plant assays that could be further developed for in situ assay of soil, water, and possibly air include the *Hordeum vulgare* (barley) chlorophyll-deficient mutation assay and barley chromosome aberration assay, *Arabidopsis* embryo mutation assay, *Vicia faba* (broad bean) root-tip cytogenetic assay, and *Allium cepa* (onion) root-tip cytogenetic assay.[24-28]

Invertebrates

Invertebrates offer a truly enormous reservoir of potential sentinel species, rivaling that of plants. They form many phyla, are found in all environments, and range in size from almost microscopic to the size reached by deep sea squid. The mid size from *Dapnia magna* to large invertebrates of the size of small squid are probably the most useful because of sufficient biomass for observation and analysis. Chromosome aberrations have been reported in the purple sea urchin and the marine polychaete *Neanthes arenaceodentata* has been proposed as a cytogenetic model for marine toxicology.[29-30] The most used of this group has been the mussel (*Mytilus edulis*) in which not only survival but a variety of chromosomal changes have been monitored.[31-33]

Fish

Fish have been used as indicators of environmental contamination. The mud minnow (*Umbra limi*) has been used to assess for sister chromatid exchange (a measure of mutagenesis) and chromosomal abnormalities have been observed in *Boleophthalmus dussumeri*.[34-36]

There is a considerable amount of information on carcinogenesis in fresh water and marine fish resulting from environmental contamination.[37-53] These include skin lesions, carcinomas, hepatomas, and papillomas as well as hyperplastic and preneoplastic lesions in a variety of freshwater and marine fish. The fish include the brown bullhead, catfish, white suckers and drum fish from the Great Lakes and their tributaries, sauger and walleye from Torch Lake, Michigan, black bullhead catfish in wastewater treatment ponds in Alabama and 17 species of fish from the Fox River, Illinois. Marine fish include English sole and other flatfish from Puget Sound, tomcod from the Hudson River in New York, and winter flounder from Boston Harbor.

Amphibians and Reptiles

Amphibians and reptiles have not been much used for bioassay. For the most part they appear to have been considered unsuitable for laboratory bioassay although the eggs and larvae of some amphibians could be used. Bullfrog tadpoles have been used as a model for genotoxicity of environmental contaminants.[40,54] The difficulty of breeding and raising reptiles may have discouraged their use. The information on reptiles is particularly bleak.

Most reptiles are carnivorous and all amphibians are carnivorous during some part of their life cycle. Amphibians are frequently herbivorous as larvae, but become carnivorous in the adult or neotenic stage. As carnivores the two groups are high in the food chain and may be at or near the top in spite of the fact that they themselves may be food to others. Their position in the food chain further enhances their potential for biological concentration and usefulness as sentinel species.

Birds

Birds are excellent organisms for biological monitoring. They are easy to observe, sensitive to many toxic compounds, have well-known life histories, and are abundant at many trophic levels and almost all ecosystems.[55] Studies of piscivorous birds, such as the great blue heron (*Ardea herodius*), have a high potential for exposure to various environmental contaminants because of their dietary habits and widespread occurrence including feeding and nesting in urban areas.[56-58] Studies of the great blue heron in western Oregon and British Columbia have provided information on the population, reproductive status, and nesting and feeding requirements of this species in the Northwest.[59,60] Raptors, such as the red-tailed hawk (*Buteo jamaicensis*), have documented the bioavailability, biomagnification, and food transfer of compounds like organophosphates, organochlorines, and PCBs.[58,61-63] The red-tailed hawk is an opportunistic hunter, feeding on rodents, birds, snakes, and insects. It is widely distributed across North America and is found in many ecological biomes.[64] Trapping methods for this species and other raptor species are documented and there is a growing database on its biochemical response to environmental contaminants such as pesticides.[65-66]

The European starling is another avian species which may be useful as a biological indicator. The starling is common from arctic Canada to the subtropics of Mexico and the present population of starlings in North America is estimated to be at least 200 million. Starlings are primarily grassland feeders, and usually forage within 200 m of their nest. During the breeding season, both the adult and nestling diet consists almost entirely of invertebrates obtained from the surface or from the upper few centimeters of soil.[67] Nest boxes are readily accepted as nest sites by starlings.[68-70]

The enhancement of indigenous populations of birds and other organisms can often be achieved by increasing a limited resource. For example, nest sites are one of the most common limiting sources for avian species. The placement of nest boxes or platforms can attract such species as European starling (*Sturnus vulgaris*), common barn owl (*Tyto alba*), great tit (*Parus major*), and eastern bluebird (*Sialia sialis*).[71-73] Such enhanced populations can provide a statistically significant number of individuals to study and facilitate observation and access to both young and adults. Resource enhancement might be considered for other major groups of organisms.

Mammals

Mammals are obviously useful as sentinel species, e.g., the white footed mouse.[4,74,75] Field trapping methods and population techniques are established for many small mammals.[76] Species candidates such as the members of the genera *Peromyscus* and *Microtus* are commonly found in many habitats over most of North America. Mice of the genus *Peromyscus* are among the most widely distributed mammals in North America. Small mammals such as mice and voles have small home ranges making it relatively easy to estimate what contaminants they have encountered.[77] Most small mammal species burrow in the soil increasing the potential for exposure to various environmental contaminants.[78] Small mammals have been used extensively in both laboratory and field experiments providing a large database about their life histories and responses to chemical exposure.[4,79-85]

Large mammals should be considered. Carnivores such as the coyote and raccoon, and the opposum, a marsupial with omnivorous food habits, may biologically accumulate xenobiotics and serve as useful sentinels. Game animals—deer, opossum, raccoon, squirrel—are also of potential value because they are used for human consumption.

Domestic mammals, e.g., the pet dog and cat, pigs, goats and cattle occur in great numbers and are often in contact with the general environment and with man's specific environment. These species should be carefully considered for sentinels. Incidences of bone sarcoma, mesothelioma, and bladder cancer have been studied in dogs related to several environmental factors.[86-89]

The human, the ultimate domestic mammal, may well serve as a sentinel species. This does not advocate using the human as a bioassay organism, but using information gathered in physicals, surgery, medical treatment, autopsies, etc. This is done already to some extent with workers employed in some industries, such as vinyl chloride and pesticides, exposure to toxicological agents and life styles such as consumption of alcohol and smoking.[90-97]

Sentinel Bioassays

Bioassay, biological endpoints, biomarkers, biological function—whatever terms are applied—can be used to assess the toxicity of individual chemicals and complex mixtures in the laboratory and to assess environmental exposure and predict ecological and human health effects.[98] Potential bioassay endpoints are numerous and depend upon the species and the category of endpoints to be measured, e.g., excretory function, mutagenesis, biomass, behavior, enzyme induction, etc. Many bioassays are covered in other chapters of this book. The discussion of bioassays will include those that fall under in situ field test bioassay endpoints as well as strictly laboratory test bioassay endpoints. In many cases the same bioassay endpoint can be used in the laboratory or in the field. In in situ field tests, while the exposure of the organism is always in situ in the field, the actual measurement of the bioassay may be in the field, or the exposed organism (or parts of it) may be taken to the laboratory for final analysis.

Chemical Analysis and Radiation Measurement

Chemical analysis of parent compounds and metabolite residues has been shown to be an extremely effective tool for monitoring chemical uptake and accumulation. There was a time that chemical analysis was much preferred, but bioassay is becoming very important in assessing the effects of xenobiotics and not just their concentration. Watson et al. found significant trophic level movement of polychlorinated biphenyls and other organochlorine compounds in a PCB-contaminated waste site.[83] Residues were found in both biota and upper trophic level consumers. Several other researchers have found similar results with voles (*Microtus pennsylvanicus*) at Love Canal, ring-billed gulls (*Larus delawarenis*), snapping turtles (*Chelydra serpentina*), and wild mink (*Mustela vison*) and otter (*Lutra canadensis*).[82,99–101] Residues of 2,3,7,8-tetrachlorodibenzo-p-dioxin have been found in deer mice (*Peromyscus maniculatus*), prairie voles (*Microtus leucogaster*), coyote (*Canis latrans*), blue racer snake (*Coluber constrictor*), and garter snake (*Thamnophis sirtalis*) from Times Beach.[102] Heavy metal contamination in small mammals has been associated with wastewater-irrigated habitats, highways and a mercury contaminated site.[103–107] Selenium residues have been found in mammals utilizing California irrigation drainwater and may have implications for other metal and metalloid-contaminated sites.[84]

Several researchers have used wildlife to assess the effects of radiation from nuclear waste disposal areas.[108–112] Halford et al. censused on-site wild waterfowl to determine species composition and residence times.[113] Specimens of waterfowl, small mammals, and passerine birds were collected and muscle tissue was used to determine radionuclide content. Commercially raised, wing-clipped mallards (*Anas platyrhynchos*) were released on the radioactive leaching ponds. Attached and implanted dosimeters were used to determine radionuclide content. Species collected enabled researchers to evaluate radionuclide exposure differences between habitat use and ecosystem component (dabbling vs. diving ducks and terrestrial vs. aquatic species) and exposure source (internal vs. external exposure). The hazard to wildlife and to hunters who hunt on-site and consume ducks shot on-site was evaluated. It was concluded that there was no threat to human health. Other studies evaluated radiation in raptor nestings and their small mammal food base, in herons, in other bird species, and in other wildlife.[109,111,114-118] Wildlife appear to be an excellent monitoring tool for tracking radiation levels on-site and for estimating possible off-site movement of contamination.

Mortality

The ultimate and often most conspicuous endpoint of adverse exposure is death. Increased mortality has been found in voles (*Microtus pennsylvanicus*) in the immediate area of the Love Canal hazardous waste site and in birds inhabiting a DDT contaminated site at Wheeler National Wildlife Refuge.[82,119] In wildlife toxicology, mortality can be determined by systematic carcass searches on-site, or more indirectly through an evaluation of species diversity, density, and age-class ratios. The on-site diversity, density, and age-class structure, when compared to a field control or reference site, can be used to identify the absence of key species whose presence on site is altered by the site contaminants. These indices can also be used to assess the health of the on-site population in terms of size and reproductive potential.

Death also occurs in plants; whole systems can die. The dying forests of parts of the U.S., Canada, and central and northern Europe due to acid environmental conditions, fields of stunted corn due to misapplication of herbicides, and purposeful defoliation of areas of southeast Asia some 20 years ago are examples.

In wildlife toxicology, due to confounding factors such as immigration, emigration, migration and predation, mortality is seldom the most sensitive measure of effect, although it is often used in pesticide application assessments. An additional factor is the difficulty in locating carcasses.[120-122] Intoxication in wildlife may lead to behaviors or physical disabilities which result in death occurring in such a way that finding the carcass is unlikely (e.g., in a burrow or under vegetation). Plants, as sessile organisms, do not present this problem within a year, but may between years.

Sudden death on-site of plants or animals is an important and often dramatic indicator and can be a guide to further investigation. Death for extended periods of time after exposure, death off-site, and delayed death are all other components of importance.

Behavior

Most avian behavioral research today is concerned with the effects of pesticide exposure, but it could be modified to assess waste site contaminant exposure. Behavioral measures include basic motor activity, homing and orientation, breeding behavior, foraging behavior, and predator-prey relationships.[123-127] If critical behaviors such as breeding and foraging are altered as a result of contaminant exposure, the survivability of the individual, and possibly the population, is affected. Behavioral measurements taken in situ, in conjunction with other endpoints, could show direct effects of environmental contaminants on wildlife.

Reproduction

Reproduction is a common and important endpoint. Many bioassays measure the concentration which decreases the population size by some measured amount. The acute algal test with *Selenastrum capricornutum* is based on cell reproduction. Reproduction parameters are commonly used in wildlife toxicology, particularly avian wildlife toxicology, with endpoints such as egg number, number of nestlings hatched, number of birds fledged, juvenile weight, adult reproduction behavior, gonadal development, etc. Reproduction is also measurable in higher plants, although it is often considered a part of crop production, e.g., grain and fruit yield. Reproduction is measured in chickens as egg production and in food animals as litter size.

Physiological, Biochemical, Mutagenic and Carcinogenic Endpoints

Differences in physiological parameters such as body weight, organ weight, body measurements, body temperature, heart rate, fat content, and tissue histopathology can be used as a measure of exposure to waste site contaminants.[82,128,129] Expected impairments and measurements will be a function of the basic toxicity of the compounds present on-site.

Biochemical changes can be used to assess toxic material exposure. With exposure, changes in metabolic enzymes and pathways may occur. For example, reductions in brain and plasma cholinesterase activity have been used extensively to assess exposure in wildlife to cholinesterase-inhibiting compounds such as organophosphate pesticides.[70,130,131] These analyses are especially indicative of exposure when integrated with fecal residue analysis of both parent compound and metabolites.[66,132]

Other biochemical endpoints are increasingly receiving attention for their ability to indicate exposure to environmental pollutants. The induction of detoxification enzymes can be used to assess exposure to many organic contaminants and many endpoint measurements are available including induction of the mixed function oxidases, cytochrome P-450, ATPase, monoamine oxidases, NADPH cytochrome c reductase activities, and other cytochromatic enzyme activities.[133–139]

Adduct formation between xenobiotics and DNA, RNA and protein, commonly investigated in animals, but little studied in plants, are biochemical changes associated with mutations and carcinogenesis, particularly the DNA and RNA adducts. Adduct formation is unequivocal proof of exposure to a xenobiotic and offers considerable use as a bioassay.

Changes in immune response may also be indicative of toxic material exposure. Recent studies with small mammals and fish offer insights into the potential use of wildlife using newly developed immunoassays for assessing the impact of contaminated sites of the immune system.[79–81,140–142]

Changes in hormone metabolism may also be used as an indicator of pollutant exposure, including changes in plasma luteninizing hormone concentrations and effects of steroidogenesis.[143-144]

Other biochemical markers may include measurement of bile metabolites for PAHs and chlorinated phenolics, glycogen and sulfhydryl concentrations; glutathione peroxidase activity; perturbations in n-glutathione metabolism; alterations in specific heme biosynthetic pathway enzymes and in heme-dependent processes; changes in δ-aminolevulinic acid; changes in δ-aminolevulinic acid dehydratase; and porphyrin profile changes.[4,145-148]

The uptake, metabolism, and elimination of toxicants are critical parameters for modeling food chain transfer and risk assessment.[149-151] Many endpoints are biomedical endpoints and traditionally have not been applied in situ, but they offer exciting new approaches in evaluating impacts to wildlife and in interpreting wildlife results for human risk assessment.

Several biochemical biomarkers are restricted to plants. The most prominent are involved in photosynthesis, particularly the electron transport system of photosynthesis of all plants, and nitrogen fixation. When electron transport is blocked in photosynthesis an increased proportion of the absorbed energy is reemitted as fluorescence. This phenomenon can be used to monitor a plant's response to any factor which directly or indirectly affects photosynthetic metabolism. Stress conditions such as those caused by toxic materials can reduce the efficiency of photosynthesis by disturbing the pigment-protein apparatus or blocking the electron transport system in chloroplasts. The increase or decrease of fluorescence compared to a control can indicate a lesion in the chain of events in photosynthesis or damage to the chloroplasts. The nitrogen fixation of legumes can be used also as a biomarker to assess environmental toxicity. Little use of these has been made to date, but several laboratories are developing the procedures for using these functions as bioassays.[152]

In addition to the presence of known carcinogens on site, the oxidative metabolism of many toxicants results in metabolic intermediates which are active mutagens/carcinogens.[153-154] Recent advances in assay development may allow for the use of these endpoints for in situ wildlife monitoring of mutagenesis and carcinogenesis. DNA and RNA adduct formation and cytochrome P-450 activation have been found to be indicative of carcinogenic activation using several species and compounds in vivo and in vitro.[129,155-159] Mutagenic responses have been found in situ in plants in association with a lead smelter and petrochemical sites and a petrochemical waste dump.[3,160]

In vivo and in vitro mutagenic assays are routinely being conducted on xenobiotics, and their application to wildlife could provide a direct evaluation of immunotoxicity, carcinogenicity, and mutagenicity of waste site contaminants and a realistic evaluation of the potential impact on man.[161,162]

SUMMARY AND CONCLUSION

The assessment of the toxic effects of environmental contamination and the development of biological early warning systems for appraisal of these toxic effects can be advanced by the use of sentinel species and sentinel bioassays. The selection of organisms and changes in biological functions as sentinels can be made from currently used species and bioassays. In addition, an effort can be made to "plan" a species and "plan" a bioassay by developing a list of desired characteristics for both, discuss these with knowledgeable colleagues and, in a purposeful manner, come up with a best-effort combination which will accomplish the desired biological monitoring. The discussion of organism and bioassays in this article can serve as a backdrop or departure point for this purpose.

Lists of candidate sentinel species of plants, invertebrates, fish, amphibians, reptiles, birds, and mammals and a list of candidate sentinel bioassays are available upon request to the senior author.

REFERENCES

1. Plewa, M. J ., E. D. Wagner, G. J. Gentile, and J. M. Gentile. "An Evaluation of the Genotoxic Properties of Herbicides Following Plant and Animal Activation," *Mutat. Res.* 136:233–245 (1984).

2. Lower, W. R., P. S. Rose, and V. K. Drobney. "*In Situ* Mutagenic and Other Effects Associated with Lead Smelting," *Mutat. Res.* 54:83–95 (1978).

3. Lower, W. R., W. A. Thompson, V. K. Drobney, and A. F. Yanders. "Mutagenicity in the Vicinity of a Lead Smelter," *Teratogen. Carcinogen. Mutagen.* 3:231–253 (1983).

4. Lower, W. R., and R. K. Tsutakawa. "Statistical Analysis of Urinary δ-Aminolevulinic Acid (ALAU) Excretion in the White-Footed Mouse Associated with Lead Smelting," *Environ. Pathol. Toxicol.* 1:551–560 (1978).

5. Lower, W. R., and W. A. Thompson. "An Indirect Test of Correlation," *Environ. Toxicol. Chem.* 7:77–80 (1988).

6. Thompson, W. A., and W. R. Lower. "Indirect Determination of Correlation," *J. Stat. Plan. Inference* 16:83–87 (1987).

7. Hopke, P. K., M. J. Plewa, J. B. Johnston, D. Weaver, S. G. Wood, R. A. Larson, and T. Hinesly. "Multitechnique Screening of Chicago Municipal Sewage Sludge for Mutagenic Activity," *Environ. Sci. Technol.* 16:140–147 (1982).

8. Lower, W. R., V. K. Drobney, B. J. Aholt, and R. Politte. "Mutagenicity of the Environments in the Vicinity of an Oil Refinery and a Petrochemical Complex," *Teratogen. Carcinogen. Mutagen.* 3:65–73 (1983).

9. Schairer, L. A., J. Van't Hof, C. G. Hayes, R. M. Burton, and F. J. de Serres. "Exploratory Monitoring of Air Pollutants for Mutagenicity Activity with the *Tradescantia* Stamen Hair System," *Environ. Health Pers.* 27:51–60 (1978).

10. Schairer, L. A., J. Van't Hof, C. G. Hayes, R. M. Burton, and F. J. de Serres. "Measurement of Biological Activity of Ambient Air Mixtures Using a Mobile Laboratory for *In Situ* Exposures: Preliminary Results from the *Tradescantia* Plant Test System," in *Applications of Short-Term Bioassays in the Fractionation of Complex Environmental Mixtures.* M. D. Waters, S. Nesnow, J. L. Hisingh, S. Sandhu, and L. Claxton, Eds. (New York: Plenum Press, 1979), pp. 421–440.

11. Lower, W. R., A. G. Underbrink, A. F. Yanders, K. Roberts, T. K. Ranney, G. T. Lombard, D. D. Hemphill, and T. Clevenger. "New Methodology for Assessing Mutagenicity of Water and Water Related Sediments," in *2nd Int. Conf. on Ground-*

water Quality Research. N. N. Durham, and A. E. Redelfs, Eds. (Stillwell, OK: Oklahoma State University Printing Services Publishers, 1984), pp. 194–196.

12. Lower, W. R., A. F. Yanders, T. R. Marrero, A. G. Underbrink, V. K. Drobney, and M. D. Collins. "Mutagenicity of Bottom Sediment from a Water Reservoir," *Environ. Toxicol. Chem.* 4:13–19 (1985).

13. Ichikawa, S. *"In Situ* Monitoring with *Tradescantia* around Nuclear Power Plants," *Environ. Health Pers.* 37:145–164 (1981).

14. Yamaguchi, H. "Variation in Mutation in *Tradescantia* Stamen Hairs Cultivated near a Power Reactor Site," in seminar on Somatic Mutations in *Tradescantia* as a Model System in Radiobiology and Environmental Studies Report in 1369/B. Institute of Nuclear Physics, Krakow, Poland, 1987, pp. 9–10.

15. Abraham, S. "Floral Abnormalities and Somatic Mutations in the Staminal Hairs of *Tradescantia* Clone 02 Induced by Effluents from a Titanium Factory," Report in 1369/B. Institute of Nuclear Physics, Krakow, Poland, 1987, pp. 33–45.

16. Cebulska-Warilewska, A., and K. Kulczykowska. "Somatic Mutations in *Tradescantia* as a Indicator of Air Pollution," Report in 1369/B. Institute of Nuclear Physics, Krakow, Poland, 1987, pp. 47–55.

17. Meier, A., and M. Wallenschus. "An *In Situ* Monitoring Experiment with *Tradescantia* KU9 around the PWR Unterweser in Northern Germany," Report in 1369/B. Institute of Nuclear Physics, Krakow, Poland, 1987, pp. 105–113.

18. Schairer, L. A., R. C. Sautkulis, and N. R. Tempel. "Mutagenicity of Smog and Diesel Emissions Implies that UV and/or Visible Light are Activating Agents," *Environ. Mutagen.* 5:466 (1983).

19. Ma, T.-H., and M. M. Harris. *"In Situ* Monitoring of Environmental Mutagens," *Hazard Asses. Chem.* 4:77–105 (1985).

20. Ma, T.-H., V. Anderson, M. M. Harris, and J. L. Bare. "Tradescantia- Micronucleus (Trad-MCN) Test on the Genotoxicity of Malathion," *Environ. Mutagen.* 5:127–137 (1983).

21. Ma, T.-H., W. R. Lower, F. D. Harris, J. Poku, V. A. Anderson, M. M. Harris, and J. L. Bare. "Evaluation by the Tradescantia-micronucleus Test of the Mutagenicity of Internal Combustion Engine Exhaust Fumes from Diesel and Diesel-soybean Oil Mixed Fuels," in *Short-Term Bioassays in the Analysis of Complex Environmental Mixtures III.* M. Waters, S. Sandhu, J. Lewtas, L. Claxton, N. Chernoff, and S. Nesnow, Eds. (New York: Plenum Press, 1983), pp. 89–99.

22. Ma, T.-H., M. M. Harris, V. A. Anderson, I. Ahmed, K. Mohammad, J. J. Bare, and G. Lin. "Tradescantia-micronucleus (Trad-MCN) Tests on 140 Health-related Agents," *Mutat. Res.* 138:157–167 (1984).

23. Klekowski, E., and D. E. Levin. "Mutagens in a River Heavily Polluted with Paper Recycling Wastes: Results of Field and Laboratory Mutagen Assays," *Environ. Mutat.* 1:209–219 (1979).

24. Constantin, M. J., and R. A. Nilan. "The Chlorophyll-deficient Mutant Assay in Barley *(Hordeum vulgare),* A Report of the U.S. EPA's GENE-TOX Program," *Mutat. Res.* 99:37–49 (1982).

25. Constantin, M. J., and R. A. Nilan. "Chromosome Aberration Assays in Barley *(Hordeum vulgare),* A Report of the U.S. EPA's GENE-TOX Program," *Mutat. Res.* 99:13–36 (1982).

26. Redei, G. P. "Mutagen Assay with Arabidopsis. A Report of the U.S. EPA's Gene-TOX Program," *Mutat. Res.* 99:243–255 (1982).

27. Ma, T.-H. "Vicia Cytogenetic Tests for Environmental Mutagens. A Report of the U.S. EPA's GENE-TOX Program," *Mutat. Res.* 99:257–271 (1982).

28. Grant, W. F., and K. D. Zura. "Plants are Sensitive *In Situ* Detectors of Atmospheric Mutagens," in *Mutagenicity: New Horizon in Genetic Toxicology*, J. A. Heddle, Ed. (New York: Academic Press, 1982), pp. 407–434.

29. Pesch, G. G., and C. E. Pesch. *"Neanthes arenaceodentata* (Polychaeta: Annelida), a Proposed Cytogenetic Model for Marine Genetic Toxicology," *Can. J. Fish Aquat. Sci.* 37:1225–1228 (1980).

30. Hose, J. E., H. W. Puffer, P. S. Osheda, and S. M. Bay. "Developmental and Cytogenetic Abnormalities Induced in the Purple Sea Urchin by Environmental Levels of Benzo(a)pyrene," *Arch. Environ. Contam. Toxicol.* 12:319–325 (1983).

31. Dixon, D. R. "Aneuploidy in Mussel Embryos (*Mytilus edulis* L.) Originating from a Polluted Dock," *Mar. Biol. Lett.* 3:155–161 (1982).

32. Dixon, D. R., and K. R. Clarke. "Sister Chromatid Exchange: A Sensitive Method for Detecting Damage Caused by Exposure to Environmental Mutagens in the Chromosomes of Adult *Mytilus edulis*," *Biol. Lett.* 3:163–172 (1982).

33. Harrison, F. L., and I. M. Jones. "An *In Vivo* Sister Chromatid Exchange Assay in the Larvae of the Mussel *Mytilus edulis*: Response to 3 Mutagens," *Mutat. Res.* 105:235–242 (1982).

34. Kligerman, A. D., and S. E. Bloom. "Sister Chromatid Differentiation and Exchanges in Adult Mudminnows (*Umbra limi*) after *In Vivo* Exposure to 5-Bromodeoxyuridine," *Chromosoma* 56:101–109 (1976).

35. Alink, G. M., E. M. H. Frederia Walters, M. A. Van der Gaag, J. F. J. Van de Kerkhoff, and C. L. M. Poels. "Induction of Sister Chromatid Exchanges in Fish Exposed to Rhine Water," *Mutat. Res.* 78:369–374 (1980).

36. Krishnaja, A. P. and M. S. Rege. "Induction of Chromosomal Aberrations in Fish *Boleophthalmus dussumieri* after Exposure *In Vivo* to Mitomycin C and Heavy Metals Mercury, Selenium and Chromium," *Mutat. Res.* 102:71–82 (1982).

37. Metcalf, C. D. "Chemical Carcinogenesis Studies with Fish," Abstr. in Society of Environmental Toxicology and Chemistry Seventh Annual Meeting, Washington, DC, 1986, p. 90.

38. Couch, J. A., and J. C. Harshbarger. "Effects of Carcinogenic Agents on Aquatic Animals: An Environmental and Experimental Overview," *Environ. Carcinogen. Rev.* 3:63–105 (1985).

39. Hose, J. E., and S. S. Smith. "Hematological Abnormalities in Fish from Highly Contaminated Areas of Southern California," Abstr. in Society of Environmental Toxicology and Chemistry Seventh Annual Meeting, Washington, DC, 1986, p. 66.

40. Knezovich, J. P., P. A. Krauter and F. L. Harrison. "Bullfrog Tadpoles as a Model for Studying the *In Vivo* Metabolism and Genotoxicity of Environmental Contaminants," Abstr. in Society of Environmental Toxicology and Chemistry Seventh Annual Meeting, Washington, DC, 1986, pg. 68.

41. Halver, J. E. "Aflatoxicosis and Rainbow Trout Hepatoma," in *Mycotoxins in Foodstuffs*, G. N. Wogan, Ed., (Cambridge: MIT Press, 1965), pp. 209–234.

42. Malins, D. C., B. B. McCain, D. W. Brown, S.-L. Chan, M. S. Myers, J. T. Landahl, P. G. Prohaska, A. J. Friedman, L. D. Rhodes, D. G. Burrows, W. D. Gronlund, and H. O. Hodgins. "Chemical Pollutants in Sediments and Diseases of

Bottom-dwelling Fish in Puget Sound, Washington," *Environ. Sci. Technol.* 18:705–713 (1984).

43. Baumann, P. C., W. D. Smith, and M. Ribeck. *"Polynuclear Aromatic Hydrocarbons, Physical and Biological Chemistry,"* (Columbus, OH: Battelle Press, 1982), p. 93.

44. Harshbarger, J. C., L. J. Cullen, M. J. Calabrese, and P. M. Spero. "Epidermal, Hepatocellular and Cholangiocellular Carcinomas in Brown Bullheads, *Ictalurus nebulosus* from Industrially Polluted Black River, Ohio," *Mar. Environ. Res.* 14:535–536 (1984).

45. Smith, I. R., H. W. Ferguson, D. A. Rokosh, and M. A. Hayes. "Papellomatosis and Cholangio-cellular Neoplasia in Fish from Lake Ontario (Hamilton) Harbor," Abstr. in Society of Environmental Toxicology and Chemistry Seventh Annual Meeting, Washington, DC, 1986, p. 124.

46. Brown, E. R., J. J. Hazdra, L. Keith, I. Greenspan, J. B. Kwapinski, and P. Beamer. "Frequency of Fish Tumors in a Polluted Watershed as Compared to Nonpolluted Canadian Waters," *Cancer Res.* 33:189–198 (1973).

47. Black, J. J., M. Holmes, P. O. Dymerski, and W. F. Zapisek. "Fish Tumor Pathology and Aromatic Hydrocarbon Pollution in a Great Lakes Estuary," in Hydrocarbons and Halogenated Hydrocarbons in the Aquatic Environment, B. K. Afgan and D. McKay, Eds. (New York: Plenum Press, 1981), p. 559.

48. Black, J. J., P. O. Dymerski, and W. F. Zapisek. "Routine Liquid Chromatographic Method for Assessing Polynuclear Aromatic Hydrocarbon Pollution in Fresh Water Environment," in *Aquatic Toxicology Hazard Assessment*, D. R. Branson and K. L. Dickson, Eds. (Philadelphia: Am. Soc. for Testing and Materials, 1981).

49. Black, J. J. "Field and Laboratory Studies of Environmental Carcinogenesis in the Niagara River," *J. Great Lakes Res.* 9:326–334 (1983).

50. Black, J. J. "Aquatic Animal Bioassays for Carcinogenesis," *Transplant. Proc.* 16:406–411 (1984).

51. Tomljanovich, D. A. "Growth Phenomena and Abnormalities of Sauger, *Stizostedion canadense* (Smith), of the Keweenaw Waterway," M.S. Thesis, Michigan Tech. Inst., Houghton, MI (1974).

52. Smith, C. E. "Hepatomas in Atlantic Tomcod from the Hudson River Estuary," Abstr. in Society of Environmental Toxicology and Chemistry Seventh Annual Meeting, Washington, DC, 1986, p. 65.

53. Murchelano, R. A. "Hepatotoxic Lesions in Boston Harbor, Massachusetts Winter Flounder, *Pseudopleuronectes americanus*," Abstr. in Society of Environmental Toxicology and Chemistry Seventh Annual Meeting, Washington, DC, 1986, p. 124.

54. Cooke, A. S. "Tadpoles as Indicators of Harmful Levels of Pollution in the Field," *Environ. Pollut.* 25:123–133 (1981).

55. Karr, J. R. "Biological Monitoring and Environmental Assessment: A Conceptual Framework," *Environ. Manage.* 11:249–256 (1987).

56. Riley, R. G., E. A. Crecelius, R. E. Fitzners, B. L. Thomas, J. M. Burtisen, and N. S. Bloom. "Organic and Inorganic Toxicants in Sediment and Marine Birds from Puget Sound," NOAA Technical Series Memorandum NOS OMS 1. National Ocean Service, National Ocean and Atmospheric Administration, Rockville, MD, 1983, p. 125.

57. Calambokidis, J., S. Speich, J. Peard, G. H. Steiger, J. C. Cubbage, D. M. Frye,

and L. J. Lowenstine. "Biology of Puget Sound Marine Mammals and Marine Birds: Population Health and Evidence of Pollution Effects," NOAA Technical Memorandum. National Oceanic and Atmospheric Administration, Rockville, MD, 1985, p. 159.

58. Henny, C. J., L. J. Blus, and C. S. Hulse. "Trends and Effects of Organochlorine Residues on Oregon and Nevada Wading Birds, 1972–1983," *Colonial Waterbirds* 8:117–128 (1985).

59. Pratt, H. M., and D. W. Winkler. "Clutch Size, Timing of Laying, and the Reproductive Success in a Colony of Great Blue Herons and Great Egrets," *Auk* 102:59–63 (1985).

60. Gibbs, J. P., S. Woodward, M. L. Hunter, and A. E. Hutchinson. "Determinants of Great Blue Heron Colony Distribution in Coastal Maine," *Auk* 104:38–47 (1987).

61. Henny, C. J. "An Analysis of the Population Dynamics of Selected Avian Species with Special Reference to Changes during the Modern Pesticide Era," Wildlife Research Report No. 1, U.S. Dept. of the Interior, Washington, DC, 1972, p. 99.

62. Henny, C. J., L. J. Blus, E. J. Kolbe, and R. E. Fitzner. "Organophosphate Insecticide (Famphur) Topically Applied to Cattle Kills Magpies and Hawks," *J. Wildl. Manage.* 49:648–658 (1985).

63. Henny, C. J., E. J. Kolbe, E. F. Hill, and L. J. Blus. "Case Histories of Bald Eagles and Other Raptors Killed by Organophosphorus Insecticides Topically Applied to Livestock," *J. Wildl. Dis.* 23:292–295 (1987).

64. Heintzelman, D. S. "Hawks and Owls of North America," (New York: Universe Books, 1979), pp. 33–37.

65. Bloom, P. H. "Capturing and Handling Raptors," in *Raptor Management Techniques Manual*, Giron Pendleton, B. A., B. A. Millsapp, K. W. Cline, and D. M. Bird, Eds. (Washington, DC: National Wildlife Federation, 1987), pp. 99–123.

66. Hooper, M. J. "Avian Cholinesterases: Their Characterization and Use in Evaluating Organophosphate Insecticide Exposure," Ph.D. Thesis, University of California, Davis, CA (1988).

67. Feare, C. *The Starling*, (New York: Oxford University Press, 1984), p. 315.

68. Powell, G. V. N. and D. Gray. "Dosing Free Living Nestling Starlings with an Organophosphate Pesticide, Famphur," *J. Wildl. Manage.* 44:918–921 (1980).

69. Grue, C. W., G. V. N. Powell, and M. J. McChesney. "Care of Nestlings by Wild Female Starlings Exposed to an Organophosphate Pesticide," *J. Appl. Ecol.* 19:327–335 (1982).

70. Robinson, S. C., C. J. Driver, R. J. Kendall, and T. E. Lacher, Jr. "Effects of Agricultural Spraying of Methyl Parathion on Reproduction and Cholinesterase Activity in Starlings (*Sturnus vulgaris*) in Skagit Valley, Washington," *Environ. Toxicol. Chem.* 7:343–349 (1988).

71. Marti, C. D., P. Wagner, and K. Denne. "Nest Boxes for the Management of Barn Owls," *Wildl. Soc. Bull.* 7:145–148 (1979).

72. Greenwood, P. J., P. Harvey, and C. M. Perrins. "Great Tit (*Parus major*) Studied in Nest Boxes," *Animal Behav.* 27:645–651 (1979).

73. Mitchell, R. T., H. Blagbrough, and R. VanEtten. "The Effects of DDT Upon the Survival and Growth of Nestling Songbirds," *J. Wildl. Manage.* 17:45–54 (1953).

74. Scanlon, P. F., R. J. Kendall, R. L. Lochmiller, and R. L. Kirkpatrick. "Lead Concentrations in Pine Voles From Two Virginia Orchards," *Environ. Pollut. (Series B)* 6:157–160 (1983).

75. Ringer, R. K. "The Future of a Mammalian Wildlife Toxicology Test: One Researcher's Opinion," *Environ. Toxicol. Chem.* 7:339–342 (1988).

76. Schemnitz, S. D. The Wildlife Management Techniques Manual, Wildlife Society, Washington, DC, 1980, pp. 686.

77. Southern, H. N. "The Stability and Instability of Small Mammal Populations," in *Ecology of Small Mammals*, D. M. Stoddart, Ed., (London: Chapman and Hall, 1979), pp. 103–134.

78. Hayward, G. F., and J. Phillipson. "Community Structure and Functional Role of Small Mammals in Ecosystems," in *Ecology of Small Mammals*, D. M. Stoddart, Ed. (London: Chapman and Hall, 1979), pp. 135–211.

79. Olson, L. J., and R. D. Hinsdill. "Immunosuppression in Deer Mice *Peromyscus maniculatus* Fed Chlorinated Biphenyls Arochlor 1254 and Chlorocholine Chloride," *Fed. Proc.* 42:Abstr. 3811 (1983).

80. Olson, L. J., and R. D. Hinsdill. "Influence of Feeding Chlorocholine Chloride and Glyphosine on Selected Immune Parameters in Deer Mice, *Peromyscus maniculatus*," *Toxicology* 30:103–114 (1984).

81. Olson, L. J., A. Fairbrother, R. D. Hinsdill, and T. M. Yuill. "Immuno Modulatory and Viral Challenge Effects of Feeding Cyclo Phosphamide, Chloro Choline Chloride and Glyphosine to Old Deer Mice *Peromyscus maniculatus*," *Fed. Proc.* 43:Abstr. 490 (1984).

82. Rowley, M. H., J. J. Christian, D. K. Basu, M. A. Pawlikowski, and B. Paigen. "Use of Small Mammals (Voles) to Assess a Hazardous Waste Site at Love Canal, Niagara Falls, New York," *Arch. Environ. Contam. Toxicol.* 12:383–397 (1983).

83. Watson, M. R., W. B. Stone, J. C. Okoniewski, and L. M. Smith. "Wildlife as Monitors of the Movement of Polychlorinated Biphenyls and Other Organochlorine Compounds from a Hazardous Waste Site," Northeast Fish and Wildlife Conference, Hartford, CT, May 5–8, 1985, pp. 91–104.

84. Clark, D. R. "Selenium Accumulation in Mammals Expposed to Contaminated California USA Irrigation Drainwater," *Sci. Total Environ.* 66:147–168 (1987).

85. Block, E. K. "The Effects of the Organophosphate Pesticide COUNTER on *Peromyscus*: An Integrated Laboratory and Field Study," M.S. Thesis, Western Washington University, Bellingham, WA, 1988, p. 110.

86. Schneider, R., C. R. Dorn, and D. O. N. Taylor "Factors Influencing Canine Mammary Cancer Development and Postsurgical Survival," *J. Natl. Cancer Inst.* 43:1249–1261 (1969).

87. Tjalma, R. A. "Canine Bone Sarcoma: Estimation of Relative Risk as a Function of Body Size," *J. Natl. Cancer Inst.* 36:1137–1150 (1966).

88. Glickman, L. T., L. M. Domanski, T. G. Maquire, R. R. Dubielzig, and A. Churg. "Mesothelioma in Pet Dogs Associated with Exposure of Their Owners to Asbestos," *Environ. Res.* 32:305–313 (1983).

89. Hayes, H. M., Jr., R. Hoover, and R. E. Tarone. "Bladder Cancer in Pet Dogs: A Sentinel for Environmental Cancer?," *Am. J. Epidemiol.* 114:229–233 (1981).

90. Lambert, B., A. Linblad, K. Holmberg, and D. Francesconi. "The Use of Sister Chromatid Exchange to Monitor Human Populations for Exposure to Toxicologically Harmful Agents," in *Sister Chromatid Exchange*, S. Wolff, Ed. (New York: Wiley-Interscience, 1982), pp. 149–182.

91. Dulout, F. N., M. C. Pastori, O. A. Olivero, M. G. Cid, D. Loria, E. Matos, N.

Sobel, E. C. DeBujan, and N. Albiano. "Sister-Chromatid Exchanges and Chromosomal Aberrations in a Population Exposed to Pesticides," *Mutat. Res.* 143:237–244 (1985).

92. Funes-Cravioto, F., B. Lambert, J. Lindsten, J. Ehrenberg, A. T. Natarajan, and S. Osterman-Geokar. "Chromosome Aberrations in Workers Exposed to Vinyl Chloride," *Lancet* 1:459 (1975).

93. Natarajan, A. T., and G. Obe. "Screening of Human Populations for Mutations Induced by Environmental Pollutants: Use of Human Lymphocyte System," *Ecotoxicol. Environ. Safety* 4:468–481 (1980).

94. Obe, G., and J. Herha. "Chromosome Aberrations in Heavy Smokers," *Hum. Genet.* 41:259–263 (1978).

95. Obe, G., D. Gobee, H. Engeln, J. Herha, and A. T. Natarajan. "Chromosomal Aberrations in Peripheral Lymphocytes of Alcoholics," *Mutat. Res.* 73:377–386 (1980).

96. Narod, S. A., G. R. Douglas, E. R. Nestmann, and D. H. Blakey. "Human Mutagens: Evidence from Paternal Exposure," *Environ. Mol. Mutagen.* 11:401–415 (1988).

97. Leonard, H., M. Duverger-Van Bogaert, A. Bernard, M. Lambaotte-Vandepaer, and R. Lauwerys. "Population Monitoring for Genetic Damage Induced by Environmental Physical and Chemical Agents," *Environ. Monitor. Assess.* 5:369–384 (1985).

98. McCarthy, J. F., L. R. Shugart, and B. D. Jimenez. "Biological Markers in Wild Animal Sentinels as Predictors of Ecological and Human Health Effects from Environmental Contamination," in 8th Life Science Symp. on Biological Indicators: Exposures and Effects, G. W. Gehrs, Ed., Knoxville, TN, March 21–23, Oak Ridge National Laboratory, Oak Ridge, TN, 1988.

99. Sileo, L., L. Karstad, R. Frank, M. V. H. Holdrinet, E. Addison, and H. E. Braun. "Organochlorine Poisoning of Ring-Billed Gulls in Southern Ontario," *J. Wildl. Dis.* 13:313–322 (1977).

100. Olafsson, P. G., A. M. Bryan, B. Bush, and W. Stone. "Snapping Turtles—A Biological Screen for PCBs," *Chemosphere* 12:1525–1532 (1983).

101. Foley, R. E., S. J. Jackson, R. L. Sloan, and M. K. Brown. "Organochlorine and Mercury Residues in Wild Mink and Otter: Comparison with Fish," *Environ. Toxicol. Chem.* 7:363–374 (1988).

102. Lower, W. R., A. F. Yanders, C. E. Orazio, R. K. Puri, J. Hancock, and S. Kapila. "A Survey of 2,3,7,8-Tetrachlorodibenzo-p-Dioxin Residues in Selected Animal Species from Times Beach, Missouri," *Chemosphere* 18:1079–1088 (1989).

103. Anthony, R. G., and R. Kozlowski. "Heavy Metals in Tissues of Small Mammals Inhabiting Waste-Water-Irrigated Habitats," *J. Environ. Qual.* 11:20–22 (1982)

104. Blair, W. F., A. L. Hiller, and P. F. Scanlon. "Heavy Metal Concentrations in Mammals Associated with Highways of Different Traffic Densities," *Va. J. Sci.* 28:61 (1977).

105. Scanlon, P. F. "Lead Contamination of Mammals and Invertebrates Near Highways with Different Traffic Volumes," in *Animals as Monitors of Environmental Pollutants* (Washington, DC: National Academy of Sciences, 1979), pp. 200–208.

106. Galluzzi, P. "Mercury Concentrations in Mammals, Reptiles, Birds, and Waterfowl Collected in the Hackensack Meadowlands," paper presented at New Jersey Academy of Science, 1981, p. 24.

107. Lipsky, D., and P. Galluzzi. "The Investigation of Mercury Contamination in the Vicinity of Berry's Creek," Management of Uncontrolled Hazardous Waste Sites National Symp., Washington, DC, 1982, p. 5.
108. Hanson, W. C., and H. A. Kornberg. "Radioactivity in Terrestrial Animals Near an Atomic Site," in *Proc. Int. Conf. Peaceful Uses of Atomic Energy*, Geneva, Switzerland, pp. 385–388.
109. Halford, D. K., and O. D. Markham. "Radiation Dosimetry of Small Mammals Inhabiting a Liquid Radioactive Waste Disposal Area," *Ecology* 59:1047–1054 (1978).
110. Halford, D. K., and J. B. Millard. "Vertebrate Fauna of a Radioactive Leaching Pond Complex in Southestern Idaho," *Great Basin Nat.* 38:64–70 (1978).
111. Craig, T. H., D. K. Halford, and O. D. Markham. "Radionuclide Concentrations in Nestling Raptors Near Nuclear Facilities," *Wilson Bull.* 91:72–77 (1979).
112. Halford, D. K., J. B. Millard, and O. D. Markham. "Radionuclide Concentrations in Waterfowl Using a Liquid Radioactive Waste Disposal Area and the Potential Radiation Dose to Man," *Health Phys.* 38:173–181 (1981).
113. Halford, D. K., O. D. Markham, and R. L. Dickson. "Radiation Doses to Waterfowl Using a Liquid Radioactive Waste Disposal Area," *J. Wildl. Manage.* 46:905–914 (1982).
114. Domby, A. H., D. Paine, and R. W. McFarlane. "Radiocesium Dynamics in Herons Inhabiting a Contaminated Reservoir System," *Health Phys.* 33:415–422 (1977).
115. Willard, W. K. "Avian Uptake of Fission Products from an Area Contaminated by Low-level Atomic Wastes," *Science* 32:148–150 (1960).
116. Cadwell, L. L., R. G. Schreckhise, and R. E. Fitzner. "Cesium-137 in Coots (*Fulica americana*) on Hanford Waste Ponds: Contribution to Population Dose and Off-Site Transport Estimates," Pacific Northwest Laboratory Report, PNL-SA-7176, Richland, WA, 1979.
117. Markham, O. D., and D. K. Halford. "Radionuclides in Mourning Doves Near a Nuclear Facility Complex in Southeastern Idaho," *Wilson Bull.* 94:185–197 (1982).
118. Arthur, W. J., and D. H. Janke. "Radionuclide Concentrations in Wildlife Occurring at a Solid Radioactive Waste Disposal Area in Southeastern Idaho," *J. Environ. Qual.* 12:117–122 (1986).
119. O'Shea, T. J., W. J. Fleming, and E. Chromartie. "DDT Contamination at Wheeler National Wildlife Refuge," *Science* 209:509–510 (1980).
120. Balcomb, R. "Songbird Carcasses Disappear Rapidly from Agricultural Fields," Office of Pesticide Programs. U.S. Environmental Protection Agency, Washington, DC, 1986.
121. Mineau, P., and D. B. Peakall. "An Evaluation of Avian Impact Assessment Techniques Following Broad-Scale Forest Insecticide Sprays," *Environ. Toxicol. Chem.* 6:781–791 (1987).
122. *Guidance Document for Conducting Terrestrial Field Studies*, Environmental Protection Agency, Ecological Effects Branch, Hazard Evaluation Division, Office of Pesticide Programs, Washington, DC, 1988, p. 67.
123. Snyder, R. L., and C. D. Cheney. "Pesticide Effects in Wild Deer Mice," *Proc. Utah Acad. Sci. Arts Lett.* 49(pt. 2):81–87 (1974).
124. Galindo, J., R. J. Kendall, C. J. Driver, and T. E. Lacher, Jr. "The Effect of Methyl Parathion on the Susceptibility of Bobwhite Quail (*Colinus virginianus*) to Domestic Cat Predation," *Behav. Neural. Biol.* 43:21–36 (1985).

125. Buerger, T. T., B. S. Mueller, R. J. Kendall, T. DeVos, and R. R. Hitchcock. "The Effect of Methyl Parathion on Activity and Survivability of Bobwhite Quail," in *Proc. Tall Timbers Game Bird Seminar*, Kyser, C., J. L. Landers and B.S. Mueller, Eds. (Tallahassee, FL: Tall Timbers Assoc., 1987), pp. 4–9.

126. Brewer, L. W., C. J. Driver, R. J. Kendall, C. Zenier, and T. E. Lacher, Jr. "Effects of Methyl Parathion in Ducks and Duck Broods," *Environ. Toxicol. Chem.* 7:375–379 (1988).

127. Bussiere, J. L., R. J. Kendall, T. E. Lacher, Jr., and R. S. Bennett. "Effect of Methyl Parathion on Food Discrimination in Bobwhite Quail (*Colinus virginianus*), *Environ. Toxicol. Chem.* 8:1125–1131 (1989).

128. Malins, D. C., B. B. McCain, D. W. Brown, M. S. Myers, M. M. Krahn, and S. L. Chan. "Toxic Chemical Including Aromatic and Chlorinated Hydrocarbons and Their Derivatives, and Liver Lesions in White Croaker (*Genyonemus lineaus*) from the Vicinity of Los Angeles," *Environ. Sci. Technol.* 21:765–770 (1987).

129. Martineau, D., A. Lagace, P. Beland, R. Higgins, D. Armstrong, and L. R. Shugart. "Pathology of Stranded Beluga Whales (*Delphinapterus leucas*) from the St. Lawrence Estuary, Quebec, Canada," *J. Comp. Pathol.* 98:287–311 (1988).

130. Grue, C. E., W. J. Flemming, D. G. Busby, and E. F. Hill. "Assessing Hazards of Organophosphate Pesticides to Wildlife," *Trans. N. Am. Wildl. Nat. Resour. Conf.* 48:200–220 (1983).

131. Grue, C. E., and B. K. Hunter. "Brain Cholinesterase Activity in Fledgling Starlings: Implications for Monitoring Exposure of Songbirds to ChE Inhibitors," *Bull. Environ. Contam. Toxicol.* 32:282–289 (1984).

132. Hooper, M. J., P. Detrich, C. Weisskopf, and B. W. Wilson. "Organophosphate Exposure in Hawks Inhabiting Orchards During Winter Dorman-spraying," *Bull. Environ. Contam. Toxicol.*, 42:651–659 (1989).

133. Burns, K. A. "Microsomal Mixed Function Oxidases in an Estuarine Fish, *Fundulus heteroclitus*, and Their Induction as a Result in Environmental Contamination," *Comp. Biochem. Physiol.* 53:443–446 (1976).

134. Lech, J. J., M. J. Vodicnik and C. R. Elcombe. "Induction of Monooxygenase Activity in Fish," in *Aquatic Toxicology*, L. J. Weber, Ed. (New York: Raven Press, 1982), pp. 107–148.

135. Payne, J. F., C. Bauld, A. C. Dey, J. W. Kiceniuk, and U. Williams. "Selectivity of Mixed-Function Oxygenase Enzyme Induction In Flounder (*Pseudopleuronectes americanus*) Collected at the Site of Baie Verte, Newfoundland, Oil Spill," *Comp. Biochem. Physiol.* 79:15–19 (1984).

136. Payne, J. F., L. L. Fancey, A. D. Rahimtula, and E. L. Porter, "Review and Perspective on the Use of Mixed-Function Oxidase Enzymes in Biological Monitoring," *Comp. Biochem. Physiol.* 86:233–245 (1987).

137. Jimenez, B. D., L. S. Burtis, G. H. Ezell, B. Z. Eagan, N. E. Lee, J. J. Beauchamp, and J. F. McCarthy. "The Mixed-Function Oxidase System of Bluegill Sunfish, *Lepomis macrochirus*: Correlation of Activities in Experimental and Wild Fish," *Environ. Toxicol. Chem.* 7:623–634 (1988).

138. Jimenez, B. D., and L. S. Burtis. "Response of the Mixed Function Oxidase 7 System to Toxicant Dose, Food and Acclimation Temperature in the Blue Gill Sunfish," *Marine Environ. Res.* 24:45–49 (1988).

139. Parke, V. D. "Cytochrome P-450 and the Detoxification of Environmental Chemicals," *Aquat. Toxicol.* 1:367–376 (1981).

140. Porter, W. P., R. Hinsdill, A. Fairbrother, L. J. Olson, J. Jaeger, T. Yuill, S. Bisgaard, W. G. Hunter, and K. Nolan. "Toxicant-Disease-Environment Interactions Associated with Suppression of Immune System, Growth and Reproduction," *Science* 224:1014–1017 (1984).

141. Fairbrother, A., T. M. Yuill, and L. J. Olson. "Effects of Three Plant Growth Regulations on the Immune Response of Young and Aged Deer Mice (*Peromyscus maniculatus*)," *Arch. Environ. Contam. Toxicol.* 15:265–275 (1986).

142. Spitsbergen, J. M., K. A. Schat, J. M. Kleeman, and R. E. Peterson. "Interactions of 2,3,7,8-Tetrachlorodibenzo-p-dioxin (TCDD) with Immune Responses of Rainbow Trout," *Vet. Immunol. Immunopathol.* 12:263–280 (1986).

143. Rattner, B. A., and S. D. Michael. "Organophosphorus Insecticide Induced Decrease in Plasma Luteinizing Hormone Concentration in White-Footed Mice," *Toxicol. Lett.* 24:655–69 (1985).

144. Freeman, H. C., G. B. Sangalang, and J. F. Uthe. "The Effects of Pollutants and Contaminants on Steroidogenesis in Fish and Marine Mammals," in *Contaminants Effects on Fisheries*, V. S. Crains, P. V. Hodson and J. O. Nraigu, Eds. (New York: John Wiley & Sons, 1988), pp. 197–212.

145. Krahn, M. M., M. S. Myers, D. G. Burrows, and D. C. Malins. "Determination of Metabolites of Xenobiotics in Bile of Fish from Polluted Waterways," *Xenobiotica* 14:633–646 (1984).

146. Oikari, A., and E. Anas. "Chlorinated Phenolics and Their Conjugates in the Bile of Trout (*Salmo gairdneri*) Exposed to Contaminated Waters, *Bull. Environ. Contam. Toxicol.* 35:802–809 (1985).

147. Ohlendorf, H. M., A. W. Kilness, J. L. Simmons, R. K. Stroud, D. J. Hoffman, and J. F. Moore. "Selenium Toxicosis in Wild Aquatic Birds," *J. Toxicol. Environ. Health* 24:67–92 (1988).

148. Kendall, R. J., and P. F. Scanlon. "Tissue Lead Concentrations and Blood Characteristics of Rock Doves (*Columba livia*) from an Urban Setting in Virginia," *Arch. Environ. Contam. Toxicol.* 11:265–268 (1982).

149. Kendall, R. J. "Wildlife Toxicology Integrated Field and Lab Studies Using Selected Model Species Might Lead to Ways of Quantifying Adverse Effects of Chemical Contaminants," *Environ. Sci. Technol.* 16:448–453 (1982).

150. Kendall, R.J. "Wildlife Toxicology: A Reflection on the Past and the Challenge of the Future," *Environ. Toxicol. Chem.* 7:337–338 (1988).

151. Breck, J. E., and C. F. Baes. "Report on the Workshop on Food Chain Modeling for Risk Analysis," Oak Ridge National Laboratory Report 6051, National Technical Information Service, Springfield, VA, 1985.

152. Judy, B. M., W. R. Lower, C. D. Miles, M. W. Thomas, and G. F. Krause. "Chlorophyll Fluorescence of a Higher Plant as an Assay for Toxicity Assessment of Soil and Water," paper presented at the First Symposium on Use of Plants for Toxicity Assessment, American Society for Testing and Materials, Atlanta, GA, April 19–20, 1989.

153. Ahokas, J. T. "Cytochrome P-450 in Fish Liver Microsomes and Carcinogen Activation," in *Pesticides and Xenobiotic Metabolism in Aquatic Organisms*, M. A. Q. Khan, J. J. Lech, and J. J. Menn, Eds. ACS Symp. Ser. 99 (Washington, DC: American Chemical Society, 1979), pp. 279–295.

154. Cummings, S. W., and R. A. Prough. "Metabolic Formation of Toxic Metabolites,"

in *Biological Basis of Detoxification*, J. Caldwell and W.B. Jakoby, Eds. (New York: Academic Press, 1983), pp. 1–30.

155. Shugart, L., and J. Kao. "Examination of Adduct Formation In Vivo in the Mouse Between Benzo[a]pyrene and DNA of Skin and Hemoglobin of Red Blood Cells," *Environ. Health Pers.* 62:223–226 (1985).

156. Shugart, L., and R. Matsunami. "Adduct Formation in Hemoglobin of the Newborn Mouse Exposed In Utero to Benzo[a]pyrene," *Toxicology* 37:241–245 (1985).

157. Shugart, L. "Quantifying Adductive Modification of Hemoglobin from Mice Exposed to Benzo[a]pyrene," *Anal. Chem.* 152:365–369 (1986).

158. Wolf, C. R. "Cytochrome P-450s: Polymorphic Multigene Families Involved in Carcinogen Activation," *Trends Genet.* 2:209–214 (1986).

159. Shugart, L., J. McCarthy, B. Jimenez, and J. Daniels. "Analysis of Adduct Formation in the Bluegill Sunfish (*Lepomis macrochirus*) Between Benzo[a]pyrene and DNA of the Liver and Hemoglobin of the Erythrocyte," *Aquatic Toxicol.* 9:319–325 (1987).

160. McBee, K. "Chromosomal Damage in Resident Small Mammals at a Petrochemical Wastes Dump Site: A Natural Model for Analysis of Environmental Mutageneisis," Ph.D. Thesis, Texas A&M University, pp. 118 (1985).

161. Stegeman, J. J. "Polynulear Aromatic Hydrocarbons and Their Metabolism in the Marine Environment," in *Polycyclic Hydrocarbons and Cancer*, H. V. Gelboin and P. O. Ts'O, Eds. (New York: Academic Press, 1981), pp. 1–58.

162. Leonard, A., M. Duverger-van Bogaert, A. Bernard, M. Lambotte-Vandepaer, and R. Lauwerys. "Population Monitoring for Genetic Damage Induced by Environmental and Chemical Agents," *Environ. Monitor. Asses.* 5:369–384 (1985).

CHAPTER 19

Application of Bioindicators in Assessing the Health of Fish Populations Experiencing Contaminant Stress

S. M. Adams, L. R. Shugart, G. R. Southworth, and D. E. Hinton

ABSTRACT

Assessment of the response of fish populations to contaminants usually involves measurement of indicators of exposure or effects, but rarely do biomonitoring studies incorporate a mixture of endpoints that include both types of indicators. Our integrated bioindicator approach involves monitoring a suite of selected exposure and effects indicators at several levels of biological organization from the biomolecular level to the community level. Measurement of these selected biological variables permits early detection of environmental problems and provides insights into causal relationships between contaminant exposure and the effects that may be ultimately manifested at the population and community levels. This approach was applied to fish populations that inhabit streams receiving chronic inputs of contaminants including PAHs, PCBs, heavy metals, and chlorine. Indicators such as detoxification enzymes and DNA damage have provided direct evidence of contaminant exposure, whereas indicators related to lipid metabolism, histopathology, and general bioenergetic condition have reflected effects at both the suborganism and organism level. This bioindicator approach is an effective technique to assess the integrated effects of contaminant stress on fish.

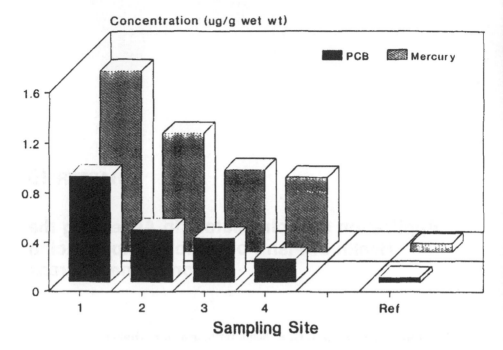

Figure 1. Concentration of PCB and mercury in redbreast sunfish from East Fork Poplar Creek (EFPC) and the reference stream demonstrating a distinct downstream gradient in the levels of these two contaminants in EFPC.

INTRODUCTION

A variety of man-induced stresses such as contaminants can interfere with the normal functioning of an organism and its ability to live in harmony with the environment. Short of death, some of the more common but serious effects of environmental stressors on aquatic organisms include changes in behavior, growth, and reproduction. Modifications of these processes in individual animals can eventually result in undesirable effects at higher levels of biological organization, such as changes in fish populations or entire ecological systems. Measurement of selected biological responses (alternatively referred to as biomarkers or bioindicators) to stress permits early detection of environmental problems (early warning indicators) and provides insights into causal mechanisms between stress and effects that may be ultimately manifested at the population and community level.

The underlying concept of the bioindicators approach in biomonitoring is that contaminant effects occur at the lower levels of biological organization (i.e., at the genetic, cell, and tissue level) before more severe disturbances are manifested at the population or ecosystem level (Chapter 1, Figure 1). Effects of pollutant stress are usually expressed first at the molecular/biochemical level (where pollutants affect the normal functioning of chemical processes in the body) (Chapter 1, Figure 1). These effects can be detected as changes in enzyme levels, cell membranes, or genetic material (DNA). Changes at these subcellular levels induce a series of structural and functional responses at the next higher level of biological organization, which can impair, for example, complex processes such as hormonal regulation, metabolism, electrolyte balance, and proper functioning of the immune system. These effects, in turn, may eventually alter the organism's ability to grow, reproduce, or survive. Ultimately, irreversible and detrimental effects may be observed at the population, community, or ecosystem levels of biological organization.

The bioindicator approach differs from more traditional methods of environmental monitoring (e.g., direct measurement of chemical pollutant levels or use of laboratory toxicity tests) by using naturally occurring organisms as sentinels or early-warning indicators of environmental damage. Because the physiological condition of an animal integrates the effects of all the environmental stressors acting on it (such as contaminants, unfavorable temperatures, water velocity, suspended sediment, oxygen, and food availability), measuring an appropriate set of biological indicators greatly strengthens assessments of the nature and extent of environmental degradation.

We have used this integrated biomonitoring method to examine effects of contaminants in streams on the Oak Ridge Reservation; on the Pigeon River (North Carolina and Tennessee), which is polluted by pulp and paper mill effluents; on the Elizabeth River (Virginia), which is contaminated by hydrocarbons and heavy metals; and on a large reservoir in South Carolina, which receives high inputs of PCBs. The purpose of this chapter is to demonstrate the use and application of bioindicators in a field situation for assessing and evaluating the effects of contaminant stress on fish populations.

APPROACH AND METHODS

As an example of the application of bioindicators in an integrated field assessment, we present here some results of our studies on a contaminated stream near Oak Ridge, Tennessee. Bioindicators measured in this study range from the biomolecular level to the organism level (see Chapter 1, Figure 1) and represent indicators of both contaminant exposure and effects.

Sampling Procedures

Adult redbreast sunfish (*Lepomis auritus*) were sampled at four sites along the length of East Fork Poplar Creek (EFPC), a third-order stream that receives point source discharges of mixed contaminants near its headwaters (further description of the stream can be found in Jimenez et al., Chapter 6). This stream has a distinct gradient in contaminant loading as evidenced by the decreasing downstream concentrations of PCBs and mercury in redbreast sunfish (Figure 1). At each of these four sites and from a reference stream (Brushy Fork Creek), 12 to 15 adult male sunfish (16-20 cm total length) were collected by electroshocking during each of four sampling periods from spring 1986 to fall 1987. Sunfish were collected for DNA integrity analysis at three sampling periods from fall 1987 to summer 1988. Within 2 min after collection, blood samples were taken by caudal puncture for biochemical analysis. Each fish was labeled with a unique identification number and transported alive to the laboratory for additional analysis.

In addition to recording total lengths and weights for each fish, observations were also made on the general condition of the fish, such as presence or absence of fin disease, body and/or mouth sores, internal and external parasites, and general condition status. Following sacrifice, livers and spleens were removed for histopathological analysis. Sections of the liver were also taken for detoxification enzyme and RNA/DNA analysis. The viscera was excised from the body cavity and the total weight recorded after all food material was removed from the stomach and intestine. Liver, visceral, and spleenic-somatic indices were calculated as the weights of these respective organs divided by the total body weight. Condition factor was calculated as $K = 10^5 W/L^3$ where W is body weight (g) and L is total length (cm).

Lipid Analysis

Following dissection and removal of the critical organs, a subsample of fish from each site was chosen for lipid analysis. Lipid biochemical analysis was performed according to a modification of the Bligh and Dyer[9] method and the Iatroscan analyzer system[20] was used for lipid class quantification. Lipid analysis for each fish included total lipids (percent of body weight); triglycerides (percent of total lipid); cholesterol; phospholipids; and the two major fractions of phospholipids, phosphatidylcholine (PC) and phosphatidylethanolamine (PE).

Serum Chemistry

Blood collected in unheparinized tubes was allowed to clot, was transferred with Pasteur pipettes to 1.5-mL conical microcentrifuge tubes labeled with the fish identification number, and was centrifuged for 2 min in a Beckman Microfuge. The clear supernatant (serum) was drawn off with clean pipettes and analyzed for glucose,[37] glutamate oxaloacetate transaminase (SGOT),[42] bilirubin,[50] cholesterol,[6] and triglycerides.[12] Total serum proteins were measured by the Biuret method.[34] All of these methods are enzymatic assays and the reagents for each assay were obtained from Roche Diagnostic Systems. Assays were performed on an automated Centrifugal Fast Analyzer System (Cobas-Fara, Roche, Inc.). Calibrations were performed using the Roche serum calibrator as the standard and monitrol Level 1 and 2 (American Dade, Miami, Florida) as internal controls.

Detoxification Enzymes

Fish hepatic microsomes were prepared by differential centrifugation.[33] The activity of 7-ethoxyresorufin O-deethylase (EROD) was measured fluorimetrically at $30°C^{14}$ and is expressed as picomoles of resorufin per minute per milligram of microsomal protein (m.p.). Proteins were measured by the Bio-Rad (Richmond, California) reagent method[10] on the Cobas-Fara using bovine serum albumin as a standard. Cytochrome P-450 and cytochrome b_5 (CB5) content were each measured by their characteristic oxidized and reduced spectra. Cytochrome P-450 samples were oxidized with carbon monoxide and reduced with sodium dithionite.[25] CB5 was reduced with NADH; CB5 assays were carried out prior to cytochrome P-450 analysis.[49] The NADPH-cytochrome c reductase (NADPH-reductase) was assayed spectrophotometrically by the reduction of the electron acceptor cytochrome c with an extinction coefficient of 21.1 cm^{-1} mM^{-1}.

DNA Integrity

DNA isolation was accomplished by homogenizing the intact liver in 1 N $NH_4OH/0.2\%$ Triton® X-100. The DNA was further purified by differential extraction with chloroform/isoamyl alcohol/phenol and passage through a molecular sieve column. DNA strand breaks were measured in the isolated DNA by an alkaline unwinding assay.[26,43,44] The technique is based on the time-dependent partial alkaline unwinding of DNA followed by determination of the duplex:total DNA ratio (F value). Since DNA unwinding occurs at single-strand breaks within the molecule, the amount of double-stranded DNA remaining after a given period of alkaline unwinding will be inversely proportional to the number of strand breaks present at the initiation of the alkaline exposure, provided that renaturation is prevented. The amounts of these two types of DNA are quantified by measuring the fluorescence that results with bisbenzimidazole-Hoechst dye No. 33258.[26,43,44] The relative number of strand breaks (N value) in DNA of sunfish from contaminated sites can be compared to those fish from a reference site as $N = (\ln F_s/\ln F_r)-1$, where F_s and F_r are the mean F values of DNA from the contaminated sites and reference site, respectively.

Histopathological Analysis

Using the quantitative morphometric approach of Hinton and Couch,[23] the following histopathological analyses were performed on sections of liver excised from the redbreast sunfish: (1) percentage of liver tissue occupied by parasites, (2) percentage of liver composed of necrotic parenchyma, (3) percentage of liver tissue composed of macrophage aggregates, and (4) percentage of liver occupied by functional parenchyma.

RNA/DNA Analysis

A 50- to 100-mg section of liver was homogenized in approximately 1 mL of distilled water for 1 min using a teflon homogenizer while keeping the sample cold. After the homogenate was brought up to a final volume of 1.5 mL with distilled water, it was transferred to 1.5-mL microcentrifuge tubes and centrifuged for 2 min in a Beckman microfuge. The RNA content of each sample was analyzed in triplicate by adding to each of three 1.5-mL centrifuge tubes 200 μL of supernatant from the liver homogenate, 1.2 mL of 95% ethanol, 0.035 mL of 2 M sodium acetate, and 0.015 mL of 1 M magnesium acetate. Samples were cooled for 20 min in a refrigerator before centrifuging for 2 min. After the supernatant was decanted, 1 mL of 0.3 M KOH was added to the tubes with the pellet and incubated at 37°C in a constant water bath until the pellet dissolved. Each tube then received 0.5 mL of 1.4 N perchloric acid before it was cooled for an additional 20 min in the refrigerator. The mixture was centrifuged for 2 min and the supernatant was recovered in 20-mL scintillation vials. The precipitate was washed once with 1 mL of 0.2 N perchloric acid, centrifuged again, and the supernatant was combined with the previous supernatant. Standards were prepared using 1100 μg of RNA and processed in exactly the same manner as the liver samples. Absorbance of the samples and standards was measured at 260 nm using a Gilford Response Spectrophotometer and a distilled water blank as a reference. Results were expressed as micrograms of RNA per milligram of liver tissue. For DNA analysis, duplicate samples were prepared by adding 3 mL of 0.2 M phosphate buffer at pH 7.0 to 100 μL of the original liver homogenate supernatant. Standards were prepared with 4.6 μg of salmon sperm DNA. The fluorescence was measured using an excitation wavelength of 360 nm and an emission wavelength of 450 nm with a Beckman LS-5 spectrofluorometer. Results were expressed as micrograms of DNA per milligram of wet weight liver tissue.

Statistical Procedures

Analysis of variance (ANOVA) was used to test for site and seasonal effects as well as their interaction effect on the individual bioindicators. If the ANOVA rejected a multisample null hypothesis of equal means, then Dunnets test[52] was used to test for significant differences in each of the East Fork Poplar Creek sites compared to the reference. Tests for homogeneity of variance of individual response variables between sites were conducted using Levenes test.[48] This is an F distribution test that compares the ratios of the variances from two independent sample populations.

To determine the integrated response of fish to the environmental conditions at each sampling site, all the bioindicator variables were considered jointly within a multivariate context by using canonical discriminant analysis available on the Statistical Analysis System.[41] This method provides a graphical representation of the similarity among sites of their integrated responses. In addition, the discriminant analysis variable selection technique of Klecka[27] was used to identify and prioritize the variables which contributed most to discriminating between integrated site responses. This variable selection procedure considered all possible combinations of the observed values and, for any specified subset size, selected those variables having the best discriminating power.

Table 1. Relative Number of DNA Strand Breaks in Redbreast Sunfish from Four
Sites in a Contaminated Stream Compared to Fish from a Reference
Stream (*N* value) for Three Sample Dates.

| Sampling Date | Sampling Sites | | | |
	1	2	3	4
	(*N* value)			
October 1987	7.10*	11.23*	11.72*	9.30*
May 1988	5.18*	5.75*	5.93*	1.19*
August 1988	3.53*	6.48*	1.11	0.26

Values indicated by an asterisk (*) are significantly different ($\alpha = 0.05$) from the reference fish.

RESULTS AND DISCUSSION

Bioindicators measured for fish in this study can be grouped into responses representative of direct contaminant exposure and those reflecting the effects of this exposure. Indicators of exposure are (1) DNA integrity and (2) detoxification enzymes; indicators of effects are (3) carbohydrate-protein metabolism, (4) lipid metabolism, (5) histopathology, and (6) overall fish health or condition. These six groups of responses reflect gradients of both ecological relevance and time-course of response to a stress such as a contaminant (Chapter 1, Figure 1). Those variables in groups 1 and 2 respond relatively rapidly to stress but have relatively low ecological relevance (exposure indicators), whereas indicators included in groups 3 to 6 are slower responding to stress and are characterized by low toxicological but high ecological relevance.

Indicators of Exposure

Levels of detoxification enzymes and DNA integrity are used to indicate the degree of exposure to contaminants. Even though these two variables are used primarily as indicators of direct contaminant exposure, effects at the biomolecular and biochemical level are manifested as damage to DNA and changes in the levels of enzymes, respectively.

DNA Integrity

DNA damage has been proposed as a useful biological parameter for assessing the genotoxic properties of environmental pollutants.[28] No clear downstream gradient in strand breaks is evident even though the highest number of breaks generally occurred in fish from the middle two sites (sites 2 and 3) (Table 1). During the first two sampling dates, DNA integrity was significantly different at all four contaminated stream sites compared to the reference stream. In the summer of 1988, however, the lower two sites were not significantly different from the reference (Table 1). Normal variations in environmental variables such

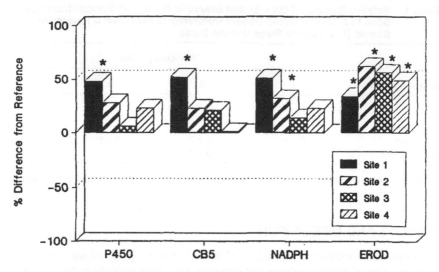

Figure 2. Relative differences in the response of the detoxification enzymes at each of the contaminated sties compared to the reference stream. Values above or below the zero line indicate that the response for fish at any particular contaminated site was higher or lower, respectively, than the same response for fish from the *reference stream*. Bars indicated by an asterisk (*) are significantly different ($\alpha \approx 0.05$) from the reference. NADPH $\approx$ NADPH-reductase.

as temperature and food availability do not significantly effect the integrity of DNA; therefore, any DNA damage observed in fish is assumed to be due to genotoxic stress.[45]

Detoxification Enzymes

The activity or levels of liver detoxification enzymes are used to indicate exposure to various xenobiotics such as PAHs, PCBs, and pesticides.[36] Many of these compounds commonly occur in industrial and municipal effluents and are accumulated within living tissues.[5,11,35]

While some differences existed in the magnitudes of response, detoxification enzymes were elevated at all the EFPC sites compared to the reference (Figure 2). Fish from site 1 nearest the outfall contained the highest levels of cytochrome P-450, CB5, and NADPH-reductase and the lowest activities of EROD compared to fish from the reference stream. For all variables excluding EROD, there appears to be a downstream gradient in activity with the highest levels immediately below the outfall and the lowest levels at site 4 indicating decreased contaminant loading downstream (Figure 2). Depressed EROD activity for fish from site 1 could be due to hepatotoxic damage inhibiting the ability of the liver cells to produce this enzyme (see Chapter 5). A wide array of organic compounds, including some identified in EFPC[31] as well as metal ions, are known to cause hepatotoxicity in fish. This phenomenon has been observed for sunfish exposed to organics in a laboratory situation (see Chapter 5).

Indicators of Effects

Effects of contaminants on fish can be manifested at various levels of biological organization (Chapter 1, Figure 1), including physiological dysfunction, structural changes in tissues and organs, behavior modifications, and impairment of growth and reproduction. Even though damage to DNA and changes in the levels or activity of enzymes can be also considered effects at the biomolecular and biochemical levels, respectively, these two variables are used in this study as indicators of direct contaminant exposure.

Carbohydrate-protein Metabolism

The indicators of carbohydrate-protein metabolism demonstrated a varied response among sites. The transaminase enzyme, SGOT, ranged from 5–70% lower at all the EFPC sites compared to the reference (Figure 3A). Changes in the transaminase enzymes are typically used to indicate tissue damage or impaired organ function;[38,39] however, due to the extreme variability of this parameter, it was not statistically different among any of the sites (Figure 3A). Serum protein is also used as an indicator of protein metabolism. Lower protein values at sites 1 and 2 compared to the reference may indicate that fish in the contaminated stream are under some type of nutritional stress, as was also found for white sucker (*Catostomus commersoni*) experiencing low food availability.[32] Even though blood glucose was slightly elevated (hyperglycimia) in fish at site 1, this variable is a generalized stress response in fish to a broad spectrum of environmental perturbations[47] and thus may reflect both direct (metabolic) and indirect (nutritional) stress responses.

Lipid Metabolism

Lipid metabolic parameters can reflect both the nutritional status and level of metabolic stress in fish. The condition or status of the lipid pool is important because the vulnerability of an organism to stress depends, in part, on this condition.[18,46] Serum triglycerides were much lower (significantly lower at sites 1 to 3) in fish from all contaminated stream sites than that of reference fish, indicating altered or impaired lipid metabolism in the former (Figure 3B). Serum cholesterol is an indicator of both nutrition and steroid metabolism in fish; however, no consistent patterns emerge between sites even though cholesterol levels in fish at sites 1 and 4 are about 10% higher than in reference fish.

Total body triglycerides reflect the energy available to an organism for mediating the effects of stress[30] and for use in critical physiological functions such as growth and gonadal development. These triglycerides also act as energy buffers during periods of food shortages.[2] Energy available for direct physiological use

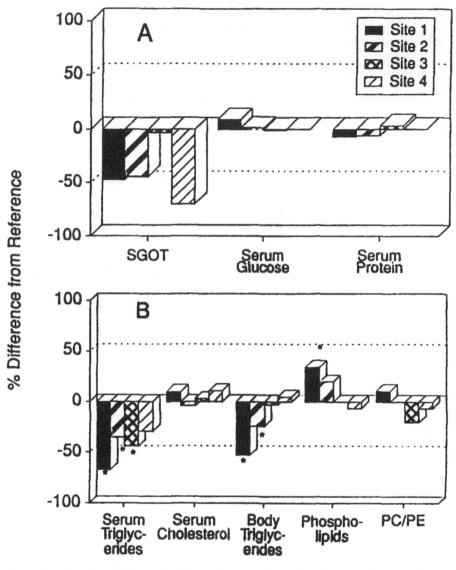

Figure 3. Relative differences in the response of the carbohydrate-protein metabolism
parameters (A) and the lipid metabolism parameters (B) at each of the contaminated sites compared to the reference stream. Values above or below the zero
line indicate that the response for fish at any particular contaminated site was
higher or lower, respectively, than the same response for fish from the reference
stream. Bars indicated by an asterisk (*) are significantly different ($\alpha = 0.05$)
from the reference. SGOT = serum glutamate oxyaloacetate transaminase; PC
= phosphatidylcholine; PE = phosphatidylethanolamine.

was 50 and 25% lower in fish from sites 1 and 2, respectively, than in fish from the reference site (Figure 3B). Levels in fish from sites 3 and 4, however, were similar to the reference.

Phospholipids, the structural components of lipids, were 35 and 20% higher in sunfish from sites 1 and 2, respectively, than in the reference stream (Figure 3B). The ratio of the two major types of phospholipids that constitute the cell wall, phosphatidylethanolamine (PE) and phosphatidylcholine (PC), can reflect cell wall membrane integrity. Membrane lipid structure influences membrane fluidity, enzyme kinetics, and electrical properties.[17] The PC/PE ratio was slightly elevated in fish at site 1 but depressed in fish from sites 3 and 4. Even though little is known about the influence of environmental variables on the PC/PE ratio, observed differences in this ratio could possibly be due to the gradient in temperatures and/or concentrations of dissolved solids between these sites.[40]

Histopathological Condition

Indicators of histopathological condition in fish from EFPC showed distinct differences compared to the reference fish (Figure 4A). The percentage of the liver composed of encysted parasites (LPARS) was about 75% higher in redbreast sunfish from sites 1 and 2 than in reference fish. Fish collected downstream of site 2 had similar or even lower percentages of parasites than reference redbreast even though this difference is not statistically significant. In addition to these encysted parasites, livers also had parasitic tracks of alteration which showed liquefactive necrosis and/or varying degrees of fibrotic changes.

An informative indicator of tissue disease or pathology is the occurrence of macrophage aggregates. These are localized centers of phagocytes (white blood cells) that invade diseased or damaged areas. Macrophage aggregates were abundant and appeared to be larger than in fish from the reference stream. The percentage of liver tissue composed of macrophage aggregates (LMACA) was consistently 40–70% higher in fish from all EFPC sites than in the reference fish and significantly higher at the upper three sites in EFPC (Figure 4A). No downstream gradient in this pathology is indicated, however.

The percentage of liver tissue actually occupied by functional parenchyma (LFPMA) demonstrated, however, a slight downstream gradient with this condition being 15% lower in fish from site 1 and only 3% lower in fish from site 3, respectively, compared to the reference (Figure 4A). These percentages may not seem dramatic (Figure 4A), but a 3–15% reduction in functional liver capacity could have a major effect on the physiological health of an organism. The reduced volume of functional liver tissue in fish from the upper sections of EFPC was probably due to the combined effect of liquefactive and coagulative necrosis along with the presence of abundant macrophages and parasites. In addition, some reduction in functional parenchma may have been due to hepatotoxic damage from contaminant exposure. These lower levels of functional liver tissue imply that fish with this condition may have reduced capacity to (1) produce

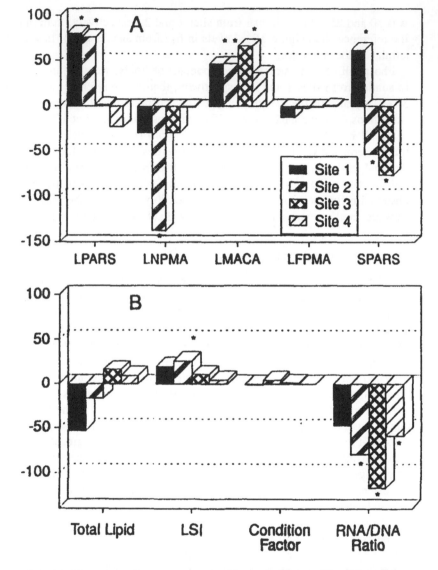

Figure 4. Relative differences in the response of the histological indices (A) and the general health indicators (B) at each of the contaminated sites compared to the reference stream. Values above or below the zero line indicate that the response for fish at any particular contaminated site was higher or lower, respectively, than the same response for fish from the reference stream. Bars indicated by an asterisk (*) are significantly different (α = 0.05) from the reference. LPARS = liver parasites, LNPMA = necrotic liver parenchyma, LMACA = macrophage aggregates in liver, LFPMA = functional parenchyma in liver, SPARS = spleen parasites, LSI = liver-somatic index.

enzymes for detoxifying contaminants, (2) store important glycogen and lipid energy reserves, (3) manufacture vitellogenin necessary for proper egg development in the female, and (4) convert and process protein and lipid compounds into physiologically useful energy.

Histological analysis of the spleen indicated that the parasite load (SPARS) of fish from site 1 was 60% higher than that of reference fish (Figure 4A). Sunfish from the lower sections of EFPC, however, had lower parasite loads than did those from the reference stream. Because the spleen is a hemopoietic organ, injury due to parasitic infestations could reduce production of both red and white blood cells, possibly resulting in anemia and increased vulnerability to disease.[7]

General Health Indicators

Four condition indices were measured as indicators of general health for fish from each site. The condition factor is a generalized indicator of overall fitness or "plumpness" and can reflect the integrated effect of both nutrition and metabolic costs induced by stress. Probably because the condition factor is fairly insensitive to changes in body condition,[1] this index was very similar between fish from all sites (Figure 4B). One possible explanation for this, considering the lower lipid levels in fish from sites 1 and 2, is that a decline in body weight resulting from a reduction in energy reserves may be counteracted by an increase in body water.[15]

Total body lipid is used to indicate overall fat storage and general nutritional status of fish. Sunfish collected at sites 1 and 2 had 50% (significantly lower than reference) and 20% less total body lipid, respectively, than did fish from the reference even though levels at the two other EFPC sites were slightly higher than the reference (Figure 4B).

The liver-somatic index (LSI) reflects both short-term nutritional status and metabolic energy demands.[1,23] The LSI is also sensitive to toxicant stress, and liver enlargement due to hyperplasia (increase in cell number) and hypertrophy (increase in cell size) has been reported in fish exposed to toxic compounds.[4,16,21] An increase in the LSI could not have been due to glycogen accumulation in the liver because most fish demonstrated enhanced basophilia (hyperplasia) of the hepatocytes. Increased LSI values in EFPC fish, therefore, could have been due to toxicant exposure. In addition, there also appears to be a downstream gradient in the LSI, which reflects a similar gradient observed for the detoxification enzymes.

The RNA/DNA ratio is used as an indicator of immediate or short-term growth in fish[13,19] as well as an indicator of exposure to sublethal concentrations of toxicants.[8] Growth is one of the ultimate indicators of fish health because it integrates all the biotic and abiotic variables acting on an organism and reflects secondary impacts of chronic stress.[29,51] Growth, as indicated by the RNA/DNA ratio, was significantly lower at all the contaminated stream sites compared to the reference (Figure 4B). There is no obvious explanation of the more depressed

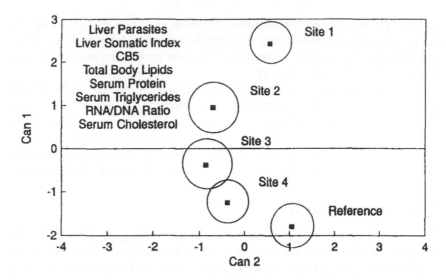

Figure 5. Segregation of integrated health responses for redbreast sunfish from four sites in the contaminated stream and the reference stream using all the bioindicators measured from spring 1986 to fall 1987. The points and circles represent site means and the 90% confidence radii of the site means. The statistically significant variables for discriminating among these sites are listed in the upper left.

growth at site 3 compared to the other EFPC sites, except to suggest that food availability was lower and/or temperatures may have been higher on a seasonal basis (particularly in the summer) at this site than at other areas in the stream.

Integrated Site Analysis

To examine the integrated response of fish to the environmental conditions (including contaminant exposure) at each site, all biochemical variables and condition indices were considered jointly within an multivariate context using canonical discrimination analysis. This method graphically represents the positions and orientations of the integrated responses of fish from each site relative to each other.

A distinct downstream gradient in integrated fish response is evident when all the bioindicators (exclusive of DNA integrity) are utilized in the discriminant analysis procedure (Figure 5). Fish from site 1 were least similar and those at site 4 most similar to reference fish. As indicated by the differences in the linear distances between the centers of the site means for each response, fish at site 2 are the most similar to those at site 1, whereas fish collected at site 4 are less similar to those immediately below the industrial outfall (Figure 5). No significant difference exists, however, in the integrated response of fish at sites 3 and 4 due to overlapping of the 95% confidence radii of these two sites.

The variables that are most significant in discriminating among sites are also

Table 2.   Percentage of Redbreast Sunfish at Each Sampling Site Appearing in the
Lowest Quartile for Each Response Variable.

Response Variable	Sampling Sites				
	1	2	3	4	Reference
Serum glucose	19	18	25	31	23
Serum cholesterol	22	28	27	20	26
Serum protein	25	30	16	24	19
SGOT	14	23	27	15	33
Total lipid	32	20	16	10	12
Body triglycerides	35	21	14	7	11
Phospholipid	3	15	29	24	19
PC/PE[a]	8	13	27	20	21
LFPMA[b]	24	10	5	2	4
EROD	22	41	16	22	0
Cytochrome P_{450}	37	26	8	25	19
NADPH-reductase	44	34	11	14	9
Condition factor	29	18	30	15	32
Visceral-somatic index	22	7	35	32	28
RNA/DNA	11	3	11	15	11

Note: A ranking in the lowest quartile represents the poorest health status.
[a]Phosphatidylcholine/phosphatidylethanolamine.
[b]Macrophage aggregates in the liver.

indicated in Figure 5. These eight variables consist of representative indicators from both the exposure and effects groups, including one indicator each from detoxification enzymes and carbohydrate-protein metabolism, two from lipid metabolism, and three from the general health indicators. In this study, representative indicators of several functional levels are required, therefore, to distinguish integrated responses of fish between sites.

Health Status Ranking

To determine the overall health status of fish at each site, all fish were first segregated into one of four quartile ranks for each response variable. For example, if a fish from site 1 had an EROD value falling within the highest 25% of all EROD values measured for all fish at all sites, then that particular fish was placed in the lowest quartile for that indicator because high EROD values are indicative of toxicant exposure or high stress. Conversely, if a fish from the reference site had an EROD value within the lowest 25% of all measured values, it was placed in the highest quartile for that parameter. For each of the bioindicator responses, then, fish from each site were segregated into quartiles and the total number of bioindicators for which a fish ranked in the lowest quartile was tabulated (Table 2). The criterion for fish health based on this approach is therefore the percentage of fish at each site that appeared in the lowest quartile for at least four of the indicators.

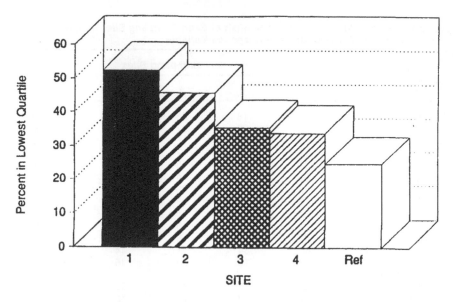

Figure 6. Integrated index of fish health status expressed as a percentage of redbreast sunfish at all the contaminated stream sites and the reference stream ranked in the lowest quartile for at least four bioindicators. Fish ranked in the lowest quartile for an indicator represent those individuals that are in poorest health for all fish at all sites.

Based on this quartile ranking procedure, there appears to be a distinct downstream gradient in overall health with the fish below the industrial outfall (site 1) demonstrating the poorest health and fish from the lower sites (sites 3 and 4) and the reference stream the highest level of health (Figure 6). For example, 52% of fish from site 1 were in the lowest quartile for at least four variables, while only 35 and 26% of the fish from site 4 and the reference stream, respectively, were in the lowest quartile for four variables.

The importance of each bioindicator in the health status ranking or the percentage of fish at each site that appear in the lowest quartile for each response is shown in Table 2. For example, 22 and 0% of the fish from site 1 and the reference, respectively, occur in the lowest quartile for EROD, while 35 and 11% occur, respectively, for triglycerides. The variables that identify fish in poor health from the upper sections of EFPC are primarily detoxification enzymes, condition indices, and indicators of lipid metabolism. Fish from the lower sections of EFPC (sites 3 and 4) are identified primarily by indicators of carbohydrate-protein and lipid metabolism. Interestingly, the particular sets of variables that are responsible for the health status ranking of fish in EFPC are also those that were identified in the integrated site analysis to be important for discriminating between sites (i.e., lipid metabolism variables and detoxification enzymes).

IMPLICATIONS AND CONCLUSIONS

Because elevated detoxification enzymes may be indicative of toxicant exposure, effects observed on fish in the upper sections of EFPC are probably due directly to contaminant loading. In addition, elevated levels of DNA damage and contaminants in fish from sites 1 to 3 provide further evidence that fish in the upper half of EFPC experience toxicant exposure. The concomitant effects on the various lipid parameters also observed for fish from the upper sections of EFPC (Figure 3) suggest that toxicant exposure has an influence on lipid metabolism either directly through metabolic processes or possibly indirectly through the food chain. Such metabolic disturbances can seriously affect the energy balance in fish, especially during periods of critical energy demands such as sexual maturation, spawning, and overwinter survival.[2]

Contaminant exposure can result in increased metabolic stress that places a negative drain on the energy and nutrient pools of organisms. With toxicological stress, additional energy is required for repair and maintenance of damaged cells, tissues, and organs. Reduction in lipid pools due to these metabolic drains can also reduce the amount and quality of energy available for maintaining immune system competence and adequate growth rates. The effects of direct toxicant stress on chemical and biological processes in organisms as they relate to immune and reproductive competence are unknown at this time but future studies will address this issue. In addition, the indirect effects of exposure on behavior and food availability to higher food-chain organisms will be investigated in future studies.

Relationships between indicators of exposure (DNA damage and detoxification enzymes) and indicators of effects in fish can also be seen by comparing the DNA and enzymes responses in Table 1 and Figure 2 to histopathological and general health indices in Figure 4. Not only does an inverse relationship exist between the enzymes and lipid metabolism (Figure 2 vs Figure 3), but also there are apparent correlations between the enzymes and both the condition indices and parasite loads in the liver (Figure 2 vs Figure 4). These correlations suggest potential causal relationships between exposure to contaminants and organism level effects. Additional evidence supporting causal relationships between contaminant exposure and biological effects is the decreasing downstream levels of mercury and PCBs in fish tissue (Figure 1). The effects of contaminant exposure at higher levels of biological organization also appear to be related to levels of toxicant exposure in EFPC as demonstrated by a downstream gradient of increasing species richness (community status indicator) and reduced fecundity of sunfish at site 1 (population level indicator).[3]

This study has demonstrated that biomarkers can be used to (1) identify early warning signs of impaired health in fish, (2) suggest causative relationships between contaminant exposure and effects at the organism level, and (3) document effects of contaminant stress on fish.

Another promising use of bioindicators is in evaluating the effectiveness of remedial actions at hazardous waste sites. Recent federal and state regulations place strict compliance standards on industry relative to the release, disposal, treatment, and recovery of hazardous wastes. Bioindicators are excellent tools to monitor the health of biological populations, to evaluate the effectiveness of remedial actions, and to determine if industry is complying with these regulations. Application of these biological indicators in a wide variety of aquatic systems should allow us to better anticipate problems and help minimize the probability of future environmental harm.

ACKNOWLEDGMENT

This work was sponsored by the Department of Environmental Management of the Health, Safety, Environment, and Accountability Divison, Oak Ridge Y-12 Plant. The Oak Ridge Y-12 Plant and Oak Ridge National Laboratory are operated by Martin Marietta Energy Systems, Inc., under contract DE-AC05-84OR21400 with the U.S. Department of Energy. Research conducted at the Oak Ridge National Environmental Research Park.

REFERENCES

1. Adams, S. M., and R. B. McLean. "Estimation of Largemouth Bass, *Micropterus salmoides* Lacepede, Growth using the Liver Somatic Index and Physiological Variables," *J. Fish Biol.* 26:111–126 (1985).
2. Adams, S. M., J. E. Breck, and R. B. McLean. "Cumulative Stress-induced Mortality of Gizzard Shad in a Southeastern U.S. Reservoir," *Environ. Biol. Fishes* 13:103–112 (1985).
3. Adams, S. M., K. L. Shepard, M. S. Greeley, Jr., B. D. Jimenez, M. G. Ryon, L. R. Shugart, and J. F. McCarthy. "The Use of Bioindicators for Assessing the Effects of Pollutant Stress on Fish," *Mar. Environ. Res.* 28:459–464 (1989).
4. Addison, R. F. "Hepatic Mixed Function Oxidase (MFO) Induction in Fish as a Possible Biological Monitoring System," in *Contaminant Effects on Fisheries*, V. W. Cairns, P. V. Hodson, and J. O. Nriagu, Eds. (New York: John Wiley & Sons, 1984), pp. 51–60.
5. Ahokas, J. T., N. T. Kärki, A. Oikari, and A. Soivio. "Mixed Function Monooxygenase of Fish as an Indicator of Aquatic Environment by Industrial Effluent," *Bull. Environ. Contam. Toxicol.* 16:270–274 (1976).
6. Allain, C. C., L. S. Poon, C. S. G. Chan, W. Richmond, and P. C. Fu. "Enzymatic Determination of Total Serum Cholesterol," *Clin. Chem.* 20:470–475 (1974).
7. Anderson, D. P. "Immunological Indicators: Effects of Environmental Stress on Immune Protection and Disease Outbreaks," in *Biological Indicators of Stress in Fish*, S. M. Adams, Ed. (Bethesda, MD: American Fisheries Society), in press.
8. Barron, M. G., and I. R. Adelman. "Nucleic Acid, Protein Content, and Growth of Larval Fish Sublethally Exposed to Various Toxicants," *Can. J. Fish. Aquat. Sci.* 41:141–150 (1984).

9. Bligh, E. G., and W. J. Dyer. "A Rapid Method of Total Lipid Extraction and Purification," *Can. J. Biochem. Physiol.* 8:911–917 (1959).

10. Bradford, M. M. "A Rapid and Sensitive Method for the Quantitation of Protein Utilizing the Principle of Protein-dye Binding," *Anal. Biochem.* 72:248–254 (1976).

11. Brown, D. A., R. W. Gossett, G. P. Hershelman, C. F. Ward, A. M. Westcott, and J. N. Cross. "Municipal Wastewater Contamination in the Southern California Bight. I. Metal and Organic Contaminants in Sediments and Organisms," *Mar. Environ. Res.* 18:291–310 (1986).

12. Bucolo, G., and H. David. "Quantitative Determination of Serum Triglycerides by the Use of Enzymes," *Clin. Chem.* 19:476–482 (1973).

13. Bulow, F. J. "RNA-DNA Ratios as Indicators of Recent Growth Rates of a Fish," *J. Fish. Res. Board Can.* 27:2343–2349 (1970).

14. Burke, M. D., and R. T. Mayer. "Ethoxyresorufin: Direct Fluorimetric Assay of a Microsomal O-dealkylation which is Preferentially Inducible by 3-methylcholanthrene," *Drug. Metab. Dispos.* 2:583–588 (1974).

15. Cunjak, R. A., and G. Power. "Seasonal Changes in the Physiology of Brook Trout, *Salvelinus fontinalis* (Mitchill), in a Sub-arctic River System," *J. Fish Biol.* 29:279–288 (1986).

16. Fletcher, G. L., M. J. King, J. W. Kiceniuk, and R. F. Addison. "Liver Hypertrophy in Winter Flounder Following Exposure to Experimentally Oiled Sediments," *Comp. Biochem. Physiol.* 73C:457–462 (1982).

17. Friedman, K. J., D. M. Easton, and M. Nash "Temperature-induced Changes in Fatty Acid Composition of Myelinated and Non-myelinated Axon Phospholipids," *Comp. Biochem. Physiol.* 83B:313–319 (1986).

18. Glebe, B. D., and W. C. Leggett. "Temporal, Intra-population Differences in Energy Allocation and Use by American Shad (*Alosa sapidissima*) during the Spawning Migration," *Can. J. Fish. Aquat. Sci.* 38:795–805 (1981).

19. Haines, T. A. "An Evaluation of RNA-DNA Ratio as a Measure of Long-term Growth in Fish Populations," *J. Fish. Res. Board Can.* 30:195–199 (1973).

20. Harvey, H. R., and J. S. Patton. "Solvent Focusing for Rapid and Sensitive Quantification of Total Lipids on Chromarods," *Anal. Biochem.* 116:312–316 (1981).

21. Heath, A. G. "Water Pollution and Fish Physiology" (Boca Raton, FL: CRC Press, 1987), 245 pp.

22. Heidinger, R. C., and S. D. Crawford. "Effect of Temperature and Feeding Rate on the Liver-somatic Index of the Largemouth Bass, *Micropterus salmoides*," *J. Fish. Res. Board Can.* 34:633–638 (1977).

23. Hinton, D. E., and J. A. Couch. "Pathological Measures of Marine Pollution Effects," in *Concepts in Marine Pollution Measurements*, H. H. Harris, Ed. (College Park, MD: Sea Grant College, University of Maryland, 1984), pp. 7–32.

24. Jimenez, B. D. et al. "Laboratory and Field Responses in the Bluegill Sunfish (*Lepomis sp.*), this volume, Chapter 6.

25. Johannesen, K. M., and J. W. DePierre. "Measurements of Cytochrome P_{450} in the Presence of Large Amounts of Contaminating Hemoglobin and Methemoglobin," *Anal. Biochem.* 86:725–32 (1978).

26. Kanter, P. M., and H. S. Schwartz. "A Hydroxylapatite Batch Assay for Quantitation of Cellular DNA Damage," *Anal. Biochem.* 97:77–84 (1982).

27. Klecka, W. R. "Discriminant Analysis." Sage University paper series on Quantitative

Applications in the Social Sciences, Series no. 07-019 (Beverly Hills, CA: Sage Publications, 1980).

28. Kohn, H. W. "The Significance of DNA-Damaging Assays in Toxicity and Carcinogenicity Assessment," *Ann. N.Y. Acad. Sci.* 407:106–118 (1983).

29. Larkin, P.A. "Fisheries Management: An Essay for Ecologists," *Ann. Rev. Ecol. System* 9:57–73 (1978).

30. Lee, R. M., S. D. Gerking, and B. Jezierska. "Electrolyte Balance and Energy Mobilization in Acid-Stressed Rainbow Trout, *Salmo gairdneri*, and their Relation to Reproductive Success," *Environ. Biol. Fish.* 8:115–123 (1983).

31. Loar, J. M., S. M. Adams, M. C. Black, H. L. Boston, A. J. Gatz, M. A. Huston, B. D. Jimenez, J. F. McCarthy, S. D. Reagan, J. G. Smith, G. R. Southworth, and A. J. Stewart. First Annual Report on the Y-12 Plant Biological Monitoring and Abatement Program. ORNL/TM, Oak Ridge National Laboratory, Oak Ridge, TN, 1988.

32. Lockhart, W. L., and D. A. Metner. "Fish Serum Chemistry as a Pathology Tool," in *Contaminant Effects on Fisheries*, V. W. Cairns, P.V. Hodson, and J. O. Nriagu, Eds. (New York: John Wiley & Sons, 1984), pp. 73–85.

33. McKee, M. J., A. C. Hendricks, and R. E. Ebel. "Effects of Naphthalene on Benzo[a]pyrene Hydroxylase and Cytochrome P-450 in *Fundulus heteroclitus*," *Aquat. Toxicol.* 3:103–114 (1983).

34. National Committee for Clinical Laboratory Standards (NCCLS). 1979. NCCLS Approved Standards: ACS-1 Specification for Standard Protein Solution (Bovine Serum Albumin), 2nd ed. National Committee for Clinical Laboratory Standards, Villanova, Pennsylvania.

35. Neff, J. M. *Polycyclic Aromatic Hydrocarbons* (London: Applied Science Publishers, 1978), p. 262.

36. Payne, J. F., and W. R. Penrose. "Induction of Aryl Hydrocarbon Benzo(a)pyrene Hydroxylase in Fish by Petroleum," *Bull. Environ. Contam. Toxicol.* 14:112–116 (1975).

37. Peterson, J. I., and D. S. Young. "Evaluation of the Hexokinase/glucose-6-phosphate Dehydrogenase Method of Determining Glucose in Urine," *Anal. Biochem.* 23:301–316 (1968).

38. Rao, K. S. P., and K. V. R. Rao. "Tissue Specific Alteration of Aminotransferases and Total ATPa$_{ses}$ in the Fish (*Tilapia mossambica*) under Methyl Parathion Impact," *Toxicol. Lett.* 20:53–57 (1984).

39. Rhodes, L., E. Casillas, B. McKnight, W. Gronlund, M. Myers, O. Olson, and B. McCain. "Interactive Effects of Cadmium, Polychlorinated Biphenyls, and Fuel Oil on Experimentally Exposed English Sole (*Parophrys vetulus*)," *Can. J. Fish. Aquat. Sci.* 42:1870–1880 (1985).

40. Roche, H., J. Jouanneteau, and G. Peres. "Effects of Adaptation to Different Salinities on the Lipids of Various Tissues in Sea Dace (*Dicentrarchus labrax*)," *Comp. Biochem. Physiol.* 74B:325–330 (1983).

41. *SAS User's Guide: Statistics*, Version 5 (Cary, NC: SAS Institute, Inc., 1982), 956 pp.

42. Scandinavian Committee on Enzymes. "Recommended Methods for the Determination of Four Enzymes in Blood," *Scand. J. Clin. Lab. Invest.* 33:291–300 (1974).

43. Shugart, L. R. "An Alkaline Unwinding Assay for Detection of DNA Damage in Aquatic Organisms," *Mar. Environ. Res.* 24:321–325 (1988a).

44. Shugart, L. R. "Quantitation of Chemically Induced Damage to DNA of Aquatic Organisms by Alkaline Unwinding Assay," *Aquat. Toxicol.* 13:43–52 (1988b).
45. Shugart, L. R. "DNA Damage as an Indicator of Pollutant-induced Genotoxicity," in 13th Symposium on Aquatic Toxicology and Fish Assessment; Sublethal Indicators of Toxic Stress," W. G. Landis and W. H. van de Schalci, Eds. ASTM, Philadelphia, PA, in press.
46. Shul'man, G. E. *Life Cycles of Fish. Physiology and Biochemistry* (New York: John Wiley & Sons, 1974), 258 pp.
47. Silbergeld, E. K. "Blood Glucose: A Sensitive Indicator of Environmental Stress in Fish," *Bull. Environ. Contam. Toxicol.* 11:20–25 (1974).
48. Sokal, R. R., and F. J. Rohlf. *Biometry* (San Francisco, CA: W.H. Freeman, 1981).
49. Stegeman, J. J., R. L. Binder, and A. Orren. "Hepatic and Extrahepatic Microsomal Electron Transport Components and Mixed Function Oxygenases in the Marine Fish *Stenotomus versicolor*," *Biochem. Pharmacol.* 28:3431–3439 (1979).
50. Tietz, N. W. *Textbook of Clinical Chemistry.* (Philadelphia: W.B. Saunders, Co., 1986), 1919 pp.
51. Waters, T. F. "Secondary Production in Inland Waters," *Adv. Ecol. Res.* 10:91–164 (1977).
52. Zar, J. H. *Biostatistical Analysis.* (Englewood Cliffs, NJ: Prentice-Hall, 1984), 718 pp.

CHAPTER 20

A Comparative Evaluation of Selected Measures of Biological Effects of Exposure of Marine Organisms to Toxic Chemicals

Edward R. Long and Michael F. Buchman

ABSTRACT

An evaluation of the relative responsiveness and sensitivity of selected measures of biological effects to a range in contaminant concentrations was performed in the San Francisco Bay area in 1987. The evaluation was performed to provide information to be used in selecting biological measures for possible inclusion in the National Status and Trends (NS&T) Program of the National Oceanic and Atmospheric Administration (NOAA). The NS&T Program analyzes three media—sediments, bottomfish, and bivalves—routinely at sites nationwide. This evaluation included biological tests of two of those media: sediments and fish.

The overall approach chosen for the evaluation involved analyses of samples collected in the field at sites that were presumed to represent a range in chemical contamination, an approach similar to that of a practical workshop in Norway.[1] The biological measures that were evaluated included five types of sediment toxicity tests; analyses of macrobenthos community composition; sediment profiling photography; induction of hepatic aryl hydrocarbon hydroxylase, induction of hepatic cytochrome P-450 enzymes, plasma steroid hormone concentration, reproductive success and incidence of micronuclei in erythrocytes of fish. The biological tests were performed with subsamples of samples that were also analyzed for chemical concentrations. In a sequence of statistical procedures, the data from the toxicity tests of sediments were compared with each other and with the respective sediment chemistry data and the biological measures in the fish were compared with each other and with the chemical data from analyses of the fish livers. It was presumed and hypothesized before the evaluation began that biological tests most applicable to the NS&T Program would be those that were able to indicate differences among sampling locations over a range in chemical contamination and/or between sampling locations and laboratory controls, had relatively large ranges in response among mean values for the sampling locations, had relatively small analytical variability, and indicated patterns in biological response that generally paralleled the pattern in chemical contamination.

INTRODUCTION

The goal of the NS&T Program is to determine the status of and trends in environmental quality of marine and estuarine areas of the United States. To satisfy that goal, NOAA has begun monitoring the concentrations of selected, potentially toxic, chemical contaminants.[2] The NS&T Program is currently analyzing sediments from about 200 sites, bivalves from about 150 sites, and fish from about 50 sites nationwide for chemical contaminants. Quantitative data are generated for a large suite of potentially toxic contaminants at each of these sites annually. The analyses, however, include only very limited measures of the biological significance of the contaminants that are found in the test media. There are no standards with which to judge the biological relevance of the contaminant data from sediments, bivalves, and biota. Until such standards are developed and accepted, additional empirical evidence is needed to determine which sites are sufficiently contaminated to be of some biological concern.

An evaluation of prospective measures of biological effects was initiated in the San Francisco Bay area in 1987 to determine the relative attributes or performance of selected tests that may be most useful in the NS&T Program. Those measures of effects that are most promising may be used on a broader scale as a part of the NS&T Program testing protocols.

Detailed descriptions of all of the methods, all of the data, and the comparative evaluations of the biological measures were reported by Long and Buchman.[3] This article briefly summarizes the results of that evaluation. In addition, an accompanying chapter in this volume describes many of the methods and some of the data from this same evaluation.[4]

In this evalution, the following tests and analyses were performed by the following investigators:

(1) Collection and chemical analyses of superficial sediments (Battelle New England, Ocean Sciences Division and Science Applications International Corporation (SAIC)); Dr. Andrew Lissner, principal investigator;
(2) Solid phase sediment toxicity test with the amphipod *Rhepoxynius abronius* (E.V.S. Consultants); Dr. Peter Chapman, principal investigator;

(3) Solid phase sediment toxicity test with the amphipod *Ampelisca abdita* (Springborn-Life Sciences and SAIC); Dr. Ronald Breteler, principal investigator;

(4) Sediment elutriate toxicity test with the larvae of the mussel *Mytilus edulis* (E.V.S. Consultants); Dr. Peter Chapman, principal investigator;

(5) Sediment elutriate toxicity test with the larvae of the sea urchin *Strongylocentrotus purpuratus* (Southern California Coastal Water Research Project (SCCWRP)); Dr. Steven Bay, principal investigator;

(6) Sediment pore water toxicity test with the polychaete *Dinophilus gyrociliatus* (Battelle New England, Ocean Sciences Division); Dr. Scott Carr, principal investigator;

(7) Taxonomic analyses of benthic community structure (SAIC and Marine Ecological Consultants (MEC)); Dr. Arthur Barnett, principal investigator;

(8) Analyses of sedimentological and biological characteristics with sediment profiling photography (SAIC); Dr. Donald Rhoads, principal investigator;

(9) Fish collection, tissue chemical analyses, hepatic AHH analyses (Lawrence Livermore National Laboratory (LLNL)), and plasma steroid hormone analyses (University of Texas); Dr. Robert Spies, principal investigator;

(10) Fish blood micronucleated erythrocyte analyses (SCCWRP); Dr. Jeffrey Cross, principal investigator; and

(11) Fish liver cytochrome P-450 enzyme and EROD analyses (Woods Hole Oceanographic Institution (WHOI)); Dr. John Stegeman, principal investigator.

The results of the benthic community analyses and the sediment profiling photography were reported by Barnett et al.[5] and Revelas et al.,[6] respectively, and are not included in the present article.

METHODS

Sampling logistics, sample handling, and data analyses were performed separately for the tests with fish and sediments, though some of the sites overlapped. Sampling sites were selected to represent a range in contamination from relatively highly contaminated to relatively uncontaminated conditions. Data from previous studies, including those by the NS&T Program, were used to select the sampling sites.[2]

Sediment Sampling

Three independent sediment sampling stations were sampled at each of five sites for a total of 15 stations. Four sites were sampled in San Francisco Bay for the toxicity tests and chemical analyses. The sites were designated as: Oakland Inner Harbor (OA, stations 1–3); Yerba Buena Island (YB, stations 4–6) Vallejo (VA, stations 7–9); and San Pablo Bay (SP, stations 10–12). The former was expected to be highly contaminated, while the latter three were expected to be slightly to moderately contaminated. One site in Tomales Bay (TB, stations 13–15), a remote embayment located northwest of San Francisco Bay, also was sampled for these tests and was expected to be uncontaminated. The locations of the sites are illustrated in Figure 1.

The sediments were sampled using the NS&T Program protocol: three stations 50 to 100 meters (m) apart sampled at each site. Sediments were collected with a 0.1 m^2 Young grab sampler (similar to a Van Veen grab sampler). Multiple deployments of the grab sampler (usually 6 to 10) were made at each station and the upper 1 cm of sediment was removed with a Teflon-lined, stainless-steel scoop. These sediments were accumulated in a stainless steel, Teflon-lined basin until about 7 liters (L) of sediment had been composited from each station. The 7 L of composited sediment constituted the sample for each station. The sediments then were homogenized for approximately 5 min with a Teflon-lined steel spoon until the composited sample appeared homogeneous. Portions of varying sizes of the composited sample from each station then were removed for each of the chemical and sedimentological analyses and toxicity tests. Care was taken to avoid contamination of the samples. Sampling was conducted in February 1987.

Tests of sediments from the respective animal collection sites were also performed concurrently with each toxicity test and treated as laboratory controls. These sediments had been used previously as controls and determined to be relatively uncontaminated and nontoxic. Since the participating laboratories were scattered, the same material was not used as controls for all the toxicity tests. This disparity is recognized as a weakness in the study design, especially since none of the control samples was analyzed for chemical concentrations. Nevertheless, the controls served as independent test media for evaluating the viability and internal consistency of the test organisms and for determining which test samples were "toxic," i.e., significantly different from respective controls.

All toxicity tests were performed with five laboratory replicates or aliquots of the composited sediments per station. The portions for chemical analyses were frozen at $-40°C$ and stored for a maximum of 60 days until the analyses were performed. The toxicity tests were performed on three phases of the sediments: solid, elutriate, and pore water. All except the pore water test were conducted on nonfrozen portions held for no more than 5 days. Sediments were collected during a 1-week period. In order to avoid holding the sediments for no more than 5 days, most of the tests were performed in two sequential batches of samples. Each batch included a different sample of the control.

Fish Sampling

The fish sampling sites were in the Oakland Outer Harbor (OK), off Berkeley (BK), in San Pablo Bay (SP), off Vallejo (VJ), off the mouth of the Russian River (RR), and off Santa Cruz (SC) in Monterey Bay (Figure 1). The former two (OK, BK) were expected to be relatively highly contaminated, VJ was expected to be moderately contaminated, and RR, SP, and SC were expected

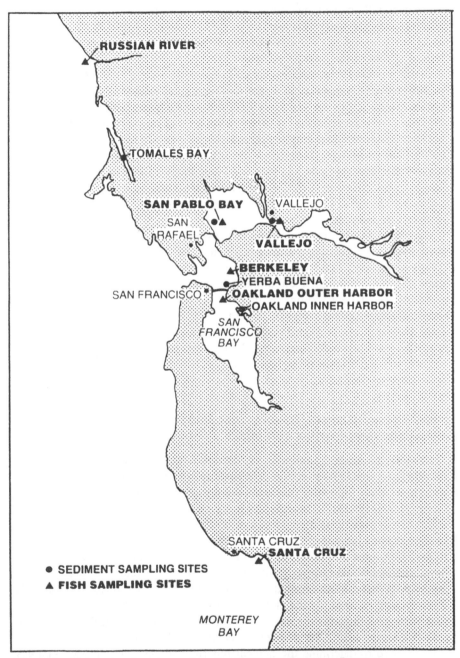

Figure 1.   Fish and sediment sampling sites in the San Francisco Bay area.

to be minimally contaminated reference sites. The species selected for the fish analyses was the starry flounder, *Platichthys stellatus*. It had been the subject of extensive research on bioavailability, contamination, and measures of biological effects in the Bay.[7]

Starry founder (*P. stellatus*) were collected with 5- and 7-m otter trawls twice: in November/December 1986 when fish were anticipated to be late in the reproductive cycle but not yet ready to spawn, and January/February 1987 when the fish were expected to be sexually mature. A target of 30 and 10 to 15 fish per site was set for each sampling period, respectively. Immediately upon capture of the fish, they were bled from the caudal vein to obtain samples for the micronuclei and hormone analyses. A 1-mL blood sample was centrifuged in the field and frozen at $-76°C$ for the future hormone analyses. A blood smear was prepared on a slide, fixed in alcohol, and kept in cold storage for the micronuclei analyses.

Captured fish were maintained on the vessel in flowing bay water until transported to the recirculating marine aquaria at LLNL. Fish were sacrificed the day following capture. Solvent-rinsed tools were used to remove livers and the gonads, which were weighed and aliquots of each stored for subsequent analyses. For each fish, the gonadosomatic index (GSI) was calculated as [gonad weight/ (body weight $-$ gonad weight)] $\times$ 1,000. The hepatosomatic index (HSI) was calculated similarly. Standard length was determined for each fish. The fish were kept alive until just before necropsy. Livers were removed and frozen at $-76°C$ for enzyme and chemical analyses. Female fish captured in January/February were taken to the LLNL to determine reproductive success.

Solid Phase Sediment Toxicity Test with the Amphipod *Rhepoxynius abronius*

Burrowing subtidal amphipods, *R. abronius*, were collected off West Beach on Whidbey Island, Washington. The toxicity of sediments was measured in 10-day static tests, following the methodology of Swartz et al.[8] as amended by Chapman and Becker.[9] This method involved placing a 2-cm layer of test sediment in 1-L glass jars, the addition of 800 mL of clean seawater, and random seeding of each jar with 20 amphipods. The avoidance endpoint was determined from daily counts of amphipods that had emerged from the sediments. After 10 days, sediments were sieved and all amphipods were removed and counted. Amphipods that did not respond to physical stimulation and missing amphipods were counted as dead. At the end of the 10-day exposure, live amphipods were transferred to a fingerbowl containing a 2-cm deep layer of control sediment and clean bioassay water. The number of individuals that had reburied within 1 hr was recorded to determine the percent reburial endpoint. All but two samples were tested in the first batch. A control sediment from the West Beach site was tested twice concurrently with the sediments from the five sites.

Solid Phase Sediment Toxicity Test with the Amphipod
Ampelisca abdita

Tube-forming amphipods, *Ampelisca abdita,* were obtained from tidal flats in Bourne Cove, a small inlet in Buzzards Bay, Massachusetts. The test animals were collected, acclimated, and used in the bioassays following the methods of Breteler et al.[52] and Scott and Redmond.[53] A volume of 200 mL (a layer about 4-cm deep) of the test sediment was placed into the exposure chambers with flowing, gently aerated seawater. Twenty amphipods were placed into each 1-L exposure chamber. The exposure chambers were monitored daily for dead and emerged amphipods. At the end of the 10-day exposures, all sediments were sieved through a 0.5-mm mesh screen. All survivors were enumerated and those unaccounted for were counted as dead. In order to minimize sample storage time, two batches of samples were tested. Sediments from the animal collection site (Bourne Cove) were tested as controls with each batch.

Sediment Elutriate Toxicity Test with Embryos of the Mussel
Mytilus edulis

Adult mussels (*M. edulis*) were collected from Deep Cove, Indian Arm, British Columbia. Spawning and fertilization of the mussels and preparation of the elutriates were accomplished following standard methods.[9] The elutriates were prepared with a 1:50 w/v, sediment/water dilution. The sediments were suspended by vigorous shaking for 10 sec and were allowed to settle for 1 hr prior to adding the embryos. Toxicity testing was conducted following the methods of Chapman and Becker.[9] Approximately 15,000 developing mussel embryos were inoculated by automatic pipette into each 1-L container. The containers were covered and incubated in a temperature-controlled room for 48 hr. After 48 hr and following gentle mixing, a 7-mL aliquot was removed from the water column by automatic pipette. Live larvae were preserved in 5% buffered formalin. The preserved samples were examined in Sedgewick-Rafter cells under 100 times magnification. Control sediments collected off Whidbey Island in Puget Sound, Washington were tested in the same manner. Percent survival was determined relative to the number surviving in the seawater control. Clean seawater drawn from Burrard Inlet, British Columbia, into the analytical laboratory was used as the seawater control. Larvae which failed to transform to the fully shelled, straight-hinged, "D" shaped prodissoconch I stage were considered abnormal.[9]

Sediment Elutriate Toxicity Test with Embryos of the Urchin
Strongylocentrotus purpuratus

Intertidal adult urchins were collected from Point Dume, in northern Santa Monica Bay, California. Spawning of gravid adult urchins was accomplished using modifications of procedures of Oshida et al.[10] A 70-mL aliquot of sediment was added to 280 mL of laboratory seawater (collected off Redondo Beach, California) to produce a sample dilution of 1:4 (v/v). This mixture then was stirred overnight, allowed to settle for 60 min and centrifuged at 2,000 times gravity (G) for 5 min to precipitate suspended particulates. The supernatant was carefully decanted and returned to its respective 400-mL beaker. From this volume, a 10-mL aliquot was removed and used in the sperm cell toxicity testing and a 220-mL aliquot was used in tests of embryos for the other endpoints.

The sperm cell test, in which the numbers of fertilized and unfertilized eggs in an aliquot of each sperm exposure sample were counted, followed the procedures of Dinnel et al.[11] The embryo toxicity tests, in which the percent of the embryos that developed abnormally was determined, were modified from the procedures of Oshida et al.[10] A duplicate sample of embryos was taken from some beakers for the cytologic/cytogenetic examinations, following the methods of Hose.[12] The embryos in 200 mL of the remaining solution were removed with a plastic screen and extracted for an echinochrome pigment measurement, following the methods of Bay et al.[13] Two batches of samples were tested to minimize storage of the sediments before testing. Each experiment included two controls; a laboratory seawater control from Redondo Beach, California and an elutriate control (laboratory seawater carried through the elutriate preparation steps, i.e., shaken and centrifuged).

Sediment Pore Water Toxicity Test with the Polychaete *Dinophilus gyrociliatus*

A population of *Dinophilus gyrociliatus* cultured in the laboratory was used in this toxicity test. A 2- to 3-L aliquot of the sample from each station was pressurized in a Teflon-lined steel cylinder with compressed air for extraction of the pore water, using the methods of Carr et al.[54] The life-cycle toxicity test with *D. gyrociliatus* was conducted to determine mortality and sublethal reproductive effects (i.e., eggs laid per female), using the procedures of Carr et al.[14,54] The tests were conducted for 7 days in 20-mL Stender dishes with ground glass lids and 10 mL of pore water per dish, using a minimum of four 1- to 2-day old animals in each dish. The tests were performed with two batches of samples. A control pore water sample from Duxbury Bay, Massachusetts was tested concurrently with each batch of samples.

Physical/Chemical Analyses of Sediments

The methods used for analyses of organic compounds were based upon the protocols of the NS&T Program of NOAA.[15] Extracts were prepared with methanol then dichloromethane, fractionation was performed by elution with appropriate solvents and identification and quantification of polynuclear aromatic hydrocarbons (PAHs) was performed by gas chromatographic-flame ionization detector

(GC-FID) and for chlorinated hydrocarbons by gas chromatographic-electron capture detector (GC-ECD).

Subsamples for analyses of trace metals, except mercury and silicon, were extracted with concentrated nitric acid and analyzed by atomic absorption spectrophotometry using graphite furnace or flame. Mercury was extracted with concentrated nitric acid and concentrated sulfuric acid and analyzed by cold vapor atomic absorption. Silicon was analyzed colorimetrically using an ultraviolet spectrophotometer.

Sediment texture (grain size) was determined as the percentage (based on dry weight) of gravel, sand, silt, and clay. Coarse and fine fractions initially were separated by wet-sieving. The silt-clay fraction was analyzed by collecting, desiccating, and weighing aqueous aliquots at timed intervals after thorough mixing of the fraction. The sand and gravel fraction, after being dried, was sieved through a 2-mm screen and the weight of gravel and sand subfractions were determined.

Total organic carbon (TOC) and total inorganic carbon (TIC) in sediments were analyzed by a modification of U.S. Environmental Protection Agency (EPA) Method 415.5 (Organic Carbon, Total) and Section 4.8 of the O.I. Corporation Model 700 TOC Analyzer Operating Procedures and Service Manual, which describes the analysis of TIC and TOC on the same aliquot of the sample.

Fish Reproductive Success

Fish captured in the January-February sampling period were taken live to the laboratory to be spawned. Spawning success of several stages in reproduction was determined. However, the sample size of gravid females collected was very small, precluding rigorous evaluation of the endpoints. Therefore, these data will not be discussed further.

Fish Plasma Hormone Analyses

Plasma samples collected from live fish were frozen at $-76°C$ for analyses of estradiol-17β and testosterone by radioimmunoassay (RIA) techniques. Radioimmunoassays for testosterone and estradiol-17β were performed on the same plasma extract. One hundred microliters of plasma were extracted with 2 mL hexane/ethyl acetate (70:30) in 12×75 mm borosilicate tubes. The extract was dried under a stream of nitrogen and reconstituted in 250 mL of gelatin assay buffer (21.2 μm phosphate buffer, pH 7.6) for the analyses.

Fish Erythrocyte Micronuclei Analyses

The number of micronuclei in peripheral erythrocytes of the fish was determined in a total of 158 fish from the two sampling periods. The micronucleus technique was applied to circulating peripheral erythrocytes in approximately 1 cc of blood removed in a heparinized syringe from the caudal vein. Blood smears were prepared immediately, allowed to air dry, and fixed in methanol for 15 min. The blood smears were kept cold (on ice in the field or in a refrigerator in the laboratory) until they were processed. The smears were stained with May-Grunwald-Giemsa procedure and examined on coded slides under a microscope at high power (1,000 ×).[16] Two replicate counts of the number of micronucleated erythrocytes per 1,000 cells were made on each smear, and the results reported as the average of the two counts. Two types of micronuclei were reported: detached and attached.

Fish Hepatic Aryl Hydrocarbon Hydroxylase Activity

Slices less than 5 mm thick were removed from the posterior liver of starry flounder (*Platichthys stellatus*), frozen and maintained at $-76°C$, and used to prepare a microsomal pellet that was assayed in vitro for microsomal enzyme activity, using the method of Nebert and Gelboin[17] as described previously.[18] A portion of the hepatic microsomes from each fish was reassayed with $10^{-4}M$ 7,8-benzflavone (BF). This concentration was determined to cause maximal suppression of aryl hydrocarbon hydroxylase (AHH) activity. An Aminco Bowman spectrofluorometer was used to quantify the 3-OH benzo[a]pyrene metabolite. Fluoresence values of assays were corrected according to quinine sulfate standards. Protein concentrations were determined by the method of Lowry et al.[19] using bovine serum albumin (BSA) as the standard. Hepatic AHH specific activities are reported as picomoles 3-OH-benzo[a]pyrene (B[a]p) per milligram protein per minute.

Fish Hepatic Cytochrome P-450 Analyses

Analyses of cytochrome P-450 enzyme activity were performed on liver microsome samples of individual fish. Hepatic microsomes were prepared from pieces of liver that had been frozen on dry ice at the time of collection. Cytochrome P-450 (extinction coefficient (e) = 91 mM^{-1}cm^{-1}) was measured by sodium dithionite difference spectra of CO-treated samples and cytochrome b_5 content (e = 185 mM^{-1}cm^{-1}) was determined from NADH difference spectra as previously described.[20] Each cuvette contained approximately 1 mg microsomal protein per milliliter.

Ethoxyresorufin O-deethylase (EROD) activity was measured by the spectrophotometric method described by Klotz et al.[21] This method directly measures product formation, like the fluorometric analysis described by Burke et al.,[22] except that the resorufin is detected by absorbance. Estradiol 2-hydroxylase activity was assayed by ^3H_2O release from (2-^3H) estradiol.[23] The cytochrome "P-450E" homologue was analyzed with an immunoblot ("Western" blot) method, using monoclonal antibody (MAb) 1-12-3 against P-450E, the major PAH-induced form isolated from the marine fish scup.[24] The characterization of this antibody and its specificity in immunoblotting has been described.[25,26]

Fish Chemical Analyses

Residues of chlorinated hydrocarbons in liver were determined using methods similar to those of Ozretich and Schroeder.[27] Macerated, dehydrated, centrifuged tissue samples were passed through disposable reverse-phase chromatography columns (Baker) with C_{18} and NH_2 solid-phase absorbents to remove interfering saturated compounds (e.g., alkanes), remaining lipids, and fatty acids. The concentrated extract was analyzed by GC (Hewlett-Packard, 5880) using a ^{65}Ni ECD and a 0.25 mm inside diameter, 30-m fused-silica capillary column internally coated with cross-linked methyl silicone. Chlorinated hydrocarbons of interest were analyzed based on retention times and response factors of authentic external standards (National Bureau of Standards/NOAA standards). Values of analytical blanks were subtracted and final concentrations were corrected for recovery of internal standards. Analyte identifications were confirmed using a Hewlett Packard Mass Selective Detector (MSD), model 5979, operating in the scan mode. The MSD was interfaced with a Hewlett Packard gas chromatograph, model 5880, equipped with a DB-1 fused-silica capillary column.

Data Analysis Methods for Sediment Toxicity Tests

Three attributes of the candidate toxicity tests were considered to be of primary importance and were evaluated: (1) sensitivity of each endpoint to test sediments relative to respective controls, (2) within-sample analytical precision (i.e., low variability among replicates), and (3) total range in biological response to the test samples relative to analytical precision (referred to hereafter as "discriminatory power"). Of secondary importance, the tests should demonstrate some concordance in response with a range in chemical contaminant concentrations. However, the evaluation of concordance between toxicity results and chemical data assumes that the etiological agents in the environmental sediments are among or co-vary with the chemical analytes that were quantified. This assumption may or may not be correct, since the most sensitive toxicity tests may identify some samples as "toxic" that otherwise would not be suspected as such based upon quantification of a limited number of chemical analytes. Other unquantified chemicals may occur in complex media such as sediments that are equally or more toxic than those that are quantified. Finally, if all tests with similar endpoints (e.g., acute toxicity) are responding to related mechanisms of toxicity, toxicity data among the bioassays should demonstrate concordance. Toxicity tests that produce results not in concordance with those of the other tests, may be responding to other mechanisms of toxicity (e.g., mutagenicity) or may be sensitive to only "nuisance" variables or may be insensitive.

All sediment toxicity test data were tested for normality by either the Kolmogorov-Smirnov test or an approximation of the Shapiro and Wilke test.[28] Since all endpoints were determined to be non-normally distributed ($0.01 < p < 0.025$), nonparametric tests were used for further data analyses. To determine the relative "sensitivities" of each test endpoint to individual samples, each composited sediment sample from each station was regarded as an independent, individual sample and the differences in results between the samples and respective controls were determined and tallied. In this statistical test, no attempts were made to determine geographic patterns in toxicity response, but, rather, to determine sensitivity of each toxicity endpoint to individual samples. Differences in results between individual samples and respective controls were tested with the Kruskal Wallis (K-W) test and Dunnett's t-tests.[29] The five replicates tested in the laboratory per sample were used as measures of within-sample variability in the K-W tests.

The standard deviations (SDs) and coefficients of variation (CVs) for the five replicate tests of each composited sample were calculated for each endpoint. Then, the averages of the CVs were determined for each of the toxicity endpoints to evaluate relative analytical "precision." The difference between the maximum and minimum mean values observed in the samples was divided by the average SD to estimate the discriminatory power, an index that was independent of the data from the controls. Correlations among toxicity endpoints normalized to respective batch controls were determined by Spearman's rank correlation.[30] Spearman's rank correlations were also determined between toxicity and bulk sediment chemical data, in which the metals data were normalized to percent fines and the organics data were normalized to TOC content.

Data Analysis Methods for Measures of Effects in Fish

The biological data were evaluated with the Kolmogorov-Smirnov test to determine if they were normally distributed. Differences among sampling sites were tested with the K-W test followed by the S-N-K test. Data from the analyses of nuclear pleomorphism in erythrocytes were tested with chi-square. Differences in chemical concentrations among sites were tested with one-way ANOVA. Correlations both among the biological measures and between biological measures and liver contaminant concentrations were determined with Spearman's rank correlation analysis. The range in response of each test among sites, the SD and the CV among the samples at each site, and the maximum range divided by the average SD were calculated as indicators of methodological sensitivity.

RESULTS

Relative Sensitivity, Precision, and Discriminatory Power of Sediment Toxicity Tests

To determine the relative sensitivity of each toxicity endpoint, the K-W test, followed by non-parametric Dunnett's t-test, was performed to determine significant differences between test samples and respective controls. The numbers and proportions of composited samples in which toxicity was significantly higher than in the respective controls are tallied in Table 1. The endpoints of *R. abronius*

TABLE 1. Results of Kruskal-Wallis Tests of Differences in Toxicity between Sediment Samples and Respective Controls with Nonparametric Dunnett's t-test.

Toxicity Endpoint	P Value[a]	No. of Batches	No of Comparisons[b]	"Toxic" Samples[c] Number	"Toxic" Samples[c] Percent
R. abronius					
Percent survival	.41 and .0001	2	15	0 + 13	87
Avoidance	.21 and .12	2	15	0 + 0	0
A. abdita					
Percent survival (flow-thru)	.22 and .03	2	15	0 + 1	7
Avoidance	.24 and .27	2	15	0 + 0	0
Percent survival (static)	.0099	1	4	1	25
M. edulis larvae					
Percent abnormal	.0001	1	15	14	93
Percent relative survival	.001	1	15	13	87
S. purpuratus larvae					
Percent abnormal	.02 and .89	2	15	2 + 0	13
Percent abnormal and retarded	.38 and .12	2	15	0 + 0	0
Echinochrome content	.11 and .09	2	15	0 + 1	7
Percent fertilization	.02 and .05	2	5	0 + 1	20
No. of mitoses per embryo	.04 and .40	2	5	3 + 0	60
Percent mitotic aberrations	.11 and .01	2	5	3 + 0	60
Micronuclei incidence	.33 and .01	2	5	0 + 1	20
Cytologic abnormality	.008 and .04	2	5	3 + 1	80
D. gyrociliatus					
Egg production	.04 and .006	2	15	2 + 3	33

[a]P value at which differences were indicated by K-W test for each batch of tests.
[b]Indicates total number of comparisons, relative to respective controls, from both batches.
[c]Tested by one-way, np Dunnett's t-test ($\alpha = 0.05$); with "toxic" defined as different (e.g., lower survival) from appropriate control.

survival, *M. edulis* abnormal development, and *M. edulis* percent survival indicated toxicity in the most samples ($\geq 87\%$). The endpoints of *A. abdita* survival, *A. abdita* avoidance, *R. abronius* avoidance, and *S. purpuratus* echinochrome content indicated the least sensitivity (0 to 7% of the samples were indicated as "toxic" relative to controls). Relative to these endpoints, those of *D. gyrociliatus* egg production and *S. purpuratus* abnormal development were intermediate in sensitivity. The endpoints of mitoses per embryo, mitotic aberrations, and cytologic abnormalities in *S. purpuratus* were recorded in tests of only five samples and indicated sensitivity to a majority of those samples relative to controls. The endpoints of reburial of *R. abronius* and survival of *D. gyrociliatus* did not indicate toxicity in any of the samples and were not evaluated further.

The averages of the SDs and CVs for each toxicity test are compared in Table

Table 2. Within-sample Precision, Range in Results, and Discriminatory Power of 15 Toxicity Endpoints Measured in 5 or 15 Sediment Samples.

Toxicity Endpoint	Number of Samples	Average of SDs	Average of CVs (%)	Sample Means		Discriminatory Power[a]
				Maximum	Minimum	
R. abronius						
Percent survival	15	10.3	21.4	91.0	28.0	6.1
Avoidance	15	2.6	83.5	7.4	1.0	2.5
A. abdita						
Percent survival	15	6.9	8.2	95.2	66.0	4.2
Avoidance	15	2.1	63.6	6.8	0.8	2.9
M. edulis						
Percent normal	15	3.2	3.9	93.4	72.7	6.5
Percent survival	15	11.3	22.6	73.0	29.1	3.9
D. gyrociliatus						
Egg production	15	2.2	31.6	10.3	2.9	3.4
S. purpuratus						
Percent normal	15	5.6	7.1	88.4	63.9	4.4
Echinochrome	15	0.007	7.4	0.109	0.082	3.9
Percent eggs fertilized[b]						
(Batch 1)	3	7.8	13.8	66.4	50.0	2.1
(Batch 2)	2	2.9	3.4	89.8	81.2	3.0
Mitoses per embryo	5	1.0	16.8	8.0	5.6	2.4
Percent mitotic aberrations	5	7.1	36.2	30.1	8.0	3.1
Number of micronuclei	5	2.2	73.9	5.0	0.6	2.0
Cytologic abnormalities	5	4.6	29.2	32.0	6.4	5.6

[a]The result of dividing the difference between the maximum and minimum mean values by the average of the SDs.
[b]Controls indicated that batches 1 and 2 behaved differently.

2, the latter as an index of precision. Given that the SDs were largely influenced by the units in which the endpoints were reported, the CVs are a better basis for comparison. Among all of the endpoints, that of percent normal development in *M. edulis* had the lowest average CV (3.9%). Among the other endpoints measured in all 15 samples, those of *A. abdita* survival and *S. purpuratus* percent normal development and echinochrome content also had relatively low average CVs. The CVs for the avoidance endpoints of *R. abronius* and *A. abdita* were the highest. The endpoints of *M. edulis* percent survival and *D. gyrociliatus* egg production were intermediate in precision. Among the endpoints in the tests with *S. purpuratus* measured in five samples, the incidence of micronuclei was highly variable, and the number of mitoses per embryo and egg fertilization success were the least variable.

The quotients obtained by dividing the total range in mean values by the average SD for each endpoint are compared in Table 2. This quotient, the discriminatory power of the test, is intended to identify those endpoints with the widest range in response and the lowest analytical variability independent of the use of and influence of the controls. The discriminatory power was highest with the *M. edulis* percent normal development endpoint and the *R. abronius* percent survival endpoint. They had 6.5 and 6.1 SDs within the range in response, respectively. Among the other endpoints measured in all 15 samples, those of avoidance of sediments by *R. abronius* and *A. abdita* had the lowest discriminatory power and those of *M. edulis* survival, *S. purpuratus* percent normal development and echinochrome content, and *D. gyrociliatus* egg production were intermediate. Among the *S. purpuratus* endpoints measured in five samples, that of cytological abnormalities had a relatively high discriminatory power, while the others had relatively low values.

Concordance Among Sediment Toxicity Endpoints

Generally, most of the toxicity endpoints indicated that the sediments from the Oakland Inner Harbor (site OA) were among the most toxic. Mean *R. abronius* survival, *A. abdita* survival, *M. edulis* survival and normal development, *S. purpuratus* echinochrome content, and *D. gyrociliatus* egg production were relatively low in the OA samples. Many of the endpoints, and others measured in the evaluation, indicated that samples from Tomales Bay (site TB) were among the least toxic. However, the percent of *M. edulis* normal development and survival, *R. abronius* survival, and *S. purpuratus* normal development were unexpectedly relatively low in those samples.

Results of a Spearman rank correlation analysis among the toxicity endpoints are summarized in Table 3. All except two of the endpoints (*R. abronius* and *A. abdita* avoidance—data shown in italics) are presented such that a high value indicates non-toxicity (e.g., percent survival, percent normal). Therefore, except for the avoidance endpoints, a positive correlation indicates concordance and a negative correlation indicates nonconcordance. Three patterns in toxicity re-

Table 3. Spearman Rank Correlations among Toxicity Test Results for Endpoints Measured in all 15 Samples.

	M. edulis (%) Survival (−)	M. edulis (%) Normal (−)	D. gyrociliatus Egg Production (−)	R. abronius Avoidance (+)	R. abronius Survival (−)	A. abdita (%) Survival (−)	A. abdita Avoidance (+)	S. purpuratus (%) Normal (−)
M. edulis % normal (−)	.950							
D. gyrociliatus egg production (−)	.264	.211						
R. abronius avoidance (+)	.314	.380	.452					
R. abronius survival (−)	.818	.761	−.132	.137				
A. abdita % survival (−)	.468	.414	.150	.166	.239			
A. abdita avoidance (+)	.354	−.273	.270	−.013	−.389	−.704		
S. purpuratus % normal (−)	.261	.307	−.239	.259	.486	−.489	.025	
S. purpuratus echinochrome (−)	−.118	−.171	−.125	−.277	−.279	.500	−.566	−.518

Note: The toxicity data were normalized to respective batch controls. Minus signs indicate endpoints that are expected to decrease with increasing contaminant concentrations and plus signs indicate endpoints (in italics) that are expected to increase.

sponses among the endpoints were apparent, based upon the Spearman rank correlation analysis. First, results for the three endpoints of *M. edulis* survival and normal development and *R. abronius* survival were relatively highly correlated with each other and not very highly correlated with any others. Second, results for the three endpoints of *A. abdita* survival and avoidance and *S. purpuratus* echinochrome content were relatively highly correlated with each other. The pattern of response with the *S. purpuratus* percent normal development endpoint contradicted that of *A. abdita* survival and *S. purpuratus* echinochrome content. Third, the results for the endpoints of *D. gyrociliatus* egg production and *R. abronius* avoidance were weakly positively correlated with each other (.452), a positive value indicating toxicity response patterns that contradicted each other. Neither was highly correlated with the results of any of the other endpoints. These three patterns in responses suggest the presence of three affinity groups of endpoints. The correlations were strongest among the endpoints in the first group and progressively weaker in the second and third groups.

Correlations Between Sediment Toxicity and Physical/Chemical Results

Results of a Spearman rank analysis of correlations between toxicity test endpoints and selected physical and chemical variables are listed in Table 4. Trace metal data were normalized to percent fines and organic chemical data were normalized to TOC content. With a total of 190 correlations, the experimental level of significance became $0.05/190 = 0.0003$ by the Bonferroni method. None of the correlations were significant at $\alpha = 0.0003$. Therefore, the results for each toxicity endpoint are treated qualitatively. Most of the endpoints in Table 4 are presented such that a high value indicates nontoxicity (e.g., percent normal, percent survival); however, high values for the endpoints of avoidance, micronuclei, cytological abnormalities and anaphase aberrations denote toxicity. Therefore, the correlations must be interpreted cautiously with regard to the sign listed at the top of each column (i.e., positive or negative correlations are expected between the biological measure and the physical/chemical variable).

Some of the endpoints that were relatively highly correlated with each other (Table 3) also indicated similar patterns in their correlations with some of the same physical/chemical parameters (Table 4). First, the toxicity endpoints of percent normal development and percent survival of *M. edulis*, percent survival of *R. abronius*, and percent normal development of *S. purpuratus* indicated a similar pattern: they were most strongly inversely correlated with sedimentological factor(s) such as percent silt, percent clay, percent fines, and/or TOC content. The *M. edulis* endpoints were also inversely correlated relatively highly with mercury concentration, whereas the *R. abronius* survival data were not very highly correlated with any of the chemical variables. Percent normal *S. purpuratus* data were also relatively highly positively correlated with the concentrations of DDE, other pesticides, zinc, and PCBs. Second, decreasing percent

Table 4. Rank Correlations among Sediment Chemistry and Toxicity Test Results for Either 15 or 5 Samples.

	M. edulis		D. gyrociliatus	R. abronius		A. abdita				S. purpuratus				
	% Survival	% Normal	Repro-duction	Avoidance	Survival	Avoidance	Survival	% Normal	% Echino-chrome	Mitoses	% Fertilized	Micro-nuclei	Cytological Abnor-malities	Anaphase Aberrations
	(−)	(−)	(−)	(+)	(−)	(+)	(−)	(−)	(−)	(−)	(−)	(+)	(+)	(+)
	15	15	15	15	15	15	15	15	15	5	5	5	5	5
Σ Pesticides	−.318	−.246	−.311	−.202	−.007	.515	−.861	.579	−.575	−.8	−.2	−.4	.7	.1
DDE	−.256	−.233	−.345	−.177	.131	.301	−.810	.691	−.567	−.9	.0	.3	.6	.9
"1-Bay" PAHs[a]	−.286	−.250	−.707	−.390	.154	.132	−.479	.364	−.418	−.7	−.2	.5	.3	.7
"No-Bay" PAHs	−.296	.218	−.175	.163	−.036	.186	−.275	.193	−.582	−.1	−.9	.8	−.1	.5
"2-Bay" PAHs	−.354	−.343	−.721	−.510	.061	.224	−.604	.371	−.368	−.7	−.2	.5	.3	.7
ΣPAHs	−.273	−.214	−.685	−.365	.148	.150	−.402	.277	−.463	−.4	−.6	.8	.1	.6
FFRI[b]	.393	.332	.575	.463	.161	−.281	.343	.075	.318	.3	.8	−.9	−.2	.7
ΣPCB	−.291	−.264	−.414	−.349	.087	.427	−.865	.587	−.506	−.9	.0	.3	.6	.9
Ag	−.366	−.284	−.316	−.133	.063	.470	−.745	.422	−.504	−.5	−.6	.7	.5	.9
Cd	−.374	−.334	.288	.352	−.270	.406	−.495	.186	−.415	−.4	−.4	.2	.1	.6
Cr	−.112	.032	.407	.335	−.239	.311	−.304	−.296	−.200	.5	−.2	−.1	.0	−.2
Cu	−.268	−.232	−.143	−.004	−.054	.404	−.554	.289	−.507	−.5	−.6	.7	.5	.9
Hg	−.443	−.490	.332	.149	−.374	.591	−.649	.084	−.386	−.7	.0	−.1	.3	.7
Ni	−.114	.011	.529	.365	−.246	.363	−.304	.189	−.289	.2	.0	−.4	−.3	−.2
Pb	−.161	−.168	−.068	.029	.036	.331	−.639	.432	−.361	−.5	−.6	.7	.5	.9
Zn	−.114	−.029	−.125	.048	.082	.343	−.625	.507	−.500	−.5	−.6	.7	.5	.9
% Silt	.368	.386	−.511	−.122	.529	−.401	.250	.146	−.064	.0	.1	.3	.5	.1

% Clay	-.496	-.661	.114	-.422	-.579	.145	.114	-.696	.357	-.2	.3	-.5	-.7	-.3
% Fines[c]	.052	-.150	.050	-.422	-.070	-.229	.472	-.629	.349	-.2	.3	-.5	-.7	-.3
TOC	-.635	-.696	.120	-.329	-.721	.279	.073	-.792	.263	.2	-.3	-.1	-.8	-.3

Note: Trace metal data have been normalized to percent fines and organic chemical data have been normalized to TOC content. The toxicity data were normalized to respective batch controls. Plus and minus signs indicate expected pattern of correlation between biological endpoint and concentrations of physical/chemical parameters.

a "Bay" refers to the stereochemistry of the ring structure in which three or more benzene rings form a bay-like indention; this structure has been associated with carcinogenic properties.

b FFRI refers to a Fossil Fuel Relative Index, a ratio measure of petroleum related PAHs; the higher the Index, the greater the proportion of PAHs normally associated with petroleum products versus combustion by-products.[31]

c Percent fines is the summation of percent clay and percent silt.

survival and increasing incidences of avoidance of sediments by *A. abdita* and decreasing echinochrome content of *S. purpuratus* also indicated a similar pattern: they were relatively highly correlated with increasing concentrations of DDE, other pesticides, and PCBs. The correlations between *A. abdita* survival and these chemicals were particularly high. Third, egg production in *D. gyrociliatus* was most inversely correlated with PAHs, and percent silt and most positively correlated with nickel and Fossil Fuel Relative Index (FFRI). Among the *S. purpuratus* endpoints measured in only five samples, the endpoints of micronuclei incidence, anaphase aberrations, mitoses per embryo, and fertilization success were relatively highly correlated with one or more classes of organic compounds. As the incidence of micronuclei and anaphase aberrations increased and as fertilization success and mitoses per embryo decreased, the concentrations of some organic compounds increased. In addition, anaphase aberrations were relatively highly correlated with several trace metals.

Relative Sensitivity of Bioeffects Measures in Fish

The data from the biological measures of the health of the fish were not normally distributed (Kolmogorov-Smirnov test). Following various transformations of the data, they remained not normally distributed. Therefore, nonparametric statistical tests generally were performed with the fish data.

Based upon the K-W test, there were no significant differences among sites in length (p = 0.106), weight (p = 0.113), GSI (p = 0.122), HSI (p = 0.708), and liver weight (p = 0.248) among females. Among the males, there were differences among sites in these measures: length (p = 0.014), weight (p = 0.010), GSI (p = 0.029), HSI (p = 0.200), and liver weight (0.0495). However, S-N-K tests could not identify which sites were different. Generally, the fish from the RR site were larger and those from the OK site were smaller than the others.

The K-W test was applied to the biological effects data to determine relative "sensitivity," i.e., ability to distinguish differences among sites (Table 5). Both EROD and cytochrome P-450E measurements indicated a similar pattern of relatively high induction of the cytochrome P-450 system in fish, especially immatures, from the BK and OK sites and low induction in fish from the SP, VJ, and RR sites. EROD activity, expressed as either units per milligram protein or units per nanomole P-450, indicated differences between sites (p = 0.0004 and p = 0.003, respectively): higher in immature fish from OK than in mature fish from SP, RR, and BK and immatures from RR (Table 5). Mean EROD activities in mature OK fish and immature BK fish were also relatively high, but not significantly different from those at other sites. Cytochrome P-450E content per milligram protein was significantly higher in immature OK fish than in mature fish from RR and SP (p = 0.0011). When expressed in proportion to total P-450 content, the P-450E content was greater in immature OK fish than in mature SP, VJ and RR fish.

Table 5. Results of Kruskal-Wallis Tests of Differences[a] in Biological Measures in *Platichthys stellatus* Collected at Five Sites in November and December 1986. Underlines Indicate Sites That Were Not Different from Each Other (Upper Case for Mature Fish, Lower Case for Immatures).

Biological Measure	Increasing Toxicity →				
EROD/mg protein	rr SP RR BK sp VJ vj OK bk ok				
EROD/nmol P-450	rr RR SP BK VJ sp vj OK bk ok				
P-450E/mg protein	SP RR VJ vj rr BK sp OK bk ok				
P-450E/nmol P-450	SP VJ RR vj BK rr sp bk OK ok				
Total P-450/mg protein	SP	RR	VJ	BK	OK
P-420/mg protein	SP	RR	BK	VJ	OK
Denatured P-450 (%)	SP	VJ	RR	OK	BK
Cytochrome b5/mg protein[b]	RR	OK			
Estradiol 2-Oh/mg protein (females)	SP	OK	RR	VJ	BK
Estradiol 2-Oh/nmol P-450 (females)	BK	OK	RR	VJ	
					SP
AHH	RR SP BK rr bk vj sp OK VJ ok				
AHH with 7,8-BF	BK RR bk rr SP vj OK sp ok VJ				
Estradiol (mature females)	RR OK BK				
Testosterone (males)	VJ SP BK RR				
Total Micronuclei	RR OK		SP	VJ	BK
Detached micronuclei	RR	OK	SP	VJ	BK
Attached micronuclei	RR	OK	VJ	BK	SP

[a] $p \leq 0.05$.
[b] Mann-Whitney test.

Mean total P-450 content and P-420 content were highest in fish from OK, but there were no significant differences ($p = .183$ and 0.786, respectively) among sites (Table 5). Also, there were no significant differences ($p = 0.385$) among sites in percent denatured P-450 content. Cytochrome b_5 content did not differ between RR and OK fish (Mann-Whitney, $p = 0.756$).

Two-way ANOVA (performed despite the fact that the data were not normally distributed) indicated there were no differences ($p = 0.09$ to 0.98) in any of the P-450 measurements between sites that were attributable to sex. However, both EROD activity and P-450E content per milligram protein in immature fish exceeded those measures in mature fish.

Estradiol 2-hydroxylase activity per milligram protein did not differ significantly ($p = 0.94$) among sites. However, estradiol 2-hydroxylase per nanomole P-450 content did ($p < 0.01$) (Table 5). The average values were lowest in mature BK fish where cytochrome P-450 induction was among the highest and highest in immature fish from SP. Consistent with this pattern was the observation that estradiol 2-hydroxylase activity in fish injected with B-naphthoflavone (BNF) was about one-half that in untreated fish. This measure was significantly lower (p = 0.007) in mature females than in immature females.

AHH activity was generally higher in immature fish than in mature fish. Significant between-site differences in AHH activity were indicated at p = 0.001, but S-N-K could not distinguish which sites were different. The mean AHH values in fish in which the samples had been exposed to 7,8-BF were generally one-third to one-fifth those of fish not exposed to this P-450 isozyme inhibitor.

Estradiol concentration in plasma was not significantly different among sites (p = 0.94) (Table 5). The overall mean value in VJ fish, which were mostly immature, was about one-fifth that in fish from OK which included more matures. Since estradiol suppresses the induction of P-450E, an inverse relationship between estradiol content and P-450E content and EROD activity would be expected. But, this relationship was not obvious with the present data. As would be expected, mature females had higher plasma estradiol concentrations than immatures. Testosterone concentrations also did not differ significantly among sites (p = 0.40) (Table 5).

The incidence of total counts of micronuclei in blood erythrocytes was significantly different among sites (K-W, p = 0.0001). S-N-K indicated that counts were lower in RR than in SP, VJ and BK fish (Table 5). Incidences among the four sites in San Francisco Bay did not differ. Counts of both attached and detached micronuclei were recorded. The incidences of attached micronuclei in all fish did not differ significantly among sites (p =. 0.061). However, the incidence of detached micronuclei did differ between sites (p = 0.001) with the same pattern as observed with total counts. The incidence of both forms did not differ among sites in males. However, total incidence and incidence of detached micronuclei were significantly lower in females from RR than in females from three other sites.

Within-Site Variability in Bioeffects Measures in Fish

Analytical variability, as evaluated with the sediment bioassays, could not be evaluated with the bioeffects measures in fish, since replicate analyses of individual fish were not performed on all the fish with all of the measures. However, within-site variability and between-site discriminatory power can be compared among measures, since all were performed on the same individual fish. Within-site variability among fish can be viewed two (opposing) ways. First, if one assumes that individual fish sampled within one area (site) had different histories of exposure to contaminants, then a bioeffects measure that has high within-site variability may simply reflect that variability if the measure is especially sensitive. If, on the other hand, one assumes that the fish sampled in one area (site) are from a relatively homogeneous population and, therefore, all the individuals have had a similar history of contaminant exposure, then a bioeffects measure with high within-site variability may indicate relatively low analytical precision. Unfortunately, *Platichthys stellatus* migrate in and out of San Francisco Bay annually and little is known of the fidelity of the returning adults to specific areas of the estuary. Although Spies et al.[7,32] have demonstrated differences in measures of contaminants in tissues and measures of bioeffects between sites in the bay, they also observed sufficient within-site variability to suggest that all the fish from any one site were not from a distinct population with identical histories of contaminant exposure.

The maximum and minimum site means, the total range in mean values, the average of the SDs, the average of the CVs, and the quotient of dividing the range by the average SD are listed in Table 6. The average CVs were highest for the endpoints of attached and total micronuclei and cytochrome P-450E and lowest for total cytochrome P-450 and estradiol-2-hydroxylase per milligram protein. The quotients of total range over the average SDs were highest for EROD per milligram protein and EROD per nanomole P-450. By comparison, these values were relatively low for total P-450, attached micronuclei, and AHH following exposure to 7,8-benzflavone.

Concordance Among Measures of Bioeffects in Fish

Results of a Spearman rank correlation analysis among the measures of effects in the fish are summarized in Table 7. As expected, many of the measures of cytochrome P-450 induction were relatively highly correlated with each other. For example, the correlation between total P-450 content and P-450E content was 0.589. Also AHH activity was most highly correlated with EROD activity. However, AHH activity was not very highly correlated with the other measures. Neither the testosterone content nor the micronuclei counts were highly correlated with other measures.

Correlations Between Bioeffects Measures and Contaminant Concentrations in Fish

Some degree of variability in contaminant levels is to be expected in a population of feral fish because of their mobility. However, it is possible that these fish were exposed to and were affected by readily metabolized and nonquantified compounds, such as aromatic hydrocarbons, that may have been partially or wholly responsible for the induction of the biological measures. If the latter case pertained, then a strong correspondence between the biological data and the quantified chemical data would not be expected.

None of the biological measures was particularly highly correlated with any of the chemical contaminants in the livers of the fish (Table 8). Many of the effects measures that were expected to be positively correlated with the chemical concentrations were unexpectedly negatively correlated with them. Total counts of micronuclei were most highly correlated (negatively) with mirex and aldrin concentrations. Cytochromes P-450 and P-450E content and EROD activity were very weakly positively correlated with tPCB concentrations. Estradiol 2-hydroxylase activity was relatively highly correlated with dieldrin concentrations;

Table 6. Maximum and Minimum Site Means, Total Ranges among Site Means, Averages of the Site SDs and CVs, and the Quotient of Dividing the Range by the Average SD for Selected Measures of Bioeffects in *Platichthys stellatus*.

	Maximum Site Mean	Minimum Site Mean	Total Range	Average SD	Average CV	Range/ Average SD
AHH (all fish)	287.1	64.7	222.4	133.8	78.4	1.7
AHH with 7,8-BF (all)	54.6	24.8	29.8	23.7	60.4	1.3
Testosterone (males)	2.2	0.5	1.7	0.8	57.3	2.1
Estradiol (females)	10.9	1.9	9.0	5.6	78.1	1.6
Total micronuclei[a]	2.4	0.1	2.3	1.5	140.5	1.5
Detached micronuclei[a]	1.5	0	1.5	0.7	82.3	2.1
Attached micronuclei[a]	1.3	0.1	1.2	1.0	149.1	1.2
Total P-450	0.267	0.168	0.099	0.095	44.0	1.0
EROD/mg protein	0.259	0.044	0.215	0.078	56.7	2.8
EROD/nmol P-450	1.47	0.32	1.15	0.47	60.8	2.4
P-450E	26.5	1.9	24.6	11.9	129.2	2.1
Estradiol 2-OH'ase/mg protein	0.038	0.016	0.022	0.012	50.8	1.8
Estradiol 2-OH'ase/nmol P-450	0.299	0.101	0.198	0.091	62.4	2.2

[a]November-December fish.

Table 7. Spearman Rank Correlations among Measures of Bioeffects in *Platichthys stellatus*.

	tP-450 + P-420 (+)*	Estradiol-2-hydroxy-lase (-)	Testos-terone (-)	P-450E/ tP-450 (+)	% Denatured P-450 (+)	Cyto-chrome P-420 (+)	Cyto-chrome P-450 (+)	EROD/ P-450 (+)	EROD/ mg Protein (+)	Cyto-chrome P-450E (+)	Total Micro-nuclei (+)
AHH (+)	.368	.031	-.061	.267	-.022	.246	.252	.448	.448	.294	.074
Total micronuclei (+)	-.119	-.006	.043	.049	.097	-.09	-.077	.191	.153	.071	
Cytochrome P-450E (+)	.375	-.195	-.073	.896	-.213	.173	.589	.52	.606		
EROD/mg protein (+)	.471	.089	-.063	.505	-.318	.24	.549	.863			
EROD/P-450 (+)	.151	.126	-.108	.529	-.094	.043	.214				
Cytochrome P-450 (+)	.9	.027	-.102	.363	-.44	.47					
Cytochrome P-420 (+)	.779	.032	-.062	-.159	.418						
% Denatured P-450 (+)	-.04	-.012	-.063	-.328							
P-450E/tP-450 (+)	-.012	-.016	-.119								
Testosterone (-)	.017	-.359									
Estradiol 2-hydroxylase (-)	.109										

*Plus signs indicate endpoints that are expected to increase with increases in contaminant concentrations and minus signs indicate endpoints (in italics) that are expected to decrease.

Table 8. Spearman Rank Correlations between Measures of Biological Effects and Concentrations of Contaminants in Livers of *Platichthys stellatus*.

		Lindane	Heptachlor	Aldrin	Heptachlor epoxide	Chlordane	Transnona-chlor	Dieldrin	Mirex	ΣDDT	ΣPCB
AHH	(+)[a]	.013	-.08	-.048	.081	.082	.133	-.197	.202	.119	-.002
Total micronuclei	(+)	-.106	.249	-.417	.107	.27	.283	.173	-.541	.215	.175
Cytochrome P-450E	(+)	.239	-.397	-.122	-.037	.062	-.008	-.207	-.011	-.019	.112
EROD/mg protein	(+)	.219	-.131	-.232	-.132	.102	.185	-.109	.002	-.009	.224
EROD/tP-450	(+)	.382	.102	-.19	.033	.108	.202	-.056	.007	.067	.233
Cytochrome P-450	(+)	-.037	-.277	-.281	.25	-.177	-.063	-.208	-.031	-.169	.089
Cytochrome P-420	(+)	.367	.258	-.155	.074	-.132	.091	.027	.246	-.05	.095
% Denatured P-450	(+)	.198	.148	-.053	.158	.007	.094	-.006	.084	.151	.007
P-450E/tP-450	(+)	.065	-.351	-.008	.026	.139	.064	-.098	-.008	.018	.137
Testosterone	(-)	-.164	-.197	-.406	-.164	.075	-.013	-.069	-.129	.038	-.082
Estradiol 2-hydroxylase	(-)	.046	.175	.136	.127	.016	.211	.416	-.142	.124	.234
tP-450 + P-420	(+)	.126	-.202	-.309	-.159	-.306	-.028	-.129	.225	-.174	.09

[a]Plus signs indicate those endpoints that were expected to increase as contaminant concentrations increase and minus signs indicate those endpoints that were expected to decrease.

whereas, a negative correlation would be expected. No single chemical or chemical class stood out as consistently being correlated with the biological measures.

In addition, very few of the correlations between the biological effects measures and measures of animal length and weight were particularly high. Notably, cytochrome P-450E was relatively highly inversely correlated with hepatosomatic index and estradiol 2-hydroxylase was relatively highly positively correlated with length, weight, and GSI.

Females predominated at all sites, except SP where a nearly equal number of males was obtained in November-December. Mature fish were most commonly examined at three of the sites; however, roughly equivalent numbers of immatures were found at the VJ and SP sites. The lengths of the fish were similar at all sites, although SP fish were slightly smaller and those at the RR site were slightly larger than those at the other three sites. The relatively high proportion of immature fish at SP was reflected in the small mean length of the fish. The liver/body weight ratios and condition factors were similar among all sites and there was relatively low variability in the data from within the sites. The gonad/body weight ratios, however, showed greater differences among the sites, and were highly variable within sites. Relative to the other sites, the gonads in SP fish, where the most immatures were captured, were smaller. The RR fish which were the largest also had the highest gonad/body weight ratios.

DISCUSSION

Sediment Toxicity

The high sensitivity of the *M. edulis* embryo test observed in this evaluation was also reported by Chapman et al.[33] in a previous study in San Francisco Bay in which six toxicity endpoints were measured. In that study, the endpoints of percent abnormal development and percent survival of *M. edulis* indicated the most samples, 5 of 9 and 8 of 9, respectively, were "toxic" relative to controls. However, Williams et al.[34] reported that a similar test performed with oyster larvae (*Crassostrea gigas*) was the least sensitive of three that were evaluated. It indicated 35% of the samples from Commencement Bay, Washington waterways were "toxic" relative to controls, compared to 39% for *R. abronius* and 63% for a Microtox™ test. In the data reported by Chapman et al.,[33] the analytical precision was somewhat lower (average CVs were 25% and 32.5%, respectively, for percent abnormal and percent survival endpoints), than in the present evaluation (average CVs of 3.9 and 22.6%, respectively). The positive correlations between the results of the *M. edulis* test and the texture and TOC content of the sediments have not been quantified through empirical experimentation. However, similar to the results observed here, Chapman et al.[33] often observed the highest toxicity in samples that had the highest percent fines and TOC content, as well as the highest concentrations of toxicants. In the present evaluation, fine-grained, organically enriched samples that were not highly contaminated with the quantified analytes were also toxic to the test organism. In summary, the endpoint of abnormal development in the *M. edulis* test was the most sensitive and had the highest precision and discriminatory power of those evaluated, but the data appeared to be more highly correlated with "nuisance variables," such as sedimentological properties, than with chemical concentrations.

Similar to the results reported here, the data from either one or both of the endpoints of the *M. edulis* test and the survival endpoint of the *R. abronius* test previously have indicated very high concordance with each other.[33,34] The relatively high sensitivity of the test of *R. abronius* survival has been previously recorded in inter-method comparisons.[33-36] With the protocols specified by Swartz et al.[8], the test is 75 percent certain of detecting statistical significance ($p <$ 0.05) if the difference in mean survival between controls and test sediments is 2.8 or greater. The average CV reported here (21.4%) was the same as that for data from nine samples reported by Chapman et al.[33] and very similar to that (22.4%) for data from seven samples reported by Mearns et al.[36] Also, the high analytical variability (average CV of 83.5%) for the avoidance endpoint observed in this evaluation has been reported elsewhere: 65.7% in nine samples;[33] and 111.2% in seven samples.[36]

The negative correlations between *R. abronius* survival and both percent clay and TOC content corroborates this relationship demonstrated quantitatively in empirical experiments conducted by DeWitt et al.[37] They observed that survival of *R. abronius* can be 15% lower in fine-grained, uncontaminated sediments than in native (muddy sand) sediments. They concluded that, together, small particle size and high organic content may sometimes cause or influence the survival of *R. abronius*, and the effects of these two properties may be especially important in relatively uncontaminated sediments. They hypothesized that sediment particle size may be a "super-variable" that is indicative of a suite of other factors that co-occur with small-grained particles and that these other factors may be the actual cause of mortality. However, the 15% effect of grain size, alone, upon survival would not explain the magnitude of the response observed in apparently' uncontaminated sediments tested in the present evaluation. In summary, the endpoint of *R. abronius* survival was among the most sensitive bioassays, had a wide range in response, relatively high discriminatory power, and intermediate precision; however, the data were more highly correlated with sedimentological variables than with chemical concentrations. The endpoints of reburial and avoidance appear to be relatively insensitive and/or highly variable.

Among ten species of fish and macroinvertebrates used to test the acute toxicity of Black Rock Harbor, Connecticut sediments, only one, *A. abdita*, indicated significant differences from controls.[38] In the present evaluation, mortalities in the control sediments exceeded those in some of the test samples and probably resulted in the underestimation of the potential sensitivity of this test. In the present evaluation, the endpoint of survival was relatively insensitive, indicating a significant difference from controls in only 1 of the 15 samples tested, a sample from the Oakland Inner Harbor that was among the most contaminated. The sensitivity of this test, as determined in the present evaluation, may have been greater if the survival of *A. abdita* in the controls had been higher. The concentrations of PCBs and many trace metals were generally more than an order of magnitude higher in Black Rock Harbor sediments[38] than in those from Oakland Inner Harbor. Dose-dependent responses of A. *abdita* to Black Rock Harbor sediments were corroborated with the negative correlations observed in the present evaluation between survival and the concentrations of several classes of organic toxicants. The organism did not indicate sensitivity to fine-grained sediments that apparently were not highly contaminated. In summary, the endpoint of A. *abdita* survival was less sensitive and had lower discriminatory power than that with R. *abronius*, but had relatively higher analytical precision than that with R. *abronius*, was not highly correlated with sedimentological variables, and was relatively highly correlated with several toxicants. The endpoint of avoidance was not sensitive and had low precision and discriminatory power.

The relative sensitivity of echinoderm sperm cells and embryos to either sewage or single dissolved chemicals has been evaluated.[11,39,40] Compared to bivalve larvae, fish, crab zooea and Microtox tests, the echinoderm tests generally were similar to the other organisms in their sensitivity to a variety of trace elements and organic compounds. We are not aware of any previously published evaluations of the relative sensitivity of this test with sediment samples. In the present evaluation, the endpoints of abnormal development and echinochrome pigment content were not as sensitive as those of abnormal development of M. *edulis* and survival of R. *abronius*. The increased incidences of cytogenetic abnormalities, micronuclei and cytological abnormalities and the decreased number of mitoses per embryo were demonstrated to be responsive to benzo[a]pyrene.[41] In the present evaluation, most of these endpoints were sensitive to many of the samples tested and were correlated with increasing concentrations of many contaminants, including hydrocarbons. However, these endpoints were evaluated in only five samples and the analytical variability was relatively high. The pattern in response among the samples indicated by the abnormal development endpoint contradicted that indicated by many of the other endpoints in the same test and in the A. *abdita* test. This contradiction may be a result of different toxicity mechanisms among these endpoints or of unintentional bias in the subjective scoring of embryos as morphologically abnormal vs normal. This contradiction in results has not been observed by the analysts at SCCWRP in previous experience with this test in assessments of complex effluents. In summary, this test facilitates the determination of many biological endpoints; most of the endpoints are intermediate in sensitivity, precision, and discriminatory power; the endpoint of normal development was negatively correlated relatively highly with sedimentological variables, but the incidences of many of the cytological/mutagenicity indicators were relatively highly correlated with chemical concentrations.

The survival of D. *gyrociliatus* appears to be relatively resistant to endosulfan and pentachlorophenol, but egg production appears to be sensitive to complex effluents and pentachlorophenol.[14] In this evaluation, survival also was insensitive. However, egg production was sensitive to about one-third of the samples tested relative to controls. The endpoint of egg production is a biologically meaningful indicator of reproductive success. As expected with a test of pore water, the results were not strongly correlated with those from the toxicity tests of the solid phase sediments or elutriates. However, they were relatively highly correlated with the concentrations of PAHs in bulk sediments. The medium that is tested, the pore water, is predicted by equilibrium-partitioning theory to be the controlling exposure medium in the toxicity of sediments to infaunal organisms.[55] However, since the borosilicate filter used in the pore water extraction method may have removed some of the potentially toxic organic compounds, this test may have underestimated the pore water toxicity. In summary, this test was intermediate in sensitivity, precision, and discriminatory power relative to the others evaluated; the data were more highly correlated with the concentrations of hydrocarbons than with other variables; and the test indicated little concordance with the data from the other tests.

Three patterns in toxicological response to the samples described above were suggested by the rank correlations among toxicity tests (Table 3) and by correlations with similar chemical or physical variables (Table 4). These patterns were not related to the medium tested (i.e., solid phase, elutriate, pore water), nor to the biological type of endpoint. Whereas the toxicity test endpoints within each affinity group indicated similar patterns in response, some indicated negative correlations with endpoints in other affinity groups. For example, the data from the *M. edulis* percent normal development endpoint (indicative of nontoxicity) and the *A. abdita* avoidance endpoint (indicative of toxicity) were positively correlated and, therefore, contradicted each other. Two endpoints of the *S. purpuratus* test, echinochrome content and percent normal development, also contradicted each other. Despite the suggestions derived from the correlations between toxicity and chemical data, the etiological agents for each test are unknown. It is possible that each endpoint responded to different physical or chemical properties, including those that were not quantified, in the very complex sediments that were tested. Therefore, until the relationships between these toxicity endpoints and specific chemicals are quantified in empirical experimentation, comprehensive assessments of sediment toxicity are best made with multiple endpoints.

Measures of Bioeffects in Fish

Measures of micronuclei incidence (particularly detached micronuclei) were sensitive; they indicated differences between the sites in San Francisco Bay and a site outside the Bay. However, there were no significant differences among sites sampled within San Francisco Bay. Relative to the other measures of bioeffects in fish, those of micronuclei incidence had relatively high within-site variability and intermediate discriminatory power. The data were not particularly highly correlated with those of the other measures of effects, and they were weakly correlated with the concentrations of selected pesticides in the fish.

Micronuclei are likely formed as a consequence of chromosome breakage and spindle dysfunction during mitosis. Micronuclei occur at background levels in uncontaminated conditions, but at very low incidences. Chromosomal aberrations and micronuclei may have a role in evolutionary changes. However, elevated incidences of micronuclei above background rates have been attributed to X-rays and specific chemicals with mutagenic properties. Therefore, as applied in the context of monitoring marine pollution, this test can be assumed to be responsive to chemicals with specific types of effects.

In this evaluation, mean micronuclei incidence in peripheral erythrocytes ranged from 0.1 per 1,000 at RR to 2.4 per 1,000 at BK. The highest incidences were among fish from the San Francisco Bay sites compared to fish from coastal reference sites. In 27 New England coastal areas, micronuclei frequencies in mature erythrocytes of winter flounder (*Pseudopleuronectes americanus*) ranged from 0.2 per 1,000 at a site at Georges Banks to 5.6 per 1,000 at a site in Long Island Sound. Overall mean frequencies in the New York Bight Apex (2.33/ 1,000, n = 39) and throughout Long Island Sound (3.94 per 1,000, n = 35) were significantly higher than in offshore areas (e.g., 0.46 per 1,000, n = 13, on the mid-Atlantic shelf).[42]

White croaker (*Genyonemus lineatus*) caught in California coastal waters off Dana Point had lower frequencies of micronuclei in peripheral erythrocytes (mean of 0.8 per 1,000, n = 28) than those caught in San Pedro Bay (mean of 3.5 per 1,000, n = 28).[43] Similar studies of kelp bass (*Paralabrax clathratus*) showed that fish from Catalina Island had lower incidences (mean of 0.6 per 1,000, n = 15) than fish caught off White Point near a major sewer discharge (mean of 6.8 per 1,000, n = 15).[43] Nuclear pleomorphism also was found in those fish with the highest micronuclei frequencies. Though differences in micronuclei frequencies between sites were significant for both species, micronuclei counts were only weakly correlated with concentrations of organic compounds in the fish livers. Highest frequencies, however, occurred in fish from areas known to have high concentrations of hydrocarbons in sediments.[43] Based upon the data from the present evaluation and from previous research, the differences in incidence of micronuclei in fish collected in contaminated and uncontaminated conditions are often significant and show a consistent pattern among species, even when differences in chemical concentrations are relatively small. The background frequencies and the degree of elevations in frequencies at sites that are contaminated do not appear to be species-specific.

In summary, it appears that the measure of the incidence of micronuclei is relatively sensitive, has intermediate discriminatory power, and the background incidences are similar among several species of marine fish, but the data are relatively variable among fish captured at the same site, and in weak concordance with the concentrations of selected organic compounds in the fish tissues.

Measures of EROD, cytochrome P-450E and estradiol 2-hydroxylase per nanomole P-450 were sensitive; they indicated differences among sampling sites. All, except the measure of cytochrome P-450E, had relatively low within-site variability and had relatively high discriminatory power. None of these measures was particularly highly correlated with tissue chemical concentrations and several were correlated with size of the fish or relative size of the gonads and/or livers. As expected, most were relatively highly correlated with each other. The measure of total cytochrome P-450 content did not indicate differences among sites, had low discriminatory power, and was not highly correlated with chemical concentrations, but had very low within-site variability

The cytochrome P-450 proteins catalyze monooxygenase reactions. They are the enzymes primarily responsible for metabolism or biotransformation of organic pollutants. This metabolism of xenobiotics can result in their inactivation and detoxification or their activation to toxic derivatives. Inactivation can lead to enhanced elimination and tolerance; activation can lead to serious organ dysfunction or pathology. Cytochrome P-450 enzymes are also responsible for both synthesis and degradation of steroid hormones. Aromatic hydrocarbons and chlorinated hydrocarbons, such as PCBs, can induce elevated cytochrome P-450 content. Cytochrome P-450 exists in different isozymes each having differing functions. One isozyme that has been isolated from fish hepatic microsomes is cytochrome P-450E, which appears to be inducible by PAH/PCB-type chemicals.[24] Cytochrome P-450E is likely an important isozyme in conducting EROD activity.

The background or uninduced levels of total cytochrome P-450 have been measured in a variety of fish and invertebrates and are similar. For example, Stegeman and Kloepper-Sams (1987)[44] reported levels ranging from 0.10 to 0.91 nmol/mg protein in mussels, crustaceans, fish, and rats. Average cytochrome P-450 content in four species of untreated fish ranged from 0.11 to 0.50 nmol/mg protein.[45] In the present study, the lowest mean level of total P-450 was 0.135 nmol/mg protein in the *P. stellatus*, very similar to results in other species and phyla. Induction of cytochrome P-450 by injection of the fish in the present study with BNF indicated a large potential for response in the *P. stellatus*. Relative to controls, BNF-injected fish showed a 2.4-fold increase in mean total P-450 content. In contrast, *Pseudopleuronectes americanus* injected with BNF showed an increase in mean total P-450 content of 1.3-fold over that of controls.[46] Mean liver microsomal cytochrome P-450 content ranged from 0.18 ± 0.05 to 0.53 ± 0.11 nmol/mg protein in *Platichthys flesus* sampled at four sites in Langesundfjord, Norway.[47] The difference in mean total P-450 content in feral fish between the sites with the highest and lowest mean values was 1.6-fold in *P. stellatus* (present study), 1.7-fold in *P. americanus*,[46] and 2.9-fold in *P. flesus*.[47]

Mean EROD activity in *P. flesus* increased roughly 14-fold (from 39 ± 19 to 547 ± 236 pmoles/min/mg protein) between the reference site and the most contaminated site in Langesundfjord, Norway.[47] The difference in mean EROD activity (units/min/nmol total P-450) between the highest and lowest sites was 4.6-fold in *P. stellatus* (present study), and 3.05-fold in *P. americanus*.[46] Mean EROD activity (units/min/ nmol total P-450) increased 4.8-fold in BNF-treated

P. stellatus relative to controls, whereas EROD activity in BNF-treated *P. americanus* changed 1.6-fold relative to controls. EROD activity in the deep-sea fish *Coryphaenoides armatus* averaged 1.175 ± 0.310 nmol/nmol P-450 in fish from the Hudson Canyon off New York and 0.178 ± 0.050 nmol/nmol P-450 in fish from the Carson Canyon off Newfoundland.[48]

The difference in mean P-450E content among sites was 13.9-fold in *P. stellatus* (present study) and 13.7-fold in *P. flesus*.[47] The difference between 3.5 ± 1.6 and 47.9 ± 18.7 pmol/min/mg protein in the activity of the P-450E isozyme in *P. flesus* from the reference and contaminated sites indicated that the fish were highly induced at the contaminated site.[47] All three of the responses (total P-450, P-450E, and EROD) paralleled the gradient in contamination in Langesundfjord, Norway reported by Addison and Edwards (1988).[49] All three appeared to be responsive to high molecular weight hydrocarbons (PAHs and PCBs) measured at the sites. None of the three was particularly responsive to the low molecular weight hydrocarbons in mesocosm exposures tested concurrently with analyses of feral fish.

In summary, it is apparent that the *P. stellatus* cytochrome P-450 system is highly responsive to exposures to organic compounds. Most of the measures indicated a significant difference between sites. Mean total P-450 and P-450E content and EROD activity were generally highest and estradiol 2-hydroxylase per nanomole P-450 was lowest in fish from the BK and/or OK sites which were located nearest known sources of potentially P-450-inducing contaminants. However, the data from all of these measures indicated only very weak correlations with contaminants accumulated in the livers of the fish. Among the endpoints evaluated in fish, within-site variability was lowest for total P-450 content and between-site discriminatory power was highest for EROD activity per mg protein. Experience with the *P. stellatus* confirmed the patterns in response observed with *P. americanus* in New England and *P. flesus* in Norway.

The analysis of AHH activity did not indicate significant differences between sampling sites, was intermediate in within-site variability, had relatively low discriminatory power, was not highly correlated with chemical concentrations in the tissues, and was weakly correlated with the EROD data. In previous studies of *P. stellatus* in the San Francisco Bay, relatively high AHH activities were observed in fish most highly contaminated with chlorinated hydrocarbons. Mean AHH activities in fish collected in 1983–1984 from the BK, OK, and SP sites were 95, 54, and 51 units (pmol 3-OH B[a]P/min/mg protein), respectively.[32] AHH activities were not suppressed in BK fish during gametogenesis (presumably, in response to exposure to organic contaminants), whereas they were suppressed in SP fish (presumably, an adaptation to avoid metabolism of steroid hormones). Hepatic AHH activity was negatively correlated with several measures of reproductive success. Some individual fish from the BK and OK sites have had AHH activities of over 300 units.[32]

Compared to *P. stellatus*, mean AHH activities in *P. americanus* from four sites in New England were relatively high (450 to 770 pmol B[a]P metabolites per minute per milligram protein) and uniform. There was a strong correlation between EROD and AHH activities in fish from the Buzzards Bay site.[46] Hepatic AHH activity in English sole (*Parophrys vetulus*) ranged from 36 to 330 pmol/mg protein/min in fish from a rural area in Puget Sound, Discovery Bay, known to be uncontaminated with aromatic hydrocarbons.[50] Fish from another area in Puget Sound with higher aromatic hydrocarbon concentrations had hepatic AHH activities that ranged from 330 to 570 pmol/mg protein/min. A strong positive correlation was observed between hepatic AHH activity and concentration of specific isozymes of cytochrome P-450. In a subsequent study, Johnson et al.[51] reported a range of 72 to 492 pmol/mg/min in mean AHH activity in *P. vetulus* from Puget Sound sites, corresponding to a 6.8-fold difference between sites. A range of 33.5 ± 21.2 to 90.2 ± 52.2 pmol B[a]P hydroxylase/min/mg protein was reported by Addison and Edwards[49] for *P. flesus* sampled along a pollution gradient in Langesundfjord, Norway. This range corresponded to a 2.7-fold difference between the least and most induced fish. Nearly a four-fold difference in response in the same species was recorded following exposure to a dilution series of oil and copper in mesocosm basins. Gender-specific and species-specific differences in AHH activities were observed in sanddabs collected in Southern California.[18] However, fish caught in areas near sources of contaminants or near natural oil seeps always had higher AHH activities than those caught in Monterey Bay, a relatively uncontaminated embayment.

In summary, it appears that AHH activity was less sensitive than some of the other measures, had relatively low discriminatory power, had moderate within-site variability, and generally can be responsive to exposure to hydrocarbons. AHH measurements complement other measures of the cytochrome P-450 system and are often correlated with them.

Fish Reproductive Success

The measures of steroid hormone content and fertilization/hatching/embryological success are important indicators of reproductive condition and success. They have been used successfully in other studies. However, in this evaluation the steroid hormone measures were relatively insensitive and the sample size available for the fertilization/hatching/embryological success tests was too small to provide interpretable data.

CONCLUSIONS

Because the contamination of marine areas near urban centers often results in complex mixtures of chemicals, biological responses to those mixtures should be expected to be equally (or more) complex. Some chemicals may be acutely lethal and a short-term bioassay of sediment could be a useful indicator of the bioavailability and toxicity of those chemicals. Other chemicals may induce subtle changes that are expressed or quantifiable over only long periods of time. Tests of mutagenicity and/or enzymatic response may be useful in evaluating exposure of biota to these chemicals. No single biological measure can be expected to suffice as the sole test or indicator of effects of complex mixtures of contaminants.

- The concentrations of many analytes in sediments differed by roughly an order of magnitude between the least and the most contaminated sampling sites.
- The toxicity test of abnormal development in *Mytilus edulis* larvae had the highest sensitivity, analytical precision, and discriminatory power of the tests that were evaluated, but the influence of sediment texture and organic carbon content should be evaluated in controlled laboratory experiments.
- Many of the whole-embryo *Strongylocentrotus purpuratus* test endpoints appeared to be less sensitive than those measured with *M. edulis* embryos, and some toxicological endpoints contradicted others in their pattern of response, but the cytogenetic endpoints appeared to indicate mutagenicity in sediment samples that had high hydrocarbon concentrations.
- The relatively highly tested and developed *Rhepoxynius abronius* test of survival had very high sensitivity, analytical precision, and discriminatory power; and the influence of fine-grained sediments upon the organism has been quantified; but, the endpoints of reburial and emergence appear to be very insensitive and highly variable, respectively.
- The relatively new test with *Ampelisca abdita* survival was less sensitive than some of the others (including that with the other amphipod *R. abronius*), but was insensitive to fine-grained sediments, had relatively high analytical precision, and moderate discriminatory power. It appears to be most useful in assessments of sediments known or suspected to be highly contaminated; but, the endpoint of emergence appears to be highly variable.
- The pore water bioassay with *Dinophilus gyrociliatus* was intermediate in sensitivity, precision and discriminatory power, the endpoint of egg production is ecologically important, and the results were correlated with hydrocarbon concentrations in the bulk sediments. Equilibrium partitioning theory suggests that the bioavailability and toxicity of chemicals in sediments is largely through exposure to the pore water, so tests of pore water toxicity should be very powerful tools. However, better methods of extracting sediment pore water that reduce the amount of organic contaminants that are removed should be developed to enhance the sensitivity of the pore water test.
- Fish from two sites in San Francisco Bay were generally more contaminated with a mixture of compounds than those from a coastal reference site and a reference site in the Bay. However, the distinction in the chemical concentrations between sites was not as clear as with the sediments.
- In general, the suite of total cytochromes P-450/EROD/P-450E measures in the livers of the fish appeared to be relatively sensitive, weakly correlated with some tissue contaminant data, relatively low in variability among fish from the same site, and inducible over a similar range in response among species. In either laboratory exposures or in feral fish, the suite of total P-450/EROD/P-450E endpoints appears to respond to contaminants similarly among the species

tested thus far. The P-450 suite of tests may be indicative of the recent exposure of feral fish to hydrocarbons, including those that may not be readily quantifiable. However, they also may indicate that the animals have successfully responded to xenobiotic exposure.

- The incidence of micronucleated erythrocytes in the fish was significantly lower in fish from the coastal reference site than in fish from most of the sites in San Francisco Bay. Incidences in fish among the sites in San Francisco Bay were not distinguishable. The determinations of micronuclei incidence in erythrocytes, particularly detached micronuclei, appeared to be relatively sensitive, intermediate in discriminatory power, but relatively highly variable among fish at a site.
- The suite of hepatic cytochrome P-450 measures and micronuclei counts, together, appeared to provide sensitive indicators of both exposure to and adverse effects of certain hydrocarbons.
- The measures of AHH activity were less sensitive than those of other measures of the cytochrome P-450 system, but provided data that complemented those from the specific P-450 analyses.
- Biochemical and other analyses that are indicative of reproductive impairment, but are quicker and less expensive than spawning studies, should be tested and evaluated further.

ACKNOWLEDGMENTS

Mr. Rick Wright and Mr. Nick Rottunda (SAIC) collected the sediment and benthos samples. Dr. Loveday Conquest (University of Washington) provided advice on statistical methods for data analyses. Mr. Kirk Van Ness (NOAA) assisted in sampling site selection. Ms. Charlene Swartzell (NOAA) typed the manuscript and revised it innumerable times. CDR Stewart McGee (NOAA) assisted in arranging vessel logistics. The following performed the chemical analyses of sediments at SAIC: Dr. Andrew Lissner, Dr. James Payne, Dr. John Clayton, Mr. Gary Farmer, Mr. Dan McNabb, Mr. Mike Guttman (organics); and Mr. R. Sims (trace metals) under the programmatic guidance of Dr. Paul Boehm and Ms. Sandra Freitas (Battelle Ocean Sciences). Dr. Peter Chapman, Ms. Roxanne Rousseau and Mr. Ian Watson (E.V.S. Consultants) performed the tests with *Rhepoxynius abronius* and *Mytilus edulis*. Dr. Steven Bay, Mr. Darrin Greenstein, Ms. Karen Rosenthal, and Ms. Valerie Raco (SCCWRP) and Jo Ellen Hose (Occidental College) conducted tests with *Strongylocentrotus purpuratus*. Dr. John Scott, Dr. Ronald Breteler, Ms. Michele Redmond, and Ms. Susan Shepard conducted the tests with *Ampelisca abdita*. Dr. R. Scott Carr, Mr. John Williams, and Mr. Carlos T. B. Frigata (Battelle Ocean Sciences Division) performed tests with *Dinophilus gyrociliatus*. Dr. Arthur Barnett, Ms. Susan Garner, Mr. Lawrence Lovell, Ms. Susan Watts, and Ms. Janice Callahan (MEC) conducted the taxonomic analyses of benthic samples. Dr. Donald Rhoads, Dr. Joseph Germano, Mr. Eugene Revelas, and Dr. Robert Merten (SAIC) performed the sediment profiling survey of San Francisco Bay. Drs. John Stegeman and Bruce Woodin and Ms. Lucia Susani (WHOI) performed the cytochrome

P-450 analyses of fish. Drs. Robert Spies and David Rice and Ms. Mari Prieta (LLNL) collected the fish, performed chemical and AHH analyses and tested reproductive success. Drs. Jeffrey Cross (SCCWRP) and Jo Ellen Hose (Occidental College) performed micronuclei analyses of fish blood. Dr. Peter Thomas (University of Texas) determined plasma steroid hormone concentrations in the fish. Drs. Howard Harris and Douglas Wolfe (NOAA) provided helpful comments on initial drafts of the manuscript.

REFERENCES

1. Bayne, B. L., K. R. Clarke, and J. S. Gray. "Background and Rationale to a Practical Workshop on Biological Effects of Pollutants," *Mar. Ecol. Prog. Ser.* 45:1–5 (1988).
2. "National Status and Trends Program: Progress Report and Preliminary Assessment of Findings of the Benthic Surveillance Project—1984," Office of Oceanography and Marine Assessment, National Oceanic and Atmospheric Administration, NOS/NOAA (1987), 81 pp.
3. Long, E.R. and M.F. Buchman. "An Evaluation of Candidate Measures of Biological Effects for the National Status and Trends Program," Office of Oceanography and Marine Assessment, National Oceanic and Atmospheric Administration, NOS OMA 45 (1989), 105 pp.
4. Spies, R. B., J. J. Stegeman, D. W. Rice, Jr., B. Woodin, P. Thomas, J. E. Hose, H. N. Cross, and M. Prieto. Chapter 5, this volume.
5. Barnett, A. M., R. G. Kanter, T. D. Johnson, and K. D. Green. "Analyses of Infaunal Community Structure from Four Sites in the San Francisco Bay Region," Report MEC05887002. NOAA 50-DGNC-6-00200 (Encinitas, CA: Marine Ecological Consultants of Southern California, 1987).
6. Revelas, E. C., D. C. Rhoads, and J. D. Germano. "San Francisco Bay Sediment Quality Survey and Analyses," NOAA Technical Memorandum NOS/OMA 35, National Oceanic and Atmospheric Administration (1987), 127 pp.
7. Spies, R. B., and D. W. Rice, Jr. "The Effects of Organic Contaminants on Reproduction of Starry Flounder, *Platichthys stellatus* (Pallas), in San Francisco Bay. II. Reproductive Success of Fish Captured in San Francisco Bay and Spawned in the Laboratory," *Mar. Biol.* 98(2):191–200 (1988).
8. Swartz, R. C., W. A. DeBen, J. K. P. Jones, J. O. Lamberson, and F. A. Cole. "Phoxocephalid Amphipod Bioassay for Marine Sediment Toxicity," in *7th ASTM Aquatic Toxicology and Hazard Assessment Symposium.* Special Technical Report 854, Milwaukee, WI, April 18, 1983. R. D. Cardwell, R. Purdy, and R. C. Bahner, Eds. (Philadelphia, PA: American Society for Testing Materials, 1985), pp. 284–307.
9. Chapman, P. M., and S. Becker. "Recommended Protocols for Conducting Laboratory Bioassays on Puget Sound Sediments," in Final Report TC-3991-04 United States Environmental Protection Agency Region 10, (1986), 55 pp.
10. Oshida, P. S., T. K. Goochey, and A. J. Mearns. "Effects of Municipal Wastewater on Fertilization, Survival, and Development of the Sea Urchin *Strongylocentrotus purpuratus*," in *Biological Monitoring of Marine Pollutants*, F. J. Vernberg, A. Calabrese, F. D. Thurberg, and W. B. Vernberg, Eds. (New York, NY: Academic Press, 1981) pp. 389–402.

11. Dinnel, P. A., J. M. Link, and Q. J. Stober. Improved Methodology for a Sea Urchin Sperm Cell Bioassay for Marine Waters. *Arch. of Environ. Contam. Toxicol.* 16: 23–32 (1987).

12. Hose, J. E. "Potential Uses of Sea Urchin Embryos for Identifying Toxic Chemicals: Description of a Bioassay Incorporating Cytologic, Cytogenetic, and Embryologic Endpoints," *J. Appl. Toxicol.* 5(3):245–254 (1985).

13. Bay, S. M., P. S. Oshida, and K. D. Jenkins. "A Simple New Bioassay Based on Echinochrome Synthesis by Larval Sea Urchins," *Mar. Environ. Res.* 8:29–39 (1983).

14. Carr, R. S., M. D. Curran, and M. Mazurkiewicz. "Evaluation of the Archiannelid *Dinophilus gyrociliatus* for Use in Short-term Life-cycle Toxicity Tests," *Environ. Toxicol. Chem.* 5:703–712 (1986).

15. MacLeod, W. D., Jr., D. W. Brown, A. J. Friedman, D. G. Burrows, O. Maynes, R. W. Pearce, C. A. Wigren and R. G. Bogar, "Standard Analytical Procedures of the NOAA National Analytical Facility, 1985–6: Extractable Toxic Organic Compounds (Second Edition)," NOAA Technical Memorandum NMFS/NWC-92, NOAA/ NMFS (1985), 121 pp.

16. Preece, A. *A Manual for Histologic Technicians.* 3rd ed. (Boston, MA: Little, Brown and Co., 1972).

17. Nebert, D. W. and H. V. Gelboin. "Substrate-inducible Microsomal Aryl Hydrocarbon Hydroxylase in Mammalian Cell Culture," *J. Biol. Chem.* 243:6242–6249 (1968).

18. Spies, R. B., J. S. Felton, and L. Dillard. "Hepatic Mixed-Function Oxidases in California Flatfishes are Increased in Contaminated Environments and by Oil and PCB Ingestion," *Mar. Biol.* 70:117–127 (1982).

19. Lowry, O. H., N. J. Rosebrough, A. L. Farr, and R. J. Randal. "Protein Measurement with the Folin Phenol Reagent," *J. Biol. Chem.* 193:265–275 (1951).

20. Stegeman, J. J., R. L. Binder, and A. Orren. "Hepatic and Extrahepatic Microsomal Electron Transport Components and Mixed-function Oxygenases in the Marine Fish *Stenotomus versicolor*," *Biochem. Pharmacol.* 28:3431–3439 (1979).

21. Klotz, A. V., J. J. Stegeman, and C. Walsh. "An Alternative 7-Ethoxyresorufin O-deethylase Activity Assay: A Continuous Visible Spectrophotometric Method for Measurement of Cytochrome P-450 Monoxygenase Activity," *Anal. Biochem.* 140:138–145 (1984).

22. Burke, M. D., S. Thompson, C. R. Elcombe, J. Halpert, T. Haaparanta, and R. T. Mayer. "Ethoxy-, Pentoxy- and Benzyloxyphenoxazones and Homologues: A Series of Substrates to Distinguish Between Different Induced Cytochromes P-540," *Biochem. Pharmacol.* 34:3337–3345 (1985) .

23. Kupfer, D., G. K. Miranda, and W. H. Bulger. "A Facile Assay for 2-hydroxylation of Estradiol by Liver Microsomes," *Anal. Biochem.* 116:27–34 (1981).

24. Klotz, A. V., J. J. Stegeman, and C. Walsh. "An Aryl Hydrocarbon Hydroxylating Hepatic Cytochrome P-450 from the Marine Fish *Stenotomus chrysops*," *Arch. Biochem. Biophys.* 226:578–592 (1983).

25. Park, S. S., H. Miller, A. V. Klotz, P. K. Kloepper-Sams, J.J. Stegeman, and H. V. Gelboin. "Monoclonal Antibodies Against Cytochrome P-450E from the Marine Teleost *Stenotomus chrysops* (scup)," *Arch. Biochem. Biophys.* 249:339–350 (1986).

26. Kloepper-Sams, P. K., S. S. Park, H. V. Gelboin, and J. J. Stegeman. "Specificity

and Cross-reactivity of Monoclonal and Polyclonal Antibodies Against Cytochrome P-450E of the Marine Fish Scup," *Arch. Biochem. Biophys.* 253:268–278 (1987).

27. Ozretich, R. J. and W. P. Schroeder. "Determination of Neutral Organic Priority Pollutants in Marine Sediment, Tissue, and Reference Materials Utilizing Bonded-Phase Sorbent," *Anal. Chem.* 58:2041–2047 (1986).

28. Zar, J. H. *Biostatistical Analysis,* (Englewood Cliffs, NJ: Prentice-Hall, Inc., Pear River, NJ: Lederle Laboratories, 1984).

29. Wilcoxon, F. and R. A. Wilcox. *Some Rapid Approximate Statistical Procedures.* Pear River, NJ: Lederle Laboratories (1964).

30. Spearman, C. "The Proof and Measurement of Association Between Two Things," *Amer. J. Psychol.* 15:72–101 (1904).

31. Boehm, P. D. and J. W. Farrington. "Aspects of the Polycyclic Aromatic Hydrocarbon Geochemistry of Recent Sediments in the Georges Bank Region," *Environ. Sci. Tech.* 13(11):840–845 (1984).

32. Spies, R. B., D. W. Rice, Jr, P. A. Montagna, R. R. Ireland, J. S. Felton, S. K. Healy, and P. R. Lewis. "Pollutant Body Burdens and Reproduction in *Platichthys stellatus* from San Francisco Bay," Report No. UCID-20386. (Livermore, CA: Lawrence Livermore National Laboratory, 1985), 95 pp.

33. Chapman, P. M., R. N. Dexter, and E. R. Long. "Synoptic Measures of Sediment Contamination, Toxicity, and Infaunal Community Composition (The Sediment Quality Triad) in San Francisco Bay," *Mar. Ecol. Prog. Ser.* 37:75–96 (1987).

34. Williams, L. G., P. M. Chapman, and T. C. Ginn. "A Comparative Evaluation of Marine Sediment Toxicity Using Bacterial Luminescence, Oyster Embryo, and Amphipod Sediment Bioassays," *Mar. Environ. Res.* 19:225–249 (1986).

35. Swartz, R. C., W. A. DeBen, and F. A. Cole. "A Bioassay for the Toxicity of Sediment to Marine Macrobenthos," *J. Water Poll. Con. Fed.* 51:944–950 (1979).

36. Mearns, A. J., R. C. Swartz, J. M. Cummins, P. A. Dinnel, P. Plesha, and P. M. Chapman. "Inter-laboratory Comparison of a Sediment Toxicity Test Using the Marine Amphipod *Rhepoxynius abronius*," *Mar. Environ. Res.* 19:13–37 (1986.).

37. DeWitt, T. H., G. R. Ditsworth, and R. C. Swartz. "Effects of Natural Sediment Features on Survival of the Phooxocephalid amphipod, *Rhepoxynius abronius*," *Mar. Environ. Res.* 25(2):99–124 (1988).

38. Rogerson, P. F., S. C. Schimmel, and G. Hoffman. "Chemical and Biological Characterization of Black Rock Harbor Dredged Material," Field Verification Program Technical Report D-35-9 (Vicksburg, MS: US Army Corps of Engineers Waterways Experiment Station, 1985), 110 pp.

39. Dinnel, P. A. and Q. J. Stober. "Application of the Sea Urchin Sperm Bioassay to Sewage Treatment Efficiency and Toxicity in Marine Waters," *Mar. Environ. Res.* 21:121–133 (1987).

40. Nacci, D., E. Jackim, and R. Walsh. "Comparative Evaluation of Three Rapid Marine Toxicity Tests; Sea Urchin Early Embryo Growth Test, Sea Urchin Sperm Cell Toxicity Test and Microtox," *Environ. Toxicol. Chem.* 5:521–525 (1986).

41. Hose, J. E., P. S. Puffer, P. S. Oshida, and S. M. Bay. "Development and Cytogenetic Abnormalities Induced in the Purple Sea Urchin by Environmental Levels of Benzo(a)pyrene," *Arch. Environ. Contam. Toxicol.* 12:319–325 (1983).

42. Longwell, A. C., D. Perry, J. B. Hughes, and A. Hebert. "Frequencies of Micronuclei in Mature and Immature Erythroeytes of Fish as an Estimate of Chromosome Mutation Rates—Results of Field Surveys on Windowpane Flounder, Winter Floun-

der, and Atlantic Mackerel," 1983/E:55 Informal manuscript, National Marine Fisheries Service, NOAA (1983), 14 pp.

43. Hose, J. E., J. F. Cross, S. G. Stevens, and D. Diehl. "Elevated Circulating Erythrocyte Micronuclei in Fishes from Contaminated Sites Off Southern California," *Mar. Environ. Res.* 22:167–176 (1987).

44. Stegeman, J. J. and P. K. Kloepper-Sams. "Cytochrome P-450 Isozymes and Monooxygenase Activity in Aquatic Animals," *Environ. Health Persp.* 71:87–95 (1987).

45. James, M. O. and J. R. Bend. "Polycyclic Aromatic Hydrocarbon Induction of Cytochrome P-450-Dependent Mixed-Function Oxidases in Marine Fish," *Toxicol. Appl. Pharmacol.* 54:117–133 (1980) .

46. Stegeman, J. J., F. Y. Teng, and E. A. Snowberger. "Induced Cytochrome P-450 in Winter Founder (*Pseudopleuronectes americanus*) from Coastal Massachusetts Evaluated by Catalytic Assay and Monoclonal Antibody Probes," *Can. J. Aquat. Sci.* 44:1270–1277 (1987).

47. Stegeman, J. J., B. R. Woodin, and A. Goksoyr. "Apparent Cytochrome P-450 Induction as an Indication of Exposure to Environmental Chemicals in the Flounder *Platichthys flesus*," *Mar. Pollut. Bull.* 18(10):359–364 (1988).

48. Stegeman, J. J., P. K. Kloepper-Sams, and J. W. Farrington. "Monooxygenase Induction and Chlorobiphenyls in the Deep-sea Fish *Coryphaenoides armatus*," *Science* 231:1287–1289 (1986).

49. Addison, R. F. and A. J. Edwards. "Hepatic Microsomal Monooxygenase Activity in Flounder (*Platichthys flesus*) from Polluted Sites in Langesundfjord and from Mesocosms Experimentally Dosed with Diesel Oil and Copper," *Mar. Ecol. Prog. Ser.* 46:51–54 (1988).

50. Varanasi, U., T. K. Collier, D. E. Williams, and D. R. Buhler. "Hepatic Cytochrome P-450 Isozymes and Aryl Hydrocarbon Hydroxylase in English sole (*Parophrys vetulus*)," *Biochem. Pharmacol.* 35(17):2967–2971 (1986).

51. Johnson, L. L., E. Casillas, T. K. Collier, B. B. McCain, and U. Varanasi. "Contaminant Effects on Ovarian Development in English sole (*Parophrys vetulus*) from Puget Sound, Washington," *Can. J. Fisheries Aquat. Sci.* 45:2133–2146 (1988).

52. Breteler, R., et al. "Application of a New Sediment Toxicity Test Using a Marine Amphipod *Ampelisca abdita*, to San Francisco Bay Sediments," in *Aquatic Toxicology and Hazard Assessment: 12th Volume*, ASTM STP 1027, U. M. Cowgill, and L. R. Williams, Eds. (Philadelphia: American Society for Testing Materials, 1989), pp. 304–314.

53. Scott, J., and Redmond, M. S. "The Effects of Contaminated Dredged Material on Laboratory Populations of the Tubiculous Amphipod *Ampelisca abdita*," in *Aquatic Toxicology and Hazard Assessment: 12th Volume*, ASTM STP 1027, U. M. Cowgill and L. R. Williams, Eds. (Philadelphia: American Society for Testing Materials, 1989), pp. 289–303.

54. Carr, R. S., J. W. Williams, and C. T. B. Frigata. "Development and Evaluation of a Novel Marine Sediment Pore Water Toxicity Test with the Polycheate *Dinophilus gyrociliatus*," *Environ. Toxicol. Chem.* 8:533–543 (1989).

55. DiToro, D. M. "A Review of the Data Supporting the Equilibrium Partitioning Approach to Establishing Sediment Quality Criteria," in *Symposium/Workshop on Contaminated Marine Sediments.* Tampa, FL., May 31–June 2, 1988 (Washington, DC: Marine Board, National Research Council, 1989), pp. 100–114.

CHAPTER 21

Disease Biomarkers in Large Whale Populations of the North Atlantic and Other Oceans

R.H. Lambertsen

ABSTRACT

Interpretation of the possible effects of pollution on the health of marine animals requires baseline information from reference populations in clean environments. This paper reviews literature on the natural diseases and infections of large whales with an emphasis on two filter-feeding species inhabiting the central North Atlantic (fin whale, *Balaenoptera physalus*, and sei whale, *B. borealis*). Despite evidence that the abundance of North Atlantic fin and sei whales is now depressed as a result of commercial exploitation, both suffer from serious endemic infections with incidences exceeding 90 percent. Pathological and biochemical findings lead to a conclusion that in severe infections the principal nematode parasite of the fin whale probably can kill its host. Recognizing a positive relationship between population density and exposure to endemic pathogens, it is cautioned that pollution-driven mortality events in expanding marine populations could be mimicked, masked, or amplified by natural phenomena.

INTRODUCTION

One pressing concern about environmental contamination is its possible effect on the health and survival of endangered marine mammals. In the past few years there has been a dramatic increase in the number of strandings and apparent mortality rates of certain species of Cetacea (whales, dolphins, porpoises) along the eastern seaboard of the United States. The possibility that pollution in coastal waters has caused immunosuppression, leading to increased mortality from disease, is one plausible explanation.

To interpret measurements of disease processes as biomarkers of environmental pollution requires historical information on the health status of a population or comparable data from unexposed individuals. This paper will review past literature on the diseases of large whales. The results of a recent international effort to analyze the endemic and epidemic disease problems of two species of baleen whales, the fin whale and the sei whale, will then be summarized. In this research American veterinary teams conducted necropsy examinations of over 150 fin whales and 42 sei whales at the Icelandic Whales Research Laboratory during the period of 1981–1987 in cooperation with the Institute of Experimental Pathology, University of Iceland. Caught in whaling operations off the west coast of Iceland during the 1981–1987 period, the whales examined represent migratory populations that inhabit the central zone of the North Atlantic. Gas chromatographic analyses of tissue contamination with organochlorine pesticides and measurements of basic epidemiologic and hematologic biomarkers of disease will be presented. From these findings cautionary tales will be told concerning the study of cetacean morbidity and mortality in relation to the possible immunosuppressive effects of pollution.

REVIEW OF HISTORICAL LITERATURE

The major previous information on the diseases of great whales comes from two independent investigations conducted in the Southern Hemisphere.[1,2] Neither provides quantitative data or frequency estimates relating to disease, and only the more recent work has provided more than a verbal account of its findings. The first of these was conducted by W. Ross Cockrill who, while a student of veterinary medicine, was Veterinary Officer in charge of meat inspection on S.S. 'Southern Venturer' during the 1948/49 whaling season, and subsequently on the S.S. 'Balaena' during 1951/52. In his veterinary thesis and three publications extending from this work, Cockrill describes the general operations on board factory ships, the methods and requirements for the inspection of whale meat, and pathologic findings relating to the blue (*Balaenoptera musculus*), fin, sei, sperm (*Physeter catodon*), and humpback (*Megaptera novaeangliae*) whales taken in the waters off Antarctica during the 1947–52 period.[1,3,4]

Surprisingly, Cockrill[1] states: "In the Antarctic pelagic seasons of 1947/48, 1948/49, 1949/50, and 1951/52, inspectorate staffs consisting, in all, of six veterinarians and eight lay meat inspectors examined some 12,000 whale carcasses without finding any evidence of specific disease." Working conditions, however, were such that this might be expected. Cockrill explains that he and his co-workers operated on the confined decks of floating factories, in the midst of intense flensing activity. They typically had 30–40 min to examine animals which, relative to their northern conspecifics, are considered giants.[1]

Some years subsequent to his first paper, however, Cockrill listed several types of parasitic infections in blue, fin, humpback, and sperm whales based on his collections from a total of 13 animals.[3] This collection had previously been given to and described in detail by Rees.[5] According to Rees, the Cestoda found included: *Priapocephalus grandis*, from the small intestine of the sperm whale; *Tetrabothrius affinis*, from the small intestine and major bile duct of the blue whale; *T. ruudi*, from the small intestine of the fin whale; *T. wilsoni*, from a sperm whale (site unrecorded); and an unidentified *Tetrabothrius* species from the intestine of the blue whale. Markowski[6] later characterized the unidentified species, naming it *T. schaeferi*. Acanthocephala included *Bolbosoma hamiltoni*, from the large intestine of a fin whale, and *B. porrigens*, from the large intestine of two humpback whales.[5] Nematodes included *Anisakis physeteris*, from the first and second gastric compartments of the sperm whale, and *Crassicauda crassicauda*, the smallest of several congeneric species of giant spirurid nematodes which commonly infect the urogenital system of cetaceans.[3,5–12]

Cockrill[3] described at some length the renal lesions produced by *Crassicauda*. The lesions were characterized by large digitate masses of adult worms encased by a fibrotic tissue reaction infiltrated with inflammatory cells and mineral deposits. Hamilton[9] had earlier described the extension of this type of lesion from the kidney to the renal vein in fin and blue whales, while Rewell and Willis[12] had described a similar lesion in the 'Discovery' collection that protruded from the renal vein into the vena cava. In addition, Cockrill[1,3,4] described a number of benign tumors, abscesses, various traumatic lesions, and malformation of a blue whale fetus; he also reported that in the course of his studies no lesions whatsoever had been observed in the reproductive or respiratory organs.

Critically evaluated, however, this appears to be due to an investigative effort which did not include thorough postmortem examination. Cockrill's primary responsibilities were the inspection of whale meat for human consumption.[1] Given the exigencies of the operational conditions under which he worked, systematic examination of the internal viscera was not performed. Cockrill does not describe the method he employed in his examinations but states that even the external surfaces of the lungs could not be inspected thoroughly due to logistical difficulties associated with the flensing operation.[1]

In the second major study of diseases in great whales, Uys and Best[2] described lesions encountered during the biologic examination of over 2000 carcasses of sperm, sei, Bryde's (*B. edeni*), and blue whales landed at the Saldanha Whaling Station, Saldanha Bay, South Africa. These lesions include: multiple uterine fibromyomata in a sperm whale, a benign melanotic tumor in a sei whale, focal muscle and fat necrosis in 11 sei whales and 1 Bryde's whale, a necrotic renal mass in a Bryde's whale, a necrotic endometrial mass in a sei whale, caseous necrosis in the testis and epididymis of a blue whale, ductal obstruction of 1 mammary gland of a sei whale, hepatic siderosis in a sei whale, and multiple perianal lymphoid masses in a sperm whale. Histopathologic findings on these lesions are presented. Again, however, the method for postmortem examinations was not described and it is unlikely that a necropsy protocol was followed. Best collected the specimens while acting as the biological inspector for the whaling station,[2] a position which entails other time-consuming responsibilities. More recently, Best and McCully[13] described one case of phycomycosis of the epaxial musculature of a southern right whale calf (*Balaena australis*).

Other accounts of great whale pathology have helped to identify, but not quantify, disease factors in great whales. MacKintosh[14] reviewed the disease findings of the early 'Discovery' investigations on whales. Because these findings have not been published by the primary observers, with the exception of the report of Rewell and Willis cited above,[12] I quote MacKintosh:[14] "Observations made by members of the 'Discovery' Committee's staff and by various whaling inspectors may be summarized as follows. Externally, callosities and apparently bony protuberances, sometimes of considerable size, have been noted, and in one Blue whale the jaw bones were deformed. Major Spencer (in the 'Southern Princess') noticed on a number of whales certain circular patches suggestive of ringworm, and some similar marks have been described on the skin of whales at South Georgia. In one or two whales an opaque lens has been noted in one eye. Healed wounds, sometimes probably caused by harpoons, are not uncommon; a flipper has occasionally been lost; and the dorsal fin, especially it seems in male Blue whales, is not infrequently truncated or deformed. Internally the pathological features most commonly noticed are various growths, tumors, nodules, cysts, etc., which may vary in size from an inch or two to two feet in diameter. Sometimes they are described as hard, bony, or calcareous, and sometimes as containing a pus-like fluid. Such bodies of one kind or another have been found loose in the abdominal cavity on several occasions, but have been noted also in the peritoneum, in the stomach, in the uterine mesentery, in the wall of the rectum, in the kidney, between the lungs and pleura, in the trachea, in connective tissue under the blubber, in the shoulder, and even between the vertebrae. Among examples of this kind, Dr. E. H. Marshall, who visited the Ross Sea in the factory ship 'C. A. Larsen', noted multiple fibromata of the trachea (one being "larger than a cricket ball"), a large abscess involving the shoulder joint, and *a very large pyonephrosis* [my emphasis]. On several occasions a form of pleurisy has been noticed in which there were extensive adhesions of the lungs to the pleura."

Clarke,[15] in a treatise on whaling in the Azores, summarized the diseases, deformities, and injuries known in the sperm whale on the basis of his examination of 37 animals and a review of earlier findings. He describes white circular lesions on the flanks of two males and suggested "ringworm" as a possible etiology. Matthews[16] had earlier described similar lesions associated with a fungus of the genus *Trichophyton*. Clarke[15] also found firm ovoid tumors on the floor of the mouth of two whales, and a fibroma on the lower jaw of another.

The occurrence of ambergris, a pathological fecal impaction of the colon, was noted. Deformities included the absence of teats in one whale, and the occurrence of a spiral deformity of the lower jaw. The latter type of deformity is known from whaling records to be common in this species. Injuries included scars on the head region from the hooks and suckers of principal prey species of the sperm whale, teuthid squids, and a scalloping of the posterior border of the flukes suggested to be due to the bites of fish. Parasitic conditions in this sperm whale population included infection of the first and second stomach compartments with *Anisakis physeteris,* and infections of the intestines with *Priapocephalus grandis.* The blubber was commonly infested with small cysticercoids identified as the cestode species *Phyllobothrium physeteris.*

Comprehensive lists of the parasites of cetaceans are given by Baylis[17] and Dailey and Brownwell.[18] Markowski[6] describes the cestodes of whales collected during the early 'Discovery' expeditions.

Recently, Albert and co-workers have undertaken a major veterinary study of the bowhead whale (*Balaena mysticetus*) in conjunction with the yearly take of small numbers of this species for subsistence by the Inupiat Eskimos of Alaska.[19] The primary objective of that effort, however, has not been a study of disease per se, but rather, the correlation of basic biologic and physiologic function to anatomic structure. Pathology has been pursued in detail only on an incidental basis largely because of the extreme adversity of the polar conditions in which the work was done. Two reports on disease in the bowhead whale have been published. In a histopathologic study, Migaki and Albert[10] described the occurrence of a small, benign lipoma in the liver of a *B. mysticetus* calf. Migaki et al.[21] described a small cyst in the wall of the first stomach compartment of a bowhead whale caused by a parasite of the genus *Contracecum.*

Heckman[22] describes the parasitologic fauna of bowhead whales taken by Inupiat whale hunters. In his report he identifies two protozoans, one species of trematode, two nematode species, and one species of amphipod from tissue and fecal samples. Localized skin damage was associated with the diatoms and amphipod. In addition, a localized but prominent inflammatory response in the forestomach of one whale was found in association with a larval nematode of the anisakid type.

Microbial investigations of the bowhead whale were conducted by Johnston and Shum[23] and Smith et al.[24] The former investigators identified 19 bacterial species from swabs of various tissue sites. The latter isolated two adenoviruses from colonic tissue samples. However, there exists no evidence from whales that these microbial agents actually produce systemic disease. Serology revealed the presence of serum neutralizing activity, presumably antibody, to five serotypes of calicivirus,[24] suggesting natural caliciviral infections of the bowhead whale and the development of an immune response. Sera from the four bowhead whales examined revealed no activity against *Leptospira interrogans* bacterial antigens or a battery of 22 different types of influenza virus.

Two recent reports provide information on disease conditions found in male sperm whales inhabiting North Atlantic waters. In a study of 31 sperm whale bulls taken in whaling operations off the west coast of Iceland in 1981–82, three whales showed proliferative epidermal lesions identified by light and electron microscopy as virus-associated genital papillomas.[25] In one of the older animals in this group an unusual combination of pathologic lesions was documented.[26] Postmortem findings included heavy combative scarring of the head, dermal fistulae scattered over much of the body surface, extensive virus-associated genital papillomatosis, apparent extension of papillomatous lesions over the skin of the lower left flank, dark macular lesions of the epidermis infested with pennate diatoms, partial duodenal obstruction with a man-made foreign body (plastic bucket), colo-rectal obstruction with ambergris, localized degeneration with hy-dronephrosis of the right kidney, and a deeply ulcerative gastric nematodiasis associated with invasive *Anisakis spp*. It was concluded that the complex of problems seen in this animal were related both to artificial degradation of its habitat and health risks naturally associated with its ecology and senescence.

A lethal condition associated with the ingestion of plastic marine debris was noted in another sperm whale bull examined at the Icelandic Whales Research Laboratory in 1982. The first and second stomach compartments of this animal were found to contain a large polypropylene fishing net (otter trawl) attached to a coil of polypropylene line about 50 m long. These foreign bodies severely distended both stomach compartments involved and completely obstructed the proximal alimentary tract. Other findings included body emaciation and a dense, severely atrophic liver about one third normal size. Together these observations indicated that this sperm whale bull was in an advanced state of starvation at the time of capture due to ingestion of a large plastic foreign body.[27] In the entire study population of 32 sperm whales systematic necropsies revealed that at least 12 (37.5 percent) had ingested some form of plastic or metallic debris of human origin.[28]

NORTH ATLANTIC FIN AND SEI WHALES

"When we experience the conflicts ever more deeply, we are living the truth. The quiet conscience is an invention of the devil."

Albert Schweitzer

Independent efforts began in 1981 to quantify disease biomarkers in fin and sei whales taken in commercial operations off the west coast of Iceland. These were initiated by a group of American veterinary scientists out of concern that conventional population models used in fisheries management were significantly overestimating the recovery rates and recovery potentials of exploited whale stocks.[29] This recognized persistent misconceptions concerning the extent and severity of disease in large whale populations and their probable contribution to overly optimistic stock projections. Reinforcing such misconceptions are generalized claims that efficient transmission of pathogens is unlikely in the marine environment.[30]

Theoretical arguments were advanced at the international level by the Institute of Biomedical Aquatic Studies, warning that intensive exploitation may paradoxically *increase*, not decrease, natural mortality rates in whale herds.[31] This possibility was related to a demographic shift in herd composition toward young age classes typically caused by heavy exploitation. It was cautioned that the immunological naiveté of immaturity could significantly increase herd-level mortality rates in depleted stocks even if food availability per capita improved. In apparent confirmation, a shipboard survey some years later found far fewer numbers of blue and fin whales in Antarctic waters than expected on the basis conventional models, despite 23 and 13 years of total protection, respectively.[32]

The obvious contribution that ignorance of disease factors could make toward excessive optimism in conventional whale population models led to an intensified veterinary effort to investigate these factors in greater detail.[31] Scientists from many nations asserted that if catches were to be continued by national governments for research purposes during an indefinite ban on commercial harvests then disease studies were important and critical.[29] The government of Iceland agreed to allow these specialized studies in parallel with its separate Whale Research Programme, 1986–1990. The Icelandic program, carried out by the Marine Research Institute (Reykjavik), primarily involved fisheries biology of the type which produced today's conventional population models. Iceland's effort proved controversial and led to well known acts of international sabotage against its whaling fleet by a militant conservation group.[33]

Ironically, interference directed by another such group[33] succeeded only in stopping those veterinary studies being attempted on the fin and sei whales taken by Iceland for its program in fisheries research. The Icelandic research catch continued virtually unhindered. The result was that a multidisciplinary conservation effort that many scientists believed necessary if Iceland's program was to be undertaken (Figure 1) was forced to a halt. Also stopped were quantitative studies aimed at establishing objective baseline data on biomarkers of population health that future generations will need to measure, and thereby understand, critical growth processes in ocean ecosystems.

It is notable that the veterinary research suppressed[33] was aimed in addition at validating new methods for acquiring health-related information from large whale populations using nonlethal sampling techniques (remote biopsy sampling systems). Building on the pioneering efforts of Winn et al.,[34] these specialized sampling methods had been further refined and field proven in population-level studies of large whale genetics by the Institute of Biomedical Aquatic Studies.[35,36] It had been hoped that completion of the essential validation studies on disease biomarkers would enable these virtually harmless methods to supplant others which necessarily relied on whaling.

With this historical perspective comes a first cautionary tale about the evaluation of disease biomarkers in whales. When working toward the same goal as

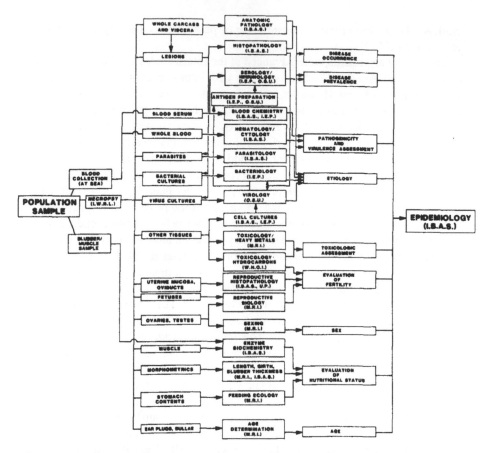

Figure 1. Multidisciplinary research plan to evaluate health status of North Atlantic fin and sei whale populations. I.W.R.L., Icelandic Whales Research Laboratory; I.B.A.S., Institute for Biomedical Aquatic Studies (University of Florida): I.E.P., Institute of Experimental Pathology (University of Iceland): M.R.I., Marine Research Institute (Reykjavik); O.S.U., Oregon State University; W.H.O.I., Woods Hole Oceanographic Institution.

politically active groups it may at times prove expedient, although certainly irresponsible, to forsake one's deeply held scientific convictions. By this means a more serene life no doubt will be led by avoiding all risk of timely or otherwise significant involvement. Yet some still should refuse to embrace that devil of reticence known (and rejected) by the thoughtful Schweitzer.

Let us hope that national governments also can understand that to choose ignorance out of political expedience is to forsake very serious human responsibilities to the future. It is the lack of information, not its acquisition, that engenders risk.

Table 1.   Spike Recoveries for Organochlorine Pesticides in Whales.

Pesticide[a]	Recovery (%) for Blank Spikes[b]			Recovery (%) for Blubber Spikes[c]	
	112783-A	112783-B	112783-C	120683-B	120683-C
HCB	91.6	102	91.8	80.4	98.7
-BHC	105	105	102	95.1	114
-BHC	99.1	104	105	98.4	115
-BHC	103	105	104	102	118
OChlor	103	109	106	90.7	113
Hepta epox	53.0	66.4	71.7	68.2	78.4
-Chlor	104	103	104	96.0	111
T-Nona	104	103	103	103	119
-Chlor	104	104	104	102	121
o,p'-DDE	98.9	108	101		
p,p'-DDE	94.6	107	102	65.4	77.2
o,p'-DDD	109	106	108	101	118
o,p'-DDT	96.3	99.3	97.5	98.2	118
p,p'-DDD	105	109	108	98.9	124
p,p'-DDT	100	102	108	101	116

[a]See Table 2 for full names of pesticides.
[b]Procedural spike into methylene chloride.
[c]Procedural spike into whale F-199 blubber.

From Dobskey, P., R. H. Lambertsen, and J. Farrington. Unpublished results (1984).

Toxicology

Preliminary gas chromatographic determinations of organochlorine pesticide concentrations were conducted on tissue samples from three fin whales and two sei whales.[37] These measurements were made by Dr. Paul Dobskey in the laboratory of Dr. John Farrington of the Woods Hole Oceanographic Institution. Analytical methods employed soxhlet extraction of either wet tissue with acetonitrile followed by methylene chloride or of freeze-dried tissue by methylene chloride to extract lipids. Depending on the moisture and lipid contents of the samples, variations on standard procedures using drying agents such as $MgSO_4$ or Na_2SO_4 were employed. Lipid extracts were concentrated by combinations of rotary evaporation and the use of Kuderna-Danish concentrators. Concentrated lipid extracts were then chromatographed on combinations of silica and fluorisil columns to achieve desired separations of pesticides into appropriate fractions (usually two fractions) devoid of most lipids except some selected biogenic hydrocarbons such as pristane, n-heptadecane, phytadienes, and squalene. Column chromatography fractions were then analyzed by gas chromotography on high resolution glass columns installed in Carlo-Erba Model 4160 GC's equipped with a Ni-63 electron detector.

Table 1 presents some of the recovery data associated with analyses of Icelandic whale samples. Tables 2 and 3 present data on organochlorine pesticide levels

Table 2. Organochlorine Pesticides in Whales $\left(\dfrac{\text{Nanogram } (10^{-9} \text{ g})}{\text{Gram wet weight}}\right)$ (Blubber, Except as Indicated).

Pesticide[a]	F-155[b]	F-155	F-155[c] Muscle	F-155[c] Liver	F-156	F-199	F199	SE-200	SE-278
HCB	302	257	2.8	11.7	296	119	92.5	25.3	40.4
α-BHC	87.9	106	1.1	4.8	88.9	30.2	35.3	34.9	48.6
γ-BHC	13.2	13.5	0.19	0.31	12.9	4.3	4.3	5.9	8.0
β-BHC	22.0	22.6	0.81	10.5	19.9	7.6	6.1	4.7	6.7
OChlor	48.8	54.5	0.85	1.8	46.5	23.3	19.9	10.4	13.5
Hepta epox	50.1	52.1	1.1	2.9	41.4	23.9	20.2	18.2	18.2
γ-Chlor					6.4	8.5	5.0		
T-Nona	232	284	4.5	7.0	214	116	80.5	34.8	51.0
α-Chlor	51.5	52.7	1.5	2.9	48.1	23.7	21.7	8.1	9.6
p,p'-DDE	1300	1540	19.2	32.6	1246	510	482	134	289
o,p'-DDD	150	167	3.7	15.4	126				
o,p'-DDT	692	743	10.0	1.4	585	204	159	60.4	110
p,p'-DDD	518	594	10.0	19.9	442	207	197	76.6	117
p,p'-DDT	437	680	5.3	5.5	494	164	139	95.2	201

[a]HCB—Hexachlorobenzene; α-BHC—Hexachlorocyclohexane, alpha isomer; γ-BHC—Gamma isomer; β-BHC—Beta isomer; Ochlor—Oxychlordane; α-Chlor—Cis-chlordane; γ-Chlor—Trans-chlordane; Hepta epox—Heptachlor epoxide; T-Nona—Trans nonachlor.
[b]See Table 3 legend for whale description.
[c]Freeze-dried weight, all other samples are blubber.

Table 3. Organochlorine Pesticides in Whales: Concentrations Normalized to Lipids $\left(\dfrac{\text{Nanogram }(10^{-9}\text{ g})}{\text{Gram Methylene}}\right)$; (Chloride Extractable Lipids).

Pesticide[a]	F-155[b]	F-155	F-155 Muscle	F-155 Liver	F-156[c]	F-199[d]	F199	SE-200[e]	SE-278[f]
HCB	396	257	55.1	68.5	366	147	115	30.9	48.8
α-BHC	115	106	21.6	28.3	110	37.4	44.1	42.5	58.7
γ-BHC	17.3	13.5	3.6	1.8	16.0	5.3	5.4	7.2	9.6
β-BHC	28.9	22.6	16.0	61.0	24.6	9.4	7.7	5.7	8.1
OChlor	64.0	54.5	16.6	10.5	57.4	28.9	24.8	12.7	16.3
Hepta epox	65.7	52.1	21.7	17.1	51.1	29.6	25.2	22.2	22.0
γ-Chlor					7.9	10.6	6.2		
T-Nona	304	284	89.0	40.7	264	144	100	42.4	61.6
α-Chlor	67.6	52.7	28.7	16.6	59.4	29.4	27.1	9.9	11.7
p,p'-DDE	1710	1542	378	190	1540	632	602	163	349
o,p'-DDD	197	167	72.3	89.6	156				
o,p'-DDT	908	743	196	39.8	722	253	198	73.7	133
p,p'-DDD	680	594	196	116	547	257	246	93.5	141
p,p'-DDT	574	680	104	32.0	610	203	173	116	242

[a]See Table 2 for full name of pesticides.
[b]Male fin whale 51 ft long.
[c]Male fin whale 53 ft long.
[d]2-year-old female fin whale 57 ft long.
[e]Male sei whale 41 ft long.
[f]8-year-old female sei whale 42 ft long.

Table 4. Yearly Incidence of *Crassicauda boopis* Infection in Fin Whales Taken Off the West Coast of Iceland, 1981–1987.

1981	1982	1983	1984	1985	1986	1987
86.7%	97.2%	91.2%	—	94.5%	100%	100%
(15)	(36)	(34)		(57)	(40)	(15)

Percent cases with adult *Crassicauda boopis* found in kidneys. Total number of whales in which kidneys were examined given in parentheses.

in various tissue samples from fin and sei whales. It is notable that these data indicate low levels of tissue contamination. Thus the available information, although limited, points to the value of pathological findings in these populations as baselines for comparison with marine mammals in contaminated coastal zones.

Pathology, Hematology, and Parasitology

Pathological investigation focused on identifying endemic infections that cause, or could cause, systemic illness. A systematic necropsy protocol was followed.[38] Infections of high incidence which appeared capable of causing systemic effects were considered to be of major importance. Fin and sei whales contrasted in the variety of internal parasitic infections, the incidence of presumptive viral diseases, and the frequency of external lesions. Sei whales exceeded fin whales in all these categories.

Only one major endemic disease was found in the fin whale. This was caused by infection of the urinary tract and vascular system with the giant spirurid nematode *Crassicauda boopis*.[38,39] Other problems seen in fin whale produced little or no evidence of systemic effects. For example, infection with the giant liver fluke, *Lecithodesmus goliath,* occurred with low incidence and caused localized liver abscesses. Infections of the colon with the small trematodes. *Ogmagaster spp.,* were common but noninvasive. Cardiovascular disease, other than that associated with crassicaudiosis, appeared benign or infrequent.

The incidence of *C. boopis* infection over the 7 year study is given in Table 3. Prominent lesions associated with adult *C. boopis* occurred in the kidneys, lung, renal veins, and vena cava. An extensive mesenteric arteritis was common and appeared to be caused by somatic migrations of *Crassicauda* larvae from the gut in the walls of arterial vessels.

Histopathological examinations revealed mixed eosinophilic, neutrophilic, and mononuclear infiltrates diffusely distributed throughout the tunica media and subendothelial connective tissue of the mesenteric arteries. In many sections dense accumulations of eosinophils mixed with smaller numbers of neutrophils and mononuclear cells occurred in these same layers. The tunica intima had areas of thickening adjacent to areas where the normal laminated pattern of the vessel wall was disrupted by hemorrhage and inflammation. Nematode larvae were found in some histological sections within the tunica media, close to the subintimal space.[38]

Respiratory disease appeared absent except for localized pulmonary lesions that surrounded necrotic pieces of adult crassicaudid nematodes. These lesions were sometimes numerous and diffusely distributed throughout the lungs, particularly in juvenile animals. They apparently originated from portions of the mineralized lesions surrounding *C. boopis* in kidneys that had fragmented in the venous circulation and been swept in the bloodstream to the lung.[38]

Large, mineralized tissue masses surrounded the tangled bodies of adult worms in the renal veins. These typically consisted of an exuberant fibrocellular tissue reaction around the worms. Less commonly the body of the parasite lay free in the vessel lumen. In advanced cases, chronic severe occlusive phlebitis was evident (Figure 2). In one immature animal bilateral occlusion of the renal veins was associated with swelling and massive abscessation of both kidneys (pyonephrosis). Lesions originating in the renal veins frequently extended into the vena cava and were enmeshed by large, obturating blood clots. Eighty percent of a sample of eighty-seven whales had some primary renal vein involvement.[38]

The tails of the adult parasites lay in the ureteral system of the kidney. Microscopic examinations revealed their eggs and larvae in the urine. This indicated a contaminative mode of transmission of the parasite through the environment and the possibility of paratenic or intermediate hosts. Histopathological examinations also revealed crassicaudidiform eggs in the pulmonary airspace in the region of the lesions in the lungs. Interpretation of the overall pattern of pathology led to a formal hypothesis concerning the mechanisms of transmission between whales and various larval migration streams of the parasite within the definitive cetacean host (Figure 3).

From the combined pathological findings it was concluded that severe infections with *Crassicauda boopis* probably could kill the host by causing congestive renal failure secondary to occlusion of the renal veins. It was also suggested that powerful laboratory tests for the evaluation of renal impairment in mammals could be used to measure systemic effects that portend mortality.[31,38]

Data from absolute and differential white blood cell counts of 47 fin whales examined in 1985 are summarized in Table 3. Methods of blood sampling from the tail flukes as described elsewhere were used.[38,40] These large animals gave an average total white cell count that was significantly lower than in terrestrial mammals. The extent to which vascular stasis and leucocyte sedimentation after death could have reduced total white blood cell counts was not clear. However, blood cell counts from samples obtained directly from the heart within 5–10 min postmortem and analyzed at sea showed no major differences from counts using blood samples obtained from the vessels of the tail flukes of the same whales.[41] Thus it seemed unlikely that the comparatively low white cell count was due to intravascular sedimentation of leucocytes.

A striking finding was a very high number of eosinophils in the blood despite the low white cell count. In other mammals an eosinophil concentration in excess of 1000/μL is generally considered indicative of a systemic disease usually associated with the migratory stages of internal parasitism. Such eosinophilia is

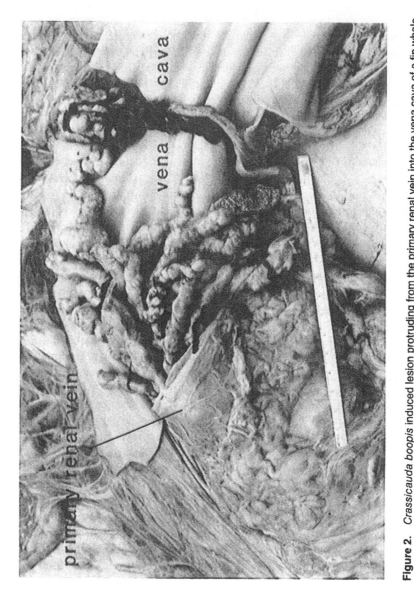

Figure 2.   *Crassicauda boopis* induced lesion protruding from the primary renal vein into the vena cava of a fin whale. The vein from the kidney is completely occluded (rule = 15 cm). (From Lambertsen, R. H., R. L. Carter, and J. Farrington, "Natural Survivorship of Fin and Sei Whales: Epidemiology." Institute of Biomedical Aquatic Studies, University of Florida. *Int. Whaling Comm.* SC/38/013, 1986.)

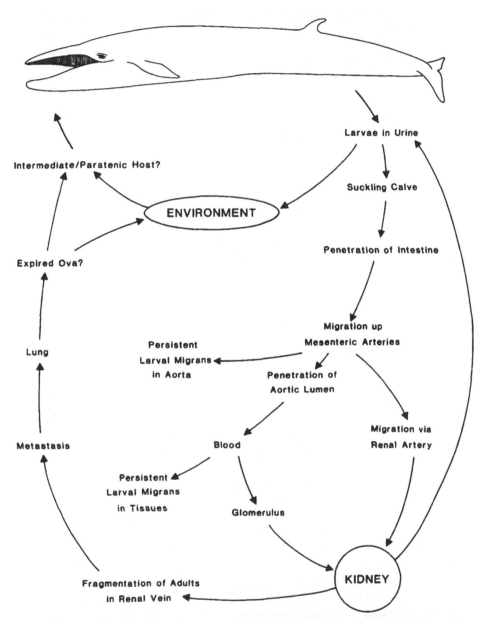

Figure 3. Hypothesized larval migration streams and transmission mechanisms of *Crassicauda boopis* infection. (From Lambertsen, R. H. "Disease of the Common Fin Whale (*Balaenoptera physalus*): Crassicaudiosis of the Urinary System," *J. Mammal.* 67:353–366 (1986). With permission.)

Table 5.   White Blood Cell Counts[a] in 47 Fin Whales Taken Off the West Coast of Iceland in 1985.

	Neutrophils	Eosinophils	Monocytes	Lymphocytes	Total WBC
		#/μL Blood			
Median	1720	1712	187	1340	5026
Mean	1822	1876	219	1533	5464
Range	432–4623	507–9477	34–651	603–3448	2770–16340
S.D.	986	1445	146	571	2498
		Percent			
Mean	32.2	32.7	4.4	29.7	
Range	8.8–62.6	14.5–58.0	5.0–20.4	15.0–56.5	

[a]Total white cell counts determined using a hemocytometer. Relative white cell counts determined using blood smears prepared from heparinized blood collected at sea from tail blood; 200–400 cells scored per determination. All counts corrected for presence of nucleated red blood cells.

a common finding in cetaceans and may reflect the almost universal infection of these species with crassicaudids or other invasive parasites.[5–12,17,18] Some fin whales had a blood eosinophil concentration of more than 9400/μL.

A survey of the serum chemical values of 31 fin whales was conducted to further investigate systemic pathologic effects.[40] This analysis found no appreciable relationship between any of the 10 serum constituents measured and chase times recorded by whaling personnel. In all cases the slopes of the linear regressions relating each measured serum constituent and chase time approximated zero (range −0.031 to 0.003), while correlation coefficients were low (−0.22 to 0.31). On this basis reasonable approximations of normal serum chemical values were calculated using statistical averages.[40]

The serum chemical and serum osmolality values of one sexually immature animal, a 17.5 m female, fit a pattern that is well known in mammals suffering from renal insufficiency or renal failure.[40] The serum concentrations of urea (114 mg/dL) and creatinine (3.5 mg/dL) in this individual were the highest found in any of the whales examined. In the same animal, deviations from normal that were consistent with renal failure, but not simple dehydration, occurred in serum electrolytes (elevated phosphate and potassium concentrations, with depressed calcium and chloride). This combination of abnormalities was not consistent with the possibility of elevated creatinine concentration due simply to capture myopathy.

It is notable that marked elevations of the concentration of creatinine and urea also occurred in a 13.8 m female fin whale, almost certainly a yearling calf, which stranded alive and then died on the coast of New England in 1977.[38] A necropsy conducted on this animal by Dr. Lawrence Dunn of the Mystic Marine-life Aquarium revealed extensive digitate masses in the renal veins and crassicaudid nematodes in the renal ducts. Serum creatinine and urea concentrations were reported as markedly higher than normal values from a variety of cetacean species. On this basis the apparent cause of death was diagnosed as renal failure caused by *Crassicauda* infection.[42] This corroborates the evidence reviewed above that crassicaudiosis can be a lethal disease in young, immunologically naive fin whales.

It is especially noteworthy that the chronic severe occlusive phlebitis which commonly affected the renal veins of fin whales has been documented in the endangered blue whale. In 1915 Hamilton[9] noted that two out of the four blue whales he examined while at the Blacksod Bay whaling station in the British Isles had large tumorous lesions in the vessels of their kidneys. His illustrated description of these lesions reveal an identical appearance to those induced by *Crassicauda boopis* in the kidney of the fin whale.

The sei whales showed a pattern of endemic disease that was distinct and far more complex than that seen in the fin whale. Although sei whales also had infections of the kidneys with *Crassicauda sp.*, the associated lesions were less severe. Large fibrocellular tumors surrounding adult *Crassicauda* were not found. Twenty-two of twenty-four (92 percent) sei whales in which the intestines were opened had invasive infections of the small bowel or colon with *Bolbosoma spp.* These acanthocephalons were rarely seen in fin whales. *Bolbosoma spp.* were attached to the muscosal lining of the intestines and caused numerous small ulcers 2–3 min in diameter. The severity of *Bolbosoma* infections varied and systemic effects were difficult to determine. Serious infections had up to three invading worms per square centimeter of the mucosal surface. These presented with a generalized secretory enteritis.

Otherwise, the sei whales were heavily scarred externally with bite wounds thought to be inflicted by small tropical sharks. Six other morphologic types of external lesions were identified, although their causes remain unknown. In addition, 4 out of 22 (18 percent) sei whales in which the respiratory tract was inspected showed an inflammation of the mucosa of the trachea and bronchi that appeared consistent with a diagnosis of virus- or mycoplasma-induced tracheobronchitis. Histopathologic study of the affected mucosa demonstrated dense mononuclear infiltrates and nodular lymphoid hyperplasia. However, limited electron microscopic studies of these muscosal lesions failed to detect virus particles or mycoplasmal agents.

Virology and Serology

One potentially significant finding in fin whales was the occurrence of small but sometimes numerous superficial lesions of the skin. These typically presented as pale gray to white epidermal maculae ranging in size from 1 to approximately 15 mm in diameter. The observation that some of these lesions were vesicular in character raised concern because the viruses which cause vesicular diseases in other mammals (e.g., Foot and Mouth Disease Virus) constitute a threat to

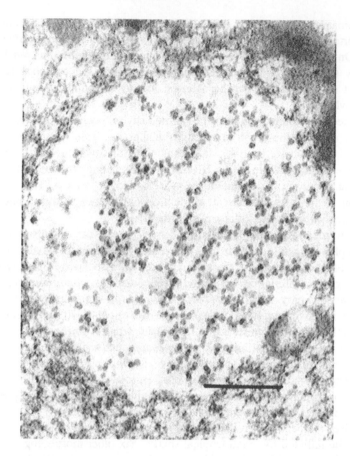

Figure 4. Transmission electron micrograph of virus particles found in skin lesions observed on fin whales in 1984. A microvesicle containing virus particles is shown. Particles were of the same size range and appearance as picornaviruses, which cause vesicular diseases in domesticated animals (bar = 0.5 μm). (From Lambertsen, R. H., R. L. Carter, and J. Farrington. "Natural Survivorship of Fin and Sei Whales: Epidemiology," Institute of Biomedical Aquatic Studies, University of Florida. *Int. Whaling Comm.* SC/38/013, 1986.)

domestic livestock. Also, if similar viral diseases occur in whales, these might cause mortality and infertility. Vesicular diseases in cattle can produce up to 50 percent mortality in young animals, although mortality in adults is seldom over 5 percent. In adult populations, mastitis, reduced lactation, and epidemic abortion are common effects.[43]

With this concern, special virological studies were begun on fin whales in 1984. Electron microscopic studies of glutaraldehyde fixed lesion material revealed virus particles similar in size and shape to the picornaviruses which cause vesicular disease, including Foot and Mouth Disease in cattle (Figure 4). Frozen lesion material was transferred to the U.S. Department of Agriculture Plum

Island Animal Disease Center for containment and attempted virus isolations. Fourteen lesion samples were cultured in duplicate on Vero and fetal bovine kidney cells. Cytopathological effects were not observed.[45] Cetacean cell cultures were thereafter established in Iceland for the attempted virus isolations. Subsequently this research had to be discontinued due to the above noted interference.

Lacking viruses cultured directly from lesions, attempts were made to continue diagnostic investigations using the indirect methods of serology. Serum samples obtained from 70 fin whales and 40 sei whales were examined for evidence of neutralizing activity against known viral pathogens in domestic animals. Virus antigen from these foreign disease threats was available at the Plum Island Animal Disease Center. Standard tests for Rinderpest, Foot and Mouth Disease, and Infectious Bovine Rhinotracheitis serum neutralizing activity were performed in microtiter plates containing fetal bovine kidney cells. The results showed toxic effects of whale serum on bovine cells but no evidence of virus neutralization. Sera were further tested for the presence of antibody to African Swine Fever (ASF) using the IEOP test. There was no evidence of ASF neutralizing activity.[45,46]

Fin and sei whale sera also were tested for calicivirus antibody by the indirect fluorescent antibody method using infected Vero cells. The positive control was rabbit- anti-calicivirus antiserum and anti-rabbit conjugate. The negative control was normal rabbit serum. Although whale serum stained infected Vero cells, the pattern of staining was not that of the positive control. Based on fluorescent microscopic examination comparisons with both positive and negative controls no calicivirus antibody could be detected.[45,46] Attempts to detect serum neutralizing activity against marine caliciviruses were repeated on serum samples from 10 fin and 10 sei whales at College of Veterinary Medicine, Oregon State University, also with negative results.[47]

CONCLUSIONS

This study allows certain conclusions and additional cautionary tales concerning the evaluation of disease burdens in whales as possible biomarkers of environmental pollution. The results show that fin and sei whale populations inhabiting the central North Atlantic suffer from a very high incidence of endemic parasitic infection. It is possible and likely that the incidence levels found are natural and not pollution-driven. Thus finding a high incidence of parasitism in a population of stranded whales in more heavily contaminated regions does not necessarily reflect pollution-driven disease events. The contrasting patterns of endemic disease found in two closely related cetacean species also reveal the danger of simple extrapolations between species when evaluating biomarkers of population health.

Finding a high incidence of parasitic infection indicates that transmission of the infection is naturally efficient. In this instance, high efficiency in transmission cycles is supported by the belief that the abundance of fin and sei whales is now low as a result of past commercial exploitation. Low abundance necessarily limits the total number of infective eggs and larvae shed into the environment from the definitive host species. As host populations increase, density-dependent increases should be expected in the rate of exposure of individuals to infectious pathogens.

For this reason it must be cautioned that pollution-driven impairments of the health status of expanding wildlife populations may be mimicked, masked, or amplified by natural phenomena. As high population densities are reached, increased rates of transmission of infectious pathogens can be expected to drive up morbidity and mortality rates in the absence of any toxic effect of pollution. If the susceptibility to pathogens is simultaneously being increased by pollution-driven immunosuppression one would predict a more marked amplification in the extent and severity of disease. Theoretically, it is possible that the recent rise in cetacean mortality in the western North Atlantic is a manifestation of this exacerbating relationship between density-dependent pathogensis and pollution-driven increases in disease susceptibility. Separating natural from pollution-driven phenomena in wild cetacean populations will require better information on trends in abundance as well as measurements of disease biomarkers in the populations exposed.

ACKNOWLEDGMENTS

Funding for aspects of this study was provided by the Committee on the Challenges of Modern Society, North Atlantic Treaty Organization, Brussels, Belgium. Views expressed by the author do not necessarily represent those of the Committee or NATO member countries. I extend my gratitude to those many workers and colleagues, with names too numerous to list, who participated in the investigation.

REFERENCES

1. Cockrill, W. R. "Antarctic Pelagic Whaling: The Role of the Veterinary Surgeon in the Whaling Industry, with Special Reference to Standards of Inspection in the Production of Whalemeat for Human Consumption, and Some Notes on the Pathology of the Baleen Whales," *Vet. Rec.* 63:111–124 (1951).
2. Uys, C. J., and P. B. Best. "Pathology of Lesions Observed in Whales Flensed at Saldanha Bay, South Africa," *J. Comp. Pathol.* 76:407–412 (1966).
3. Cockrill, W. R. "Pathology of the Cetacea: A Veterinary Study of Whales—Part I," *Brit. Vet. J.* 116:133–174 (1960).
4. Cockrill, W. R. "Pathology of the Cetacea: A Veterinary Study of Whales—Part II," *Brit. Vet. J.* 116:175–189 (1960).

5. Rees, G. "A Record of Some Parasitic Worms from Whales in the Ross Sea Area," *Parasitology* 43:127–134 (1953).

6. Markowski, S. "Cestodes of Whales and Dolphins from the Discovery Collections," *Discov. Rep.* 27:377–395 (1955).

7. Baylis, H. A. "Observations on the Genus *Crassicauda*," *Mag. Nat. Hist.* 5(9):410–419 (1920).

8. Baylis, H. A. "On *Crassicauda crassicauda* (Crepl.) Nematoda and its Hosts," *Ann. Nat. Hist.* 27(8):144–148 (1916).

9. Hamilton, J. E. "Report on Belmullet Whaling Station," *Brit. Assoc. Adv. Sci. Rep.* 1915:124–146 (1915).

10. Dailey, M. D., and W. F. Perrin. "Helminth Parasites of Porpoises of the Genus Stenella in the Eastern Tropical Pacific, with Descriptions of two New Species: *Mastigonema stenellae* gen. sp. n. (Nematoda: Spiruroidea) and *Zalophotrema pacificum* sp. n. (Trematoda: Digenea)," *Fish. Bull.* 71:455–471 (1973).

11. Geraci, J. R., M. D. Dailey, and D. J. St. Aubin. "Parasitic Mastitis in the Atlantic White-sided Dolphin, *Lagenorhynchus acutus*, as a Probable Factor in Herd Productivity," *J. Fish. Res. Bd. Can.* 35:1350–1355 (1978).

12. Rewell, R. E., and R. A. Willis. "Some Tumors Found in Whales," *J. Pathol. Bacteriol.* 61:454–456 (1949).

13. Best, P. B., and R. M. McCully. "Zygomycosis (Phycomycosis) in a Right Whale (*Eubalaena australis*)," *J. Comp. Pathol.* 89:341–348 (1979).

14. MacKintosh, N. A. "The Southern Stocks of Whalebone Whales," *Discov. Rep.* 22:199–298 (1942).

15. Clarke, R. "Sperm Whales of the Azores," *Discov. Rep.* 28:237–298 (1956).

16. Matthews, L. H. "The Sperm Whale, *Physeter catodon*," *Discov. Rep.* 17:195–263 (1938).

17. Baylis, H. A. "A List of Worms Parasitic in Cetacea," *Discov. Rep.* 6:393–418 (1932).

18. Dailey, M. D., and R. L. Brownwell, Jr. "A Checklist of Marine Mammal Parasites," in *Mammals of the Sea, Biology and Medicine*, S. H. Ridgway, Ed. (Springfield, IL: Charles C Thomas, Inc., 1972), pp. 528–576.

19. Albert, T. F., Ed. *Tissue Structural and Other Investigations on the Biology of Endangered Whales in the Beaufort Sea*, Report to the Bureau of Land Management from the Department of Veterinary Science, University of Maryland, College Park, MD (1981), 953 pp.

20. Migaki, G., and T. F. Albert. "Lipoma in the Liver of a Bowhead Whale (*Balaena mysticetus*)," *Vet. Pathol.* 19:329–331 (1982).

21. Migaki, G., R. Heckmann, and T. F. Albert. "Gastric Nodules Caused by "*Anisakis* Type" Larvae in the Bowhead Whale (*Balaena mysticetus*)," *J. Wildlife Dis.* 18:353–357 (1982).

22. Heckman, R. A. "Parasitological Study of the Bowhead Whale, *Balaena mysticetus*," in *Tissue Structural and Other Investigations on the Biology of Endangered Whales in the Beaufort Sea*, T. F. Albert, Ed. Report to the Bureau of Land Management from the Department of Veterinary Sciences, University of Maryland, College Park, MD (1981), pp. 275–304.

23. Johnston, D. G., and A. Shum. "Bacteriological Study of the Bowhead Whale, *Balaena mysticetus*," in *Tissue Structural and Other Investigations on the Biology of Endangered Whales in the Beaufort Sea*, T. F. Albert, Ed. Report to the Bureau

of Land Management from the Department of Veterinary Sciences, University of Maryland, College Park, MD (1981), pp. 255–274.

24. Smith, A. W., D. L. Skilling, and K. Benirschke. "Investigations of the Serum Antibodies and Viruses of the Bowhead whale, *Balaena mysticetus*," in *Tissue Structural and Other Investigations on the Biology of Endangered Whales in the Beaufort Sea*, T. F. Albert, Ed. Report to the Bureau of Land Management from the Department of Veterinary Sciences, University of Maryland, College Park, MD (1981), pp. 233–254.

25. Lambertsen, R. H., B. A. Kohn, J. Sundberg, and C. D. Buergelt. "Genital Papillomatosis in Sperm Whale Bulls," *J. Wild. Dis.* 23:510–514 (1987).

26. Lambertsen, R. H., and B. A. Kohn. "Unusual Multisystemic Pathology in a Sperm Whale Bull," *J. Wild. Dis.* 23:361–367 (1987).

27. Sigurjonsson, J., R. H. Lambertsen, and R. Demuth. Unpublished observation (1982).

28. Lambertsen, R. H. Unpublished results (1982).

29. "Report of the Scientific Committee," *Int. Whaling Comm.* 37:28–55 (1986).

30. Geraci, J. R. "Clinical Investigation of the 1987–1988 Mass Mortality of Bottlenose Dolphins along the U.S. Central and South Atlantic Coast," Final Report to the National Marine Fisheries Service, Office of Naval Research, and Marine Mammal Commission from the Department of Pathology, Ontario Veterinary College (1989), 63 pp.

31. Lambertsen, R. H., R. L. Carter, and J. Farrington. "Natural Survivorship of Fin and Sei Whales: Epidemiology," Institute of Biomedical Aquatic Studies, University of Florida. *Int. Whaling Comm.* SC/38/013 (1986).

32. "Report of the Scientific Committee," *Int. Whaling Comm.* SC/41 (1989).

33. Hansen, K. "[Endeavor to Stop Icelandic Whaling by Any Means: Greenpeace Adopts Attitude Away from Sabotage by Environmental Groups of Whale Catchers, but Appeals for Professional Ban on Scientist Engaged in Icelandic Whaling Project,]" translated from the Danish by H. Foss, *Kaskelot* 78:22–31 (1988).

34. Winn, H. E., Bischoff, and A. G. Taruski. "Cytological Sexing of Cetacea," *Mar. Biol.* 23:343–346 (1973).

35. Lambertsen, R. H. "A Biopsy System for Large Whales and Its Use for Cytogenetics," *J. Mammal.* 68:443–445 (1987).

36. Lambertsen, R. H., C. S. Baker, D. A. Duffield, and J. Chamberlin-Lea. "Cytogenetic Determination of Sex among Individually Identified Humpback Whales, *Megaptera novaeangliae*," *Can. J. Zool.* 66:1243–1248 (1988).

37. Dobskey, P., R. H. Lambertsen, and J. Farrington. Unpublished results (1984).

38. Lambertsen, R. H. "Disease of the Common Fin Whale (*Balaenoptera physalus*): Crassicaudiosis of the Urinary System," *J. Mammal.* 67:353–366 (1986).

39. Lambertsen, R. H. "Taxonomy and Distribution of a *Crassicauda* sp. (Nematoda: Spirurida) Infecting the Kidney of the Common Fin Whale (*Balaenoptera physalus* Linne, 1758)," *J. Parasitol.* 71:485–488 (1985).

40. Lambertsen, R. H., B. Birnir, and J. E. Bauer, "Serum Chemistry and Evidence of Renal Failure in the North Atlantic Fin Whale Population," *J. Wildl. Dis.* 22:389–396 (1986).

41. Lambertsen, R. H. Unpublished results (1984).

42. Dunn, L. Personal communication (1984).

43. Bruner, D. W., and J. H. Gillespie. *Hagan's Diseases of Domestic Animals* (Ithaca, NY: Cornell University Press, 1973), 1386 pp.

44. Lambertsen, R. H. and W. Smith. Unpublished results (1984).
45. Mebus, C. Personal communication (1986).
46. Lambertsen, R. H. "Diseases of North Atlantic Whales, Potential Threat to Domestic Livestock," Animal Health Report to the U.S. Department of Agriculture, CRIS Access. No. 009093 (1988).
47. Smith, A. F. Personal communication (1986).

CHAPTER 22

Use of Biomarkers in Ecological Risk Assessment

Glenn W. Suter II

Ecological risk assessors analyze the effects of human actions on the natural environment. Like other risk assessors, they have been concerned primarily with predictive assessments, those that estimate the nature, probability, and magnitude of effects of proposed actions such as the marketing of a new chemical product or the release of a new effluent. However, risk assessors must increasingly perform retrospective assessments, assessments of human actions that were initiated in the past and may have ongoing consequences. Examples include dumping of hazardous waste or the policy of building tall stacks on coal-fired power plants. Biomarkers are most useful in retrospective risk assessments.

Retrospective risk assessments differ from predictive assessments principally in that the environment is already affected and can be observed and measured. The components of the environment that can potentially be measured include sources, indicators of exposure, and indicators of effects (Figure 1). In many cases, one or more of these components of the contaminated environment have not been measured and cannot be measured either for technical reasons (e.g., in pelagic marine spills both the source and the affected organisms are often dispersed or dissipated before measurements can be made) or for practical reasons (e.g., nobody was available or funded to perform measurements). If the affected community cannot be measured, the effects and their relationship to the source must be simulated. However, if the affected community can be sampled and

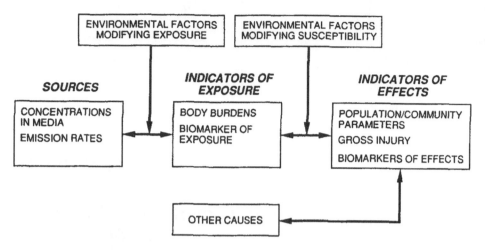

Figure 1. A diagram of the process of ecological epidemiology.

measured while effects are still detectable, then techniques of ecological epidemiology can be employed in the assessment.

Assessments fall into two categories with respect to the direction of inference (Figure 1). Source-driven assessments begin with an existing pollution source, such as a spill or an effluent, and attempt to determine the nature of effects. Effects-driven assessments begin with an observed effect such as a declining animal population or a fishless lake, and attempt to determine a cause. In both cases, the logical link between sources and effects is exposure.

Whether the problem is to establish a cause for an observed effect (effects driven) or to identify an effect associated with a source (source driven), four questions arise in ecological epidemiology: (1) is there a real effect as opposed to natural variation in ecological properties, (2) what is its cause, (3) what is its magnitude, and (4) what are its ultimate consequences?

IS THERE A REAL EFFECT?

In human health epidemiology, the effects are clearly negative (death, injury, disease), and the reality of an epidemiological effect is established by comparing the frequency of occurrence of a pathology in the study population with the expected frequency. In ecological studies, the effects are often differences in the magnitude of a variable such as biomass of fish or rate of tree growth for which there is no clearly pathological level (i.e., it is not possible to say that x kg of fish in a lake constitutes a healthy community but that x-1 kg is pathological). Even in cases such as tree mortality for which frequencies of discrete negative effects can be determined, there is no source of data comparable to human death certificates from which to determine an expected frequency. In other words, it is often not clear that there is an effect that needs a causal explanation.

One way to be sure that the effect being assessed is real is to choose as endpoints effects that are clearly abnormal, such as fish kills or fishless streams. These endpoints may not be caused by the pollutant of concern or by any other human action, but they very likely have a cause and are not simply the result of random fluctuations in the fish mortality rate or the distribution of fish populations. The obvious disadvantage of such endpoints is that the effects are more severe than is often desirable.

Another way to be sure of assessing a real effect is to establish the normal state of the community type under study. This approach is analogous to establishing an expected frequency in human health epidemiology. For this purpose, it is not sufficient to compare the subject community with a reference community (e.g., the stream reaches below and above an outfall). Two communities may differ for a variety of reasons unrelated to pollution (e.g., substrate, flow, and history of physical disturbance). Rather, it is desirable to measure the properties of several reference sites, sites that are similar to the subject site but that are not polluted or otherwise anomalously disturbed (disturbances common to the class of communities, such as regional air pollution, are ignored). The n measured properties of this set of reference sites can be used to establish an n-dimensional state space.[1] The n measurements from the subject community can then be used to estimate the probability that the subject community falls outside the reference community state space. This probability is the risk that the community is affected by some factor not affecting other communities of the same type.

If a clearly pathological level of a biomarker can be established with respect to organism-level effects, then occurrences of the marker exceeding that level are evidence of a real effect. For example, observation in moribund birds of brain cholinesterase levels that are half the normal level is sufficient to indicate that one is not simply observing natural mortality.[2] However, most biomarkers do not have clearly pathological levels, so the state-space approach is appropriate. The state space may be one-dimensional, such as a distribution of a mixed-function oxidase measure in bluegill sunfish, or biomarkers may be components of a multidimensional state space that establishes the condition of a subject population or community relative to the state-space of reference populations or communities.

WHAT IS THE CAUSE?

Attempts to prove causation in ecological epidemiology often fall prey to a logical fallacy known to epidemiologists as the "ecological fallacy"—the idea that occurrence of an effect in conjunction with a plausible environmental factor proves that factor to be the cause. It is tempting to believe that differences between sites upstream and downstream from an outfall are caused by toxicity of the effluent, but they could also be caused by the natural gradient of stream communities, enhanced flow due to augmentation by the effluent, physical disturbance of the substrate, or other factors.

Inappropriate use of hypothesis-testing statistics often compounds the "ecological fallacy." Hypothesis tests were developed for experimental studies in which it is possible to establish true replicates and randomly assign them to treatments. When that is the case, it is possible to use hypothesis tests to prove with some level of confidence that the treatments cause an effect. In ecological epidemiology there are usually no true replicates. The multiple samples taken below an outfall are pseudoreplicates, that is, samples from a single treated community. True replicates would be multiple receiving communities that are independently exposed. The nonrandom assignment is obvious; investigators do not have the power to assign effluents to stream segments. When investigators test the null hypothesis that a set of samples from a stream segment upstream of an outfall is the same as a set from a downstream segment, they are not testing a hypothesis about effluent effects; they are simply comparing sets of samples from two sites. Hypothesis tests about the relationship among sites may have value as exploratory statistics in the early stages of a study but they do not address the risks posed by pollutants.

Rather than testing a hypothesis about the difference between two sites, a risk assessor must try to describe the effects of a pollutant on an environment. This can be done by establishing a statistical model and estimating its power to explain the data.[3] The model might be a simple dichotomous one (e.g., lakes with pH less than x have no fish, or birds with brain cholinesterase concentrations less than y have been exposed to cholinesterase inhibitors). More common, however, are exposure-response models, similar to the dose-response models developed from laboratory toxicity tests. An example would be a model relating levels of cytochrome induction to percentage dilution of an effluent or total dissolved organics concentration. If environmental conditions control the occurrence or magnitude of a toxic effect, they can be included in the model. An example is Hakanson's model, which uses mercury loading, lake water pH, and lake productivity to predict the methyl mercury content of predatory fish in lakes.[4] Finally, the model may simply be a statistically derived map. For example, one might use kriging to create maps of biological properties, physical properties, and toxicant concentrations at waste sites to reveal potentially causal relations in the patterns.[5] These statistical models do not prevent the ecological fallacy. However, because they are clearly correlations and do not appear to be testing for a pollutant effect, they are less likely to induce an assessor to jump to fallacious conclusions. In addition, these models, unlike hypothesis tests, can provide information about the relationship between the degree of exposure and the magnitude of effects and about environmental factors that contribute to effects.

To actually prove causality in ecological epidemiology, one must employ the same standards as Koch's postulates for proving that a particular pathogen causes a disease.[6] They are as follows (quoted from Reference 7):

1. It must be shown that the microorganism in question is always present in diseased hosts.
2. The microorganism must be isolated from the diseased host and grown in pure culture.
3. Microorganisms obtained from the pure culture, when injected into a healthy susceptible host, must produce the disease in that host.
4. Microorganisms must be isolated from the experimentally infected host, grown in pure culture, and compared to the microorganisms in the original culture.

The first rule clearly corresponds to demonstrating a regular association between the pollutant and the effect. For effects-driven assessments, this means that every time the effect is observed, exposure to the pollutant must have preceded it. For source-driven assessments, it means that every time and everywhere the source occurs, a consistent effect occurs if susceptible organisms are present. Establishing a regular association is more difficult for source-driven assessments because the nature of the effect is unknown and the characteristics of organisms and environments that result in susceptibility are unknown. A particular pollutant may have different effects in various communities because of differences in the physical environment, such as pH or the presence of sorptive organic matter, or because of differences in the biota, such as less-sensitive species or greater functional redundancy in the community. In other words, the symptoms may be different because the target communities are different. Biomarkers may be useful in establishing a regular association of pollutants and effects. Biomarkers of effects are more likely to be consistent among species and are more likely to be characteristic of a particular pollutant or class of pollutants than are higher-level effects. For example, low cholinesterase levels are likely caused by cholinesterase-inhibiting pesticides, but low species diversity may be caused by any pollutant, physical disturbance, or natural environmental variable.

Koch's third rule clearly corresponds to toxicity testing. Tests can expose susceptible organisms to a concentration gradient of the chemical, complex effluent, or contaminated medium that is associated with effects in the field. Ideally, the tests would be conducted with the species and the life stages that are affected in the field, and the mode and duration of exposure would match the exposure in the field. More often, toxicity tests conducted for retrospective risk assessments use standard test species and exposure procedures.[8]

The analogy for rules 2 and 4 is less obvious. However, consistent indicators of exposure that are diagnostic of exposure to the pollutant or are at least characteristic of exposure to the pollutant are functionally equivalent to isolation of a pathogen. Most often, the indicator of exposure would be a body burden of the pollutant, but biochemical biomarkers of exposure would also serve. Even characteristic histopathological damage may serve as an exposure indicator. The important point is that consistent indicators of exposure in the field study and in the toxicity tests establish that the organisms in the field are exposed to the pollutant and that the degree of exposure is consistent with the degree of exposure that causes the effect in the laboratory. If these conditions are satisfied, no other cause need be invoked. Thus, exposure indicators provide the link that proves that the effects observed in the field and in the tests have a common cause.

Studies that satisfy all four of Koch's rules are the gold standard of ecological epidemiology. They truly prove that a particular source has a particular effect. Such studies are rare because (1) ecological risk assessments have not been held to the same high standards as human health studies, (2) most ecological risk assessments are not given the time or resources to complete such studies, and (3) technical limitations often prevent satisfaction of all four requirements. The obvious exceptions are studies of large-scale problems such as acidification of surface waters and the decline of forests in central Europe and the eastern United States, where the standards of proof have been high and resources are being devoted to developing and applying the necessary investigative techniques.

In the absence of a complete study, an investigator must build the best possible case with the resources available. I believe that this process is aided by the logical scheme presented in Figure 1. The arrows in the figure represent logical linkages that provide support for attribution of a cause to an effect. Quantitative or even qualitative results indicating that such linkages exist and have a consistent functional form constitute contributory evidence.

In many cases, the existence of appropriate linkages could be demonstrated by basic research, but for biomarkers that research has not been done. For example, we do not know for many biomarkers whether there is a consistent relationship between the magnitude of the source of exposure and the level of the marker, or between the level of the marker and the level of overt effects. That such relationships exist is not a given. For example, chlorinated phenols induce glutathione S-transferase in *Daphnia magna*, but the level of induction is not correlated with toxicity.[9]

Another type of evidence for causation is study of alternate possible causes. If it can be shown that natural processes, physical disturbances, or even toxicants other than the pollutant under study (in the case of source-driven assessments) are likely or proven causes, then the subject pollutant is effectively absolved. For example, if the benthic invertebrate fauna is depauperate below an outfall, habitat evaluation may show that the missing species are intolerant of the change in substrate resulting from dredging barge access to the facility. On the other hand, demonstration that a hypothesized alternate cause is not operating constitutes *de facto* evidence that the pollutant is.

WHAT IS THE MAGNITUDE OF EFFECTS?

Although for some assessments it is sufficient to demonstrate that a real effect has occurred and that the effect has a particular cause, it is often necessary to establish the magnitude of the effect so that the need for remediation can be evaluated and the benefits and costs can be balanced. For the same reason that comparison of a polluted and a clean site cannot prove causation, it cannot establish the magnitude of effects. The effects may be larger or smaller than the relative magnitudes of the effects parameter at the sites because of differences between the sites other than the degree of pollution. However, if a statistical model has been developed relating pollution levels and levels of any environmental cofactors to effects, that model can be used to estimate the extent of effects and the magnitude of effects at a site. In the absence of such a model, one can only attempt to account for the underlying differences between the sites and assume that they act independently of the pollutant. Biomarkers may contribute to the development of such models or to their verification.

WHAT ARE THE ULTIMATE CONSEQUENCES?

An assessor may wish to estimate ultimate consequences if the existing effects are increasing (e.g., due to accumulation of the pollutant or accumulation of effects on long-lived species) or if remedial actions are planned. To the extent that they capture the relationship between pollution levels, environmental characteristics, and effects, statistical models developed to describe the existing situation can be used to extrapolate into the future. However, if the statistical model is not believed to reflect mechanisms by which effects are induced or if new mechanisms of effects induction or remediation will be introduced, then it may be necessary to create a mathematical model to simulate the ultimate effects of remedial actions or of inaction.

CONCLUSIONS

If biomarkers are to become useful in ecological risk assessment, they must be linked to the other measures and indicators that contribute to retrospective assessments. It is not sufficient to show that biomarker levels differ among sites or even that a biomarker level is abnormally high at a site. Assessments are concerned with sources that may be regulated or remediated and with the organism, population, and community level effects that are to be prevented or mitigated. The biomarkers are intermediates between sources and higher-level effects; they have no importance in themselves and must be linked to sources or overt effects. Specifically, it must be shown that biomarkers increase in a regular and predictable manner with increasing exposure and that higher-level effects are predictable from biomarker levels. Otherwise, biomarker data will be suggestive of risks, but risk assessments will be based on more conventional data.

ACKNOWLEDGMENT

Publication No. 3456, Environmental Sciences Division, ORNL. Operated by Martin Marietta Energy Systems, Inc., under contract DE-AC05-84OR21400 with the U.S. Department of Energy.

REFERENCES

1. Johnson, A. R. "Diagnostic Variables as Predictors of Ecological Risk," *Environ. Manage.* 12:515–523 (1988).
2. Fite, E. C., L. W. Turner, N. J. Cook, and C. Stunkard. "Guidance Document for Conducting Terrestrial Field Studies," U.S. Environmental Protection Agency, Washington, D.C. (1988).
3. Stevens, D. "Field Sampling Design," in *Ecological Assessment of Hazardous Waste Sites*, W. Warren-Hicks, B. R. Parkhurst, and S. S. Baker, Jr., Eds. (U.S. EPA Report–EPA/600/3-89/013, 1989), pp. 4–1 to 4–13.
4. Hakanson, L. "Aquatic Contamination and Ecological Risk: An Attempt to a Conceptual Framework," *Water Res.* 18:1107–1118 (1984).
5. Stevens, D., G. Linder, and W. Warren-Hicks. "Data Interpretation," in *Ecological Assessment of Hazardous Waste Sites*, W. Warren-Hicks, B. R. Parkhurst, and S. S. Baker, Jr., Eds. (U.S. EPA Report–EPA/600/3-89/013, 1989), pp. 9–1 to 9–25.
6. Adams, D. F. "Recognition of Effects of Fluorides on Vegetation," *J. Air Pollut. Control Assoc.* 13:360–362 (1963).
7. Keeton, W. T. *Biological Science, 2nd ed.* (New York: W. W. Norton & Co., 1972).
8. Weber, C. I., W. B. Horning II, D. J. Klemm, T. W. Neiheisel, P. A. Lewis, E. L. Robinson, J. Menkedick, and F. Kessler. "Short-Term Methods for Estimating the Chronic Toxicity of Effluents and Receiving Waters to Marine and Estuarine Organisms," (U.S. EPA Report, EPA-600/4-87/028, 1988).
9. LeBlanc, G. A., B. Hilgenberg, and B. J. Cochrane. "Relationships between Structures of Chlorinated Phenols, Their Toxicity, and Their Ability to Induce Glutathione S-transferase Activity in *Daphnia magna*," *Aquat. Toxicol.* 12:147–156 (1988).

Concluding Remarks

CHAPTER 23

Concluding Remarks: Implementation of a Biomarker-Based Environmental Monitoring Program

John F. McCarthy

The preceding chapters have discussed a wide array of biomarkers that can and are being used to demonstrate exposure to and effects of environmental contaminants. From this review, there can be little doubt that measurement of biomarker responses in organisms from contaminated sites offers great promise for providing information that can contribute to monitoring programs designed for surveillance, hazard assessment, regulatory compliance or for documenting remediation. The challenge is to develop and implement a research program that will permit this promise to be fully realized. Current understanding and application of biomarkers justifies their immediate implementation in an environmental monitoring program at a pilot-scale. However, the full potential of this methodology will be realized only after a larger data base of field and laboratory studies can be accumulated and analyzed. It is the purpose of this concluding chapter to present some thoughts on a research strategy that is needed to develop the data needed to validate biomarkers and provide the scientific understanding necesssary to interpret biomarker responses of environmental species. What is proposed is an evolving monitoring program that focuses broadly on evaluation of contamination in an array of ecosystem types. The challenges and obstacles to be addressed in such a program include the following:

1. The quantitative and qualitative relationships between chemical exposure, biomarker responses and adverse effects must be established.

2. Responses due to chemical exposure must be able to be distinguished from natural sources of variability (ecological and physiological variables, species-specific differences, and individual variability) if biomarkers are to be useful in evaluating contamination.
3. The validity of extrapolating between biomarker responses measured in individual organisms and some higher-level effect at a population of community level must be established.
4. The use of exposure biomarkers in animal surrogates to evaluate the potential for human exposure should be explored.

What is outlined here is an ambitious multi-year research and development program. Ideally, such a program would be implemented as part of a long-term, inter-agency, interdisciplinary activity such as the Environmental Monitoring and Assessment Program (EMAP) being formulated by the U.S. Environmental Protection Agency in cooperation with other federal agencies or the National Status and Trends Program (NSTP) of the National Oceanic and Atmospheric Administration (see Chapter 20). EMAP is envisioned as a broadly based program to evaluate the status of environmental resources for a range of ecosystem types, identify causes for sub-nominal environmental quality, and track the changes in environmental status over time. Biomarkers could clearly contribute to EMAP or NSTP by demonstrating whether organisms are exposed to chemical contaminants, and in assessing the contribution of the contaminants to environmental deterioration. Conversely, these programs can provide an umbrella for organizing and coordinating biomarker research being conducted as more focused, smaller-scale studies, e.g., at hazardous waste sites and industrial discharges.

STATUS OF CURRENT CAPABILITIES

As is clear from the individual chapters in this volume, each biomarker is a sophisticated technical speciality in itself. It is also clear that no one biomarker is the magic panacea that can, by itself, evaluate chemical exposure and effects. Advancing the potential application of biomarkers must, therefore, be a collaborative enterprise integrating the skills of a variety of specialists. Although a number of individual researchers have evaluated specific biomarkers in laboratory and field experiments, much of this research has been limited to either (1) laboratory exposures of animals to a limited number of well-described model contaminants or (2) measurements of a single biomarker response in field-collected animals. More recently, however, a handful of research groups, mostly at government agencies and national laboratories, have begun to evaluate the responses of a suite of biomarkers in animals from polluted environments. The results have been encouraging; biomarker responses have correlated with the perceived degree of contamination, and the relative ranking of sites on the basis of molecular and biochemical responses agrees well with community level measures of ecosystem integrity. However, the same biomarkers have not been used in all the studies, repeated monitoring at the same sites is rare, and large-scale field studies have been limited almost exclusively to marine or aquatic environments.

Nevertheless, core capabilities for measuring a fairly wide array of candidate biomarkers do exist at federal agencies, national laboratories, and universities and sufficient experience exists for making rational choices about selection and sampling of animal species. The primary impediments to major progress in applying this approach to environmental monitoring is the lack of a unifying mandate and the need for stable long-term funding.

PROPOSED IMPLEMENTATION OF BIOMARKER MONITORING PROGRAM

This research plan consists of five tasks that lead from preliminary proof-of-principle demonstration of the use of biomarkers to indicate exposure, to eventual linking of biomarker responses to individual and ecosystem effects. Although tasks are presented in roughly chronological order of development, it is recognized that many components of each task are interdependent.

TASK I

Preliminary Survey: Proof-of-Principle

Research question.   Can biomarkers distinguish the presence or absence of different types of pollutants in marine, aquatic, and terrestrial environments?

Research description.   The adequacy of current understanding of, and approaches to, analysis of biomarkers in environmental species needs to be tested. Existing capabilities need to be consolidated and organized to test key questions at a very limited number of selected field sites. Results of this preliminary survey will demonstrate the strengths and weaknesses of current approaches, provide an initial screening of promising biomarkers, and establish QA/QC protocols, data base logistics, and statistical analyses that can form the basis of a long-term biomarker monitoring program.

Research objective.   The objective of this research is to compare the qualitative pattern and quantitative responses of a suite of biomarkers in sentinel species from sites polluted with specific types of contaminants, compared to the responses of organisms from pristine reference sites.

Approach.   A core group of researchers from federal agencies, national laboratories, and universities will be identified to evaluate and compare a suite of the most promising biomarkers. Biomarkers will be selected by a panel active in biomarker research. A limited number of field sites, representing aquatic, marine, and terrestrial environments, will be identified for this initial study. Within each environment, selected field sites should present relatively simple exposure patterns, typified by predominance of a single type of contaminant (e.g., heavy metals, pesticides, or carcinogens such as polycyclic aromatic hydrocarbons), and simple exposure pathways. For each type of environment, one or more reference sites will be identified that is ecologically similar to the contaminated sites, but that is free of any known sources of pollutant inputs. Sentinel species will be selected based on their habitat and trophic level to test the power of the biomarkers to identify the contribution of different routes of exposure.

Each site will be sampled two to three times a year for at least 2 years. Each member of the core group will measure the response of the biomarkers that is their speciality in animals collected at each of the different sites. The final product of this task will be a data base describing the responses of a suite of biomarkers at different levels of biological organization in the same species sampled at the same time from polluted and reference sites. Statistical analyses will evaluate: (1) how well different biomarkers distinguished the presence or absence of known contaminants (i.e., were they successful indicators of exposure); (2) whether different types of contaminants could be distinguished from the pattern of biomarker responses; and (3) how consistent the biomarkers were from season to season and year to year at the same sites (i.e., can they distinguish toxicant-related stress from natural ecological variability).

TASK II

Development, Standardization, and Validation of Key Biomarkers

Research question. Can improved biomarker methodology and fundamental understanding of biomarker responses aid in interpreting environmental monitoring data to permit a more accurate assessment of the extent of exposure to environmental contaminants and the long-term significance of that exposure?

Research description. This task provides a framework for most of the fundamental laboratory research required to develop and hone the most powerful and informative biomarkers, and to acquire the fundamental understanding necessary to interpret the significance of, and interactions among, biomarker responses. The principal thrust of this task will be directed at laboratory research with environmental species being used as sentinels. However, research focused on human health or basic biomedical studies can also contribute to advancing the objectives of this task. For example, advances in development of sophisticated biomarker methodology for estimating exposure of humans in the workplace, including development of monoclonal antibodies or DNA probes for adducts, or advanced analytical or clinical technologies that increase the sensitivity, selectivity, or speed of biomarker measurements cancontribute significantly to the application of biomarkers in an environmental monitoring program. Likewise, basic biomedical research that increases understanding of toxicant action at a molecular and cellular level will enhance capabilities to interpret biomarker responses because most of the biochemical processes in environmental species are comparable to those of laboratory animals used in biomedical research. While research focused on human health issues would not be part of the monitoring program, limited funds may permit biomedical researchers to extend their studies to make their results relevant to understanding biomarker responses in animal sentinels.

Research objective.   The objective of this research is to standardize protocols for existing biomarker measurements, develop and modify new biomarkers as needed, and acquire the fundamental understanding of the relationships between exposure, biomarker responses, and adverse effects in sentinel species at the cellular, organismic, or population level.

Approach.   Laboratory exposures with single contaminants or rationalized mixtures of contaminants are needed to establish dose-dependent relationships between exposure, biomarker responses, and toxic effects in sentinel species. Promising biomarker methods developed for certain species (including humans) may need to be modified to extend their application to a wider range of environmental species, including plants. Biomarkers that are selected for long-term monitoring programs need to be optimized for individual species, and standardized protocols need to be tested by inter-laboratory comparisons. The role of natural environmental, ecological, and physiological variables (such as seasonal changes in temperature, food availability, and reproductive status) on the magnitude and pattern of biomarker responses must be determined. The time course of biomarker responses will be an important variable in interpreting the extent and significance of exposure, and must be evaluated in different sentinel species.

This task will involve an ongoing interaction with field monitoring programs and research described in other tasks. Difficulties in interpreting biomarker responses in the field will be resolved through laboratory research in this task. Valuable and informative biomarkers that require laborious (and expensive) procedures will be automated or otherwise improved to make them more cost-effective for a routine monitoring program. New insights into biochemical aspects of toxicology will be applied and tested in species used as environmental sentinels to improve understanding and interpretation of monitoring data.

TASK III

Environmental Monitoring: Biomarkers of Exposure

Research question.   Can biomarkers measured in sentinel species distinguish quantitative and qualitative differences in exposure to contaminants in the environment?

Research description.   This task builds on and extends the preliminary survey described in Task I. That task sought to establish that the biomarkers could distinguish the presence or absence of exposure. This task begins the process of distinguishing relative degrees of exposure (for example, differences in the extent of exposure at different sites within a polluted estuary, or at increasing distances from a hazardous waste site). Likewise, while the first task selected sites characterized by a predominance of a single type of contaminant and simple and clearly delineated exposure pathways, this task will begin to include sites with more complex mixtures of contaminants and greater complexities in exposure pathways and trophic dynamics. The first task sought to establish that biomarkers were useful in aquatic, terrestrial, and marine environments, while this task will begin to explore the application of biomarkers in environments with wider geographic diversity (for example, forests, prairies, and agroecosystems are diverse examples of terrestrial environments).

This task is not a single activity, but rather a gradual evolution from the simple proof-of-principle to a comprehensive environmental monitoring program using biomarkers. Experience in different types of environments and selection of sensitive and informative biomarkers derived in the first task will provide a foundation for expanding to more complex scenarios. Successful implementation of the biomarker approach to environmental monitoring requires this gradual building of capabilities and understanding. The same attributes that make the biomarker approach a powerful tool for environmental monitoring also caution against its rapid and indiscriminate application without the benefit of carefully accumulated experience. Biomarker responses are powerful because they integrate a wide array of environmental, toxicological, and ecological factors that control and modulate exposure to, and effects of, environmental contaminants. However, these same factors may also complicate interpretation of the significance of biomarker responses in ways that may not always be anticipated. Phased increases in the extent and complexity of monitoring scenarios will test and confirm previous understanding and gradually expand that understanding until the biomarkers become a routine, well-characterized, and scientifically and legally defensible tool for monitoring and assessing environmental pollution.

Research objective.   The objective of this research is to increase understanding of the qualitative pattern and quantitative response of a suite of biomarkers in sentinel species until a capability is developed to identify the extent of exposure and the nature of the contaminant to which the organisms are being exposed. The emphasis of this task is on developing a capability to assess *exposure* rather than to evaluate the adverse effect or long-term biological significance of that exposure.

Approach.   Experiences gained from Task I will be the starting point for this task. The suite of the most effective biomarkers selected from the first task will be applied to additional sampling locations in the same areas studied in that task. The number and complexity of field sites and contamination scenarios will be gradually increased to include more complex suites of contaminants, more complex exposure pathways, and greater geographic diversity. New biomarkers should be tested as field sampling increases, and their value compared to those already in use. Improved statistical analyses are needed to describe the pattern and geographic distribution of biomarker responses. With each increase in complexity, results need to be analyzed to evaluate: (1) how well the biomarkers distinguish quantitative differences in the extent of exposure; (2) whether the pattern of responses of the suite of biomarkers can distinguish the nature of the contaminants to which the organisms were exposed (e.g., are metals or carcinogens the primary toxic agent to which the organisms are responding); (3) how effectively the biomarkers in selected sentinel species distinguish the significance of different routes of exposure (e.g., air, water, soil, sediment, and/or foodchain); and (4) whether the quantitative and qualitative responses of the biomarkers can distinguish toxicant exposure in the presence of natural environmental, ecological, and physiological variables.

TASK IV

Linking Biomarker Responses to Community Level Effects

Research question.   Can biomarkers measured at the suborganismal level predict the effects (i.e., the long-term biological significance) of exposure at the organismal, population, and community level?

Research description.   While estimating differences between, or changes in, patterns of exposure is clearly a critical objective of an environmental monitoring program, questions concerning the significance of that exposure to the long-term health and well-being of the organism or its ecosystem must ultimately be addressed. If there is some detectable level of exposure, is it "bad"? What level of exposure will lead to unacceptable or undesirable deterioration of the environment, measured, for example, in terms of the loss of commercially important species or by some more subjective aesthetic criteria. Conversely, "how clean is clean"? Can we determine when pollutant inputs are low enough to cause no long-term adverse effects? This is a critical question in decisions about remediation of existing pollution or in permitting discharges of chemicals into the environment.

The effects of contaminant exposure are difficult to address unequivocally for two reasons: (1) there is often a long latent period between exposure and the measurable expression of an often irreversible adverse effect; and (2) it is often very difficult to attribute effects observed in field studies to toxicant exposure because of biological variability among individual organisms and confounding influences of natural environmental and ecological factors such as food and habitat availability.

Implementation of a long-term monitoring program that measures biological markers at different levels of biological organization (see Figure 1 in the introductory chapter of this symposium) offers the potential for acquiring the information necessary to link sensitive and specific responses to toxicant exposure (measured at the molecular, biochemical, or physiological level) with long-term responses at the organismic, population, and community levels. Many of the suborganismal biomarkers respond quickly to changes in toxicant exposure and can be more clearly linked to specific actions of pollutants. The data base developed in a monitoring program will demonstrate the relationships between quantitative and qualitative responses in these biomarkers and long-term adverse effects at the organism level (e.g., increased susceptibility to disease or reduced fecundity), and the population and community level (changes in the structure of function of the system).

While this is a complex task, and needs to include research components involving experimental validation and ecological modeling, the dividends are very significant in terms of environmental protection and cost-savings: (1) the environmental significance of changes in pollutant inputs to an ecosystem can be rapidly assessed so that intervention can be justified before irreversible damage occurs, or to guide cost-effective remediation or input reductions; (2) chronic bioassays can be devised whose endpoints are changes in rapidly responding biochemical or molecular biomarkers, but which provide information on the long-term, population level consequences of exposure to the test chemicals; and (3) the costs of the environmental monitoring program will be reduced because measurements of biomarkers at the higher levels of biological organization can be reduced or eliminated.

Research objective. The objective of this task is to establish and verify the relationships between exposure, changes in rapidly-responding suborganismal biomarkers, and long-term adverse effects at the population or community level. The successful completion of this objective will validate the use of molecular, biochemical, and physiological level biomarkers as short-term predictors of long-term adverse effects.

Approach. Task III will develop a long-term data base on responses of biomarkers at many levels of biological organization in a diversity of environments impacted to varying extents by an array of contaminants. These data need to be analyzed to establish correlations between suborganismal biomarker responses and population and community structure and function. These correlations will permit hypotheses to be formulated which must then be tested in long-term microcosm or mesocosm experiments where exposure can be controlled. The time course of biomarker responses can be monitored and evaluated with respect to the ability of the biomarkers to forecast community level responses. Relationships between molecular and biochemical markers and population level effects may not be straightforward because of compensatory mechanisms that help regulate population dynamics in natural systems. Ecosystem models, incorporating complex trophic structures and describing higher level compensatory mechanisms, need to be evaluated as possible ways of describing and predicting links between these different levels of organization and complexity. Since fecundity is a key parameter linking individual responses and ecosystem processes, particular attention should be directed at evaluating the molecular, biochemical, or endocrinological factors regulating the reproductive capacity of an individual organism. These reproductive parameters, coupled with ecosystem models, may be useful as tools in ecological risk assessment.

TASK V

Linking Biomarker Responses in Sentinel Species to Human Epidemiology

Research question.   Can information about the nature, extent, and geographic distribution of contaminant exposure in environmental sentinels be useful in identifying the potential for human exposure and health effects attributable to environmental pollution?

Research description.   The presence of toxic and carcinogenic pollutants in the environment raises obvious concerns about the potential for human health effects due to environmental exposures. However, it is very difficult, from an epidemiological perspective, to attribute any increase in health risk to environmental sources. Measurements of biomarkers in environmental sentinels may be useful in evaluating the human health effects associated with exposure to environmental contaminants. A health effect evaluation has four fundamental phases:

1. Documentation of the nature and extent of exposure;
2. Definition of exposed and unexposed populations;
3. Diagnosis of disease in exposed populations; and
4. Establishing relationships between exposure and disease.

An environmental monitoring plan using biomarkers in environmental species can contribute directly to documenting the nature and extent of bioavailable contaminants, while the geographic pattern of animal sentinel responses can identify areas of human habitation that show evidence of bioavailable environmental contaminants. The animal sentinels provide a method for attributing risks due to *environmental* sources of exposure since animals do not share human exposure from life style or workplace pollutants. Furthermore, observations on the toxicologic effects of realistic chronic exposure in animals in contaminated environments can contribute to understanding the relationship between exposure and disease in humans.

While this task may not be central motivation for an environmental monitoring program, the potential benefits to human health assessment should not be over-looked.

Research objective.   The objective of this research is to determine if there are correlations between patterns in the extent, nature, and geographic distribution of biomarker responses in sentinel species, and patterns of epidemiological evidence of increased risk to human health.

Approach.   As Task III is implemented and expanded, information will become available on biomarker responses in sentinel species over relatively large geographic regions, including areas of human habitation. The pattern of biomarker responses (quantitative and qualitative) can be represented on a geographic map with isopleths defining areas of higher or lower levels of bioavailable contaminants. Superimposition of such a map of sentinel responses on an analogous map of human morbidity or mortality data may provide indications of correlations between environmental contaminant levels (biomarker responses of sentinels) and risks of human health effects.

SCHEDULE FOR IMPLEMENTATION OF BIOMARKER MONITORING PROGRAM

The environmental monitoring program that emerges from this proposed research plan will be the final evolution of the field monitoring in Task III, illuminated by laboratory research and development described in Task II, with ecological insights provided by Task IV. Our capabilities to interpret the data base from this program will continue to grow as field experience and experimental validation accumulate over time. In terms of budget and effort, three phases are anticipated:

1. *Initial phase*—Starting with initial workshops and planning sessions, Task I will begin to be implemented. Sampling and analyses at the selected field sites should be phased in slowly to gain experience in, and overcome initial problems with, logistics, data base management, QA/QC protocols, and other complications expected in multi-institutional collaborations. Budgets will increase in the second and third years as the full matrix of field sites, sentinel species, and suites of biomarkers for Task I are implemented and data begin to be analyzed.
2. *Learning phase*—Most of the activities described inTasks II-IV will occur during this phase of implementation. Budgets and manpower commitments will rise substantially and peak in the latter half of this phase, which is expected to last for several years. Field sampling (Task III) and laboratory research and development (Task II) will be heaviest early in this phase, then will level off as experience is gained and understanding is achieved. Tasks IV and V require an accumulation of field data, and activity in these components will begin to rise at about the middle of this learning phase.

3. *Routine monitoring phase*—This is the mature phase of the monitoring program. Most of the field monitoring sites will have been established, biomarkers and sentinel species selected, and protocols for data analysis and interpretation standardized. Testing of new biomarkers, laboratory research to understand biomarker responses, improvements in methods for data analysis and interpretation, and other components from the Phase 2 (learning phase) will always continue at some level so that the power and utility of the monitoring program will continue to grow. Budgets and manpower requirements will decrease gradually during this phase and level off as monitoring and interpretation become more routine, and as new technologies reduce the cost of biomarker measurements.

CONCLUSIONS

Biomarkers measured in environmental species have been demonstrated to be useful indicators of exposure and effects of contaminants in research studies and in a limited number of field evaluations. Because of the interdisciplinary nature of biomarker studies and the need for integration of numerous research specialities, long-term progress will be accelerated by general agreement on a common research strategy. The proposal described here is intended as a "strawman" to serve as a basis for continuing dialog toward defining an acceptable umbrella strategy. Planning and implementation of such a strategy would be facilitated if it were part of a larger mandate such as EMAP or NSTP. However, all that is really needed is for the community of researchers involved in biomarker research, in the U.S. and abroad, to decide on their common strategy and keep in communication with each other to integrate results and help each other interpret their observations. This could be accomplished by totally informal gatherings at regularly attended scientific meetings, by organization of Gordon Conference-type meetings at regular intervals, or by more formal workshops funded by consortia of interested government agencies. Hopefully, this symposium has helped to advance this agenda by gathering many of the key members of this biomarker research community, and enuciating the need for integrated collaborative research. In the spirit of pioneer communities, we invite interested readers to contact the editors of this volume to contribute ideas for establishing and implementing a research strategy for developing and validating biological markers of environmental contamination.

ACKNOWLEDGMENTS

This work was supported by the Exploratory Studies Program, Oak Ridge National Laboratory. The Oak Ridge National Laboratory is operated by Marietta Energy Systems, Inc., under contract DE-AC05-84OR21400 with the U.S. Department of Energy. Publication No. 3519. Environmental Sciences Division, Oak Ridge National Laboratory.

List of Authors

S. M. Adams, Environmental Sciences Division, Oak Ridge National Laboratory, Oak Ridge, Tennessee 37831

Kevin N. Baer, DuPont Company, Haskell Laboratory for Toxicology and Industrial Medicine, P.O. Box 50, Elkton Road, Newark, Delaware 19714

William H. Benson, Department of Pharmacology, School of Pharmacy, The University of Mississippi, University, Mississippi 38677

Michael F. Buchman, Pacific Office, Ocean Assessments Division, National Oceanic and Atmospheric Administration, 7600 Sand Point Way Northeast, Seattle, Washington 98115

Susan M. Cormier, Department of Biology, University of Louisville, Louisville, Kentucky 40292

Jeffrey N. Cross, Southern California Coastal Water Research Project, Long Beach, California 90806

S. Focardi, Dipartimento Biologia Ambientale, Università di Siena, 53100 Siena, Italy

C. Fossi, Dipartimento Biologia Ambientale, Università di Siena, 53100 Siena, Italy

A. Garg, Department of Preventive Medicine and Environmental Health, University of Kentucky Medical Center, Lexington, Kentucky 40536

Justine S. Garvey, Division of Biology, California Institute of Technology, Pasadena, California 91125

Mark Goodrich, Department of Chemistry and Marine and Freshwater Biomedical Core Center, University of Wisconsin-Milwaukee, Milwaukee, Wisconsin 53201

R. C. Gupta, Department of Preventive Medicine and Environmental Health, University of Kentucky Medical Center, Lexington, Kentucky 40536

Richard S. Halbrook, Environmental Sciences Division, Building 1505, MS 6038, P.O. Box 2008, Oak Ridge National Laboratory, Oak Ridge, Tennessee 37831

David E. Hinton, Department of Medicine, School of Veterinary Medicine, University of California-Davis, Davis, California 95616

William Hodgman, Department of Chemistry and Marine and Freshwater Biomedical Core Center, University of Wisconsin-Milwaukee, Milwaukee, Wisconsin 53201

D. A. Holwerda, Department of Experimental Zoology, University of Utrecht, The Netherlands

Jo Ellen Hose, VANTUNA Research Group, Occidental College, Los Angeles, California 90041

L. Julie Huber, Division of Toxicology, Whitaker College, and Department of Chemistry, Massachusetts Institute of Technology, Cambridge, Massachusetts 02139

R. J. Huggett, Virginia Institute of Marine Science, School of Marine Science, The College of William and Mary, Gloucester Point, Virginia 23062

Hiroji Ishida, Laboratory of Viral Oncology, Aichi Cancer Center Research Institute, Nagoya 464, Japan

Braulio D. Jimenez, Environmental Sciences Division, Oak Ridge National Laboratory, Oak Ridge, Tennessee 37831

R. J. Kendall, Institute of Wildlife and Environmental Toxicology, Clemson University, Clemson, South Carolina 29632

Ikuo Kimura, Laboratory of Viral Oncology, Aichi Cancer Center Research Institute, Nagoya 464, Japan

Naohide Kinae, Laboratory of Food Hygiene, School of Food and Nutritional Sciences, University of Shizuoka, Shizuoka, 422 Japan

Roy L. Kirkpatrick, Department of Fisheries and Wildlife Sciences, Virginia Polytechnic Institute and State University, Blacksburg, Virginia 24061

S. Krča, Center for Marine Research Zagreb, Ruder Bošković Institute, 41001 Zagreb, Croatia, Yugoslavia

Susan Krezoski, Department of Chemistry and Marine and Freshwater Biomedical Core Center, University of Wisconsin-Milwaukee, Milwaukee, Wisconsin 53201

Hidemi Kumai, Fishery Institute, Kinki University, Wakayama 649-51, Japan

B. Kurelec, Center for Marine Research Zagreb, Ruder Bošković Institute, 41001 Zagreb, Croatia, Yugoslavia

R. H. Lambertsen, Institute of Biomedical Aquatic Studies, University of Florida, Gainesville, Florida 32610-144

Darrel J. Laurén, Department of Medicine, School of Veterinary Medicine, University of California-Davis, Davis, California 95616

C. Leonzio, Dipartimento Biologia Ambientale, Università di Siena, 53100 Siena, Italy

Edward R. Long, Pacific Office, Ocean Assessments Division, National Oceanic and Atmospheric Administration, 7600 Sand Point Way Northeast, Seattle, Washington 98115

W. R. Lower, Environmental Trace Substances Research Center, University of Missouri, Columbia, Missouri 65203

E. S. Mathews, Virginia Institute of Marine Science, School of Marine Science, The College of William and Mary, Gloucester Point, Virginia 23062

Gerald McMahon, Division of Toxicology, Whitaker College, and Department of Chemistry, Massachusetts Institute of Technology, Cambridge, Massachusetts 02139

Michael J. Moore, Department of Biology, Woods Hole Oceanographic Institution, Woods Hole, Massachusetts 02543

Genji Nakamura, Fishery Institute, Kinki University, Wakayama 649-51, Japan

A. Oikari, Department of Biology, University of Joensuu, Joensuu, Finland

D. B. Peakall, Canadian Wildlife Service, Ottawa, Canada, K1A 0H3

David H. Petering, Department of Chemistry and Marine and Freshwater Biomedical Core Center, University of Wisconsin-Milwaukee, Milwaukee, Wisconsin 53201

Mari Prieto, Environmental Sciences Division, Lawrence Livermore National Laboratory, University of California, Livermore, California 94550

Richard N. Racine, Department of Biology, University of Louisville, Louisville, Kentucky 40292

David W. Rice, Jr., Environmental Sciences Division, Lawrence Livermore National Laboratory, University of California, Livermore, California 94550

Brenda Sanders, Molecular Ecology Institute and Department of Anatomy and Physiology, California State University, Long Beach, California

C.F. Shaw III, Department of Chemistry and Marine and Freshwater Biomedical Core Center, University of Wisconsin-Milwaukee, Milwaukee, Wisconsin 53201

G. R. Southworth, Environmental Sciences Division, Oak Ridge National Laboratory, Oak Ridge, Tennessee 37831

Richard Spieler, Division of Vertebrate Zoology, Milwaukee Public Museum, Milwaukee, Wisconsin 53233

Robert B. Spies, Environmental Sciences Division, Lawrence Livermore National Laboratory, University of California, Livermore, California 94550

John J. Stegeman, Department of Biology, Woods Hole Oceanographic Institution, Woods Hole, Massachusetts 02543

Glenn W. Suter II, Environmental Sciences Division, Oak Ridge National Laboratory, Oak Ridge, Tennessee 37831-6038

Peter Thomas, Marine Science Institute, University of Texas at Austin, Port Aransas, Texas 78373-1267

C. A. van der Mast, Department of Molecular Cell Biology, University of Utrecht, The Netherlands

M. B. Veldhuizen-Tsoerkan, Department of Experimental Zoology, University of Utrecht, The Netherlands

J. E. Warinner, Virginia Institute of Marine Science, School of Marine Science, The College of William and Mary, Gloucester Point, Virginia 23062

Carl F. Watson, Clement Associates Incorporated, K. S. Crump Division, 1201 Gaines Street, Ruston, Louisiana 71270

Daniel Weber, Department of Chemistry and Marine and Freshwater Biomedical Core Center, University of Wisconsin-Milwaukee, Milwaukee, Wisconsin 53201

B. A. Weeks, Virginia Institute of Marine Science, School of Marine Science, The College of William and Mary, Gloucester Point, Virginia 23062

Gerald N. Wogan, Division of Toxicology, Whitaker College, and Department of Chemistry, Massachusetts Institute of Technology, Cambridge, Massachusetts 02139

Bruce Woodin, Department of Biology, Woods Hole Oceanographic Institution, Woods Hole, Massachusetts 02543

Mitsuko Yamashita, Laboratory of Food Hygiene, School of Food and Nutritional Sciences, University of Shizuoka, Shizuoka 422, Japan

Leslie Zettergren, Department of Biology, Carroll College, Waukesha, Wisconsin 53186

Index